Health Monitoring of Aerospace Structures

Health Monitoring of Aerospace Structures

Smart Sensor Technologies and Signal Processing

EDITED BY

W.J. Staszewski, C. Boller* and G.R. Tomlinson
Department of Mechanical Engineering, Sheffield University, UK

John Wiley & Sons, Ltd

*Formerly with European Aeronautic Defence and Space Company – EADS, Munich, Germany

Telephone (+44) 1243 779777

Email (for orders and customer service enquiries): cs-books@wiley.co.uk
Visit our Home Page on www.wileyeurope.com or www.wiley.com

Other Wiley Editorial Offices

John Wiley & Sons Inc., 111 River Street, Hoboken, NJ 07030, USA

Jossey-Bass, 989 Market Street, San Francisco, CA 94103-1741, USA

Wiley-VCH Verlag GmbH, Boschstr. 12, D-69469 Weinheim, Germany

John Wiley & Sons Australia Ltd, 33 Park Road, Milton, Queensland 4064, Australia

John Wiley & Sons (Asia) Pte Ltd, 2 Clementi Loop #02-01, Jin Xing Distripark, Singapore 129809

John Wiley & Sons Canada Ltd, 22 Worcester Road, Etobicoke, Ontario, Canada M9W 1L1

Wiley also publishes its books in a variety of electronic formats. Some content that appears in print may not be available in electronic books.

Library of Congress Cataloging-in-Publication Data

Health monitoring of aerospace structures : smart sensor technologies and signal processing / edited by W.J. Staszewski, C. Boller, G.R. Tomlinson.
p. cm.
Includes bibliographical references and index.
ISBN 0-470-84340-3 (alk. paper)
1. Airplanes – Inspection. 2. Airframes – Deterioration. 3. Space vehicles – Inspection. 4. Detectors. I. Staszewski, W. J. II. boller, Chr. III. Tomlinson, Geoffrey R.

TL671.7.H43 2003
629.134′6 – dc22

2003057607

British Library Cataloguing in Publication Data

A catalogue record for this book is available from the British Library

ISBN 0-470-84340-3

Typeset in 10/12pt Times by Laserwords Private Limited, Chennai, India
Printed and bound in Great Britain by TJ International, Padstow, Cornwall
This book is printed on acid-free paper responsibly manufactured from sustainable forestry in which at least two trees are planted for each one used for paper production.

Contents

List of Contributors

EDITORS

W.J. Staszewski
Department of Mechanical Engineering
Sheffield University
Mappin Street
Sheffield S1 3JD
UK

C. Boller
Department of Mechanical Engineering
Sheffield University
Mappin Street
Sheffield S1 3JD
UK

G.R. Tomlinson
Department of Mechanical Engineering
Sheffield University
Mappin Street
Sheffield S1 3JD
UK

AUTHORS

G. Bartelds
Structures and Materials Division
National Aerospace Laboratory (NLR)
Voorsterweg 31
8316 PR Marknesse
The Netherlands

C. Biemans
DaimlerChrysler Sales Germany
Salzufer 6,
10587 Berlin
Germany

C. Boller
Department of Mechanical Engineering
University of Sheffield
Mappin Street
Sheffield S1 3JD
United Kingdom

M. Breidne
Royal Technical University
IMIT OVR
Electrum 229
16440 Kista
Sweden

Å. Claesson
Acreo AB
Electrum 236
16440 Kista
Sweden

C. Delebarre
Institut D'Electronique et de Microélectronique du Nord (IEMN)
Départment Opto-Acousto-Electronique
Université de Valenciennes
Le Mont Houy
Valenciennes Cedex F-59304
France

J.P. Dupuis
EADS
Corporate Research Centre, France
P.O. Box 76
12 Rue Pasteur
92152 Suresnes
France

P. Foote
Sowerby Research Centre
BAE SYSTEMS
FPC 267, PO Box 5
Filton
Bristol BS12 7QW
UK

S. Grondel
Institut D'Electronique et de Microélectronique du Nord (IEMN)
Départment Opto-Acousto-Electronique
Université de Valenciennes
Le Mont Houy
Valenciennes Cedex F-59304
France

J.H. Heida
Structures and Materials Division
National Aerospace Laboratory (NLR)
P.O Box 153
8300 AD Emmeloord
The Netherlands

K. Levin
Structures Department
The Aeronautical Research Institute of Sweden (FOI/FFA)
P.O. Box 11021
172 90 Stockholm
Sweden

P.A. Lloyd
DSTL
Room 1052 A2 Building
Farnborough
Hampshire GU14 0LX
UK

J. McFeat
BAE SYSTEMS
Airframe Engineering, Military Aircraft
Warton Aerodrome W427C
Preston, Lancashire
PR4 1AX
UK

L.K. Nilsson
Institute of Optical Research (IOF)
S-100 44 Stockholm
Sweden

E. O'Brien
Experimental Stress Analysis
Airbus UK
P.O. Box 77
Bristol BS99 7AR
UK

P. Papadopolous
Association for Research Technology and Training (ARTT)
P.O. Box 1527
Malikouti 1
711 10 Heraklion
Crete
Greece

R. Pressland
A380 Landing Gear
Airbus UK
P.O. Box 77
Bristol BS99 7AR
UK

I. Reed
Sowerby Research Centre
BAE SYSTEMS
FPC 267, PO Box 5
Filton
Bristol BS12 7QW
UK

L. Reithler
EADS
Corporate Research Centre
P.O. Box 76
12 Rue Pasteur
92152 Suresnes
France

M. Signorazzi
Alenia, Un'Azienda Finmeccanica Spa
Research Department
Piazza Monte Grappa 4
00195 Roma
Italy

W.J. Staszewski
Department of Mechanical Engineering
Sheffield University
Mappin Street
Sheffield S1 3JD
UK

R. Stubbe
Acreo AB
Electrum 236
16440 Kista
Sweden

G.R. Tomlinson
Department of Mechanical Engineering
Sheffield University
Mappin Street
Sheffield S1 3JD
UK

K. Worden
Department of Mechanical Engineering
Sheffield University
Mappin Street
Sheffield S1 3JD
UK

Preface

The impact that the aerospace industry has had on our lives and the world economy over the last fifty years is difficult to exaggerate. Aside from the obvious advantages relating to quick, affordable travel to far-off locations there are many more ways in which this revolution has changed our lives. The effect the aerospace industry has on the world's economic position is even more pronounced. The current global market is unimaginable without the existence of aircraft. Also, research and development for aerospace applications is at the forefront on engineering achievements and many new technologies have been transferred to other fields in recent years. However, current economic, technical and social demands have resulted in challenges for aircraft designers and operators. New large capacity aircraft are being developed and will be used widely in the future. Many of these structures will make greater use of composite materials. At the same time the current aircraft fleet is ageing continually. All these developments are a major challenge to inspection and maintenance.

Aerospace structures are currently inspected using traditional nondestructive techniques such as visual inspection, radiography and eddy current. Recent years have shown a range of different technologies and sensing techniques developed for damage detection in metallic and composite materials. This includes Acousto-Ultrasonics and guided Ultrasonic waves. Both technologies utilise optical fibre and piezoceramic sensors for damage detection. New techniques are capable of achieving continuous monitoring, integrated and on-line damage detection systems for aircraft maintenance. Recent developments in advanced signal processing, such as neural networks or wavelets, also offer the potential for more reliable and robust damage detection and prediction. This is a great opportunity for aircraft designers, manufacturers and operators. It is now clear, that sooner or later, new damage detection techniques, combined with advanced signal processing, are destined to become one of the core monitoring elements of aircraft structures. The integration of sensors, actuators, signal processors and controllers is associated with a new design philosophy leading to multifunctional and adaptable structures. The attractive potential of such technologies arise from a number of elements such as: reduced life cycle costs, reduced inspection/maintenance effort, improved performance, improved high rate operator availability, extended life of structures and improved safety. This leads to more efficient and economically attractive aircraft. All these elements are important to both manufacturers and operators of civil and military aircraft.

A vast amount of literature is available on emerging, smart technologies for damage detection. Although most of these techniques are still in a development stage, the maturity is sufficient to focus on monitoring systems that could be used in aircraft. The purpose of this book is to bring together recent damage detection technologies in the context of aerospace applications. The focus is not on details related to sensor technologies, damage monitoring techniques or signal processing procedures. The book is designed to demonstrate how all these elements need to be developed for an efficient and reliable damage monitoring system in aircraft.

Chapter 1 discusses the smart scenario for a health and usage monitoring system in aircraft structures. A review of aircraft end-user requirements is presented. This is translated into specifications for the monitoring system. Important procedures relating to the assessment of new monitoring technologies are also discussed. The material shows an example of how to develop qualification routes, test and validation standards, design and manufacturing guidelines regarding damage detection systems. Chapter 2 gives a brief introduction to health and usage monitoring of aircraft structures. This includes a description of aircraft design concepts, the most common types of failures and current inspection approaches used in practice. The first two chapters establish the need for health and usage monitoring systems based on smart technologies for aircraft structures.

Chapter 3 briefly introduces optical fibre sensors and describes load monitoring technology based on optical Bragg grating sensors. The focus is on various technological aspects which need to be addressed for effective structural usage monitoring and assessment. This includes: specifications and reliability of sensors, fibre coating technology and an optical signal processor. Chapter 4 presents various damage detection techniques which are visible for aircraft structures. Damage detection considered here is not based on monitoring loads and estimated incidents for damage to occur but on structural integration and adaptation of sensors which can directly detect damage. The material presented briefly describes: Acoustic Emission, Ultrasonics, guided wave Ultrasonics and Acousto-Ultrasonics. The first two methods are the techniques where the longest experience exists. The other methods are more in the development stage. Altogether, Chapters 3 and 4 establish the smart technology available for damage detection in aerospace structures.

Chapter 5 discusses recent developments in signal processing for multi-sensor architectures used for damage detection. Recent work in the area of structural health monitoring shows that signal processing is one of the most important elements of any damage detection systems. Various aspects related to sensor data processing, extraction and selection of features for damage detection, pattern recognition and data fusion techniques are discussed.

Chapter 6 evaluates the performance of health and usage monitoring systems during ground and in-flight tests. The feasibility of load and damage detection is demonstrated in representative tests, under realistic operational loads and environmental conditions. The work described in this chapter brings together various damage detection technologies validated under one platform of testing conditions.

We believe that the structure and content of this book is unique. Firstly, it brings together experts in the field from industrial, research and academic institutions. Secondly, it discusses the most important aspects related to smart technologies for damage detection. This includes not only monitoring techniques but also aspects related to specifications, design parameters, assessment and qualification routes. Thirdly, it demonstrates the feasibility of smart technologies for health and usage monitoring structures.

Although, the book is addressed to students, researchers and engineers working in the field of damage detection in aerospace structures, we hope that it will be also a valuable source of information in other areas of health monitoring. We hope that the book will serve as a reference for understanding the challenge behind future generation health and usage monitoring systems in aircraft structures. Since this field is advancing rapidly, we would like to steer potential readers towards the future and stimulate new developments and applications.

ACKNOWLEDGEMENTS

Most of the experimental work described in this book was supported by the European Commission under the research project called MONITOR (**M**onitoring **ON**-line **I**ntegrated **T**echnologies for **O**perational **R**eliability). MONITOR has brought together major European aircraft manufacturers with research and academic institutions in order to provide the tools by which the damage detection or the prognosis of impending damage can be given for airframe structures. The partners involved in the project were:

- *Aeronautical Research Institute (FFA)*, in Sweden
- *Aerospatiale Corporate* Research (now part of the *EADS* Group), in France
- *Airbus*, in the UK
- *Alenia Aeronautica – a Finmeccanica Company*, in Italy
- *Association for Research Technology and Training (ARTT)*, in Greece
- *BAE SYSTEMS (Military Aircraft and Sowerby Research Centre)*, in the UK
- *Daimler-Benz and Daimler-Benz Aerospace Military Aircraft* (the latter now part of the *EADS* Group), in Germany
- *Defence Evaluation and Research Agency – DERA* (now separated into two organisations: *QinetiQ* and *Defence Science and Technology Laboratory – DSTL*), in the UK
- *Institute of Optical Research (IOF),* in Sweden
- *National Aerospace Laboratory (NLR),* in the Netherlands
- *University of Sheffield, Department of Mechanical Engineering,* in the UK
- *University of Valenciennes, IEMN,* in France

The comments and conclusions related to the work presented in this book reflect the opinions of the authors not their employers and institutions involved in the research.

The editors and authors are grateful to the European Commission and all industrial partners involved in the project for their financial support.

We would like to express our thanks to the authors who have contributed to the chapters of the book. Our task was to put together the material they have provided. Unfortunately, the conversion of the material to uniform book chapters has taken much more time than we have expected. We are very grateful for their understanding and patience. The help and involvement of many other anonymous colleagues, technical staff and students involved in the MONITOR project is also greatly appreciated.

We would like to acknowledge the help of Mrs Chitra Bhattacharya who has typed some parts of the manuscript.

Finally we would like to thank our families and friends for their support in the preparation of this edited book.

Wieslaw J. Staszewski
Christian Boller
Geof R. Tomlinson
Sheffield, April 2003

1

Introduction

G. Bartelds[1], J.H. Heida[1], J. McFeat[2] and C. Boller[3]

[1] *National Aerospace Laboratory (NLR), Amsterdam, The Netherlands*
[2] *BAE SYSTEMS, Military Aircraft, Warton, UK*
[3] *Department of Mechanical Engineering, Sheffield University, UK*

1.1 HEALTH AND USAGE MONITORING IN AIRCRAFT STRUCTURES – WHY AND HOW?

To ensure structural integrity and hence maintain safety, in-service health and usage monitoring techniques are employed in many engineering areas. Structural health is directly related to structural performance and in this respect it is one of the major parameters with regard to safety of operation. This aspect of structural health is particularly relevant to transportation systems including various elements of transportation infrastructure. In this context structural health monitoring is a safety issue.

At the same time a change in structural health may affect structural performance to a degree that remedial maintenance actions become necessary. Structural repairs increase the cost of transportation in at least two ways. First, the design and implementation of repairs implies direct costs. Second, the execution of repairs generally requires the transportation system to be temporarily taken out of service and this induces indirect costs due to the loss of production volume or as a result of leasing a substitute system. To reduce repair and maintenance cost an attempt to repair can be undertaken at a very early stage of damage development to limit direct costs. Alternatively, it might be decided to postpone repair until the transportation system has to be taken out of service for scheduled major overhauls to reduce indirect costs. In this context structural health monitoring becomes an issue of cost savings. In excess, structural health monitoring may be considerable with regard to monitoring advanced repair methodologies, which have so far not received approval due to lack of knowledge in their performance and where this lack could be overcome by the application of structural health monitoring.

Health Monitoring of Aerospace Structures – Smart Sensor Technologies and Signal Processing.
Edited by W.J. Staszewski, C. Boller and G.R. Tomlinson © 2004 John Wiley & Sons, Ltd ISBN: 0-470-84340-3

In case of the option relying on the delay measure it may be necessary to adapt operational usage to limit or even stop damage growth. If sufficient knowledge exists to relate damage rates to mission types this can be achieved by usage monitoring. In general usage monitoring can be viewed as a valuable addition to structural health monitoring. Prescribed maintenance schedules are based on an estimated usage pattern. Knowledge of the actual utilisation can be translated into a severity parameter that can be compared to the value corresponding to the estimated loading spectrum. In this manner prescribed inspection intervals and times between overhauls can be tuned to actual needs.

It is worthy to note that there are substantial differences in damage development and as a consequence in the manner structural health will deteriorate with time between metal and composite structures. Whereas in metallic components cracking is a gradual and predictable process with a high probability of occurrence, the wear-out of a composite component as a result after loading environment is much less pronounced but composites may suffer from discrete traumas due to accidental damage of a nonpredictable random nature. The situation suggests that different health monitoring philosophies should be applied to the two families of structural components.

Structural health, or equivalently, the state of damage can be established either directly or indirectly. The first approach checks for the damage type (e.g. cracks, corrosion or delaminations) by applying an appropriate inspection technique. These techniques, based on physical phenomena, in fact sometimes also amount to response measurements but in this case they have a very local and direct character. The established inspection techniques vary from visual inspection by the naked eye to passing the structure through a fully automated inspection gantry. In the indirect approach structural performance or rather structural behaviour is measured and compared with the supposedly known global response characteristics of the undamaged structure. If the effect of certain damages on structural response characteristics is known, this approach provides an indirect measure of damage and of structural health. Obviously in both the direct and indirect approaches the sensitivity and the reliability of inspection are important quantitative performance measures. They are determined on the one hand by the laws of physics but on the other in practice also by the hardware and software quality of the inspection equipment and, last but not least, by the equipment operator: the inspector. In this connection human factors such as the loss of alertness in case of rare occurrences of damage and inspector fatigue in case of long and tedious inspections are important reasons to consider a smarter solution to inspection as an element of structural health monitoring.

Safety, costs and performance issues of the structural health and usage monitoring are particularly important in the aircraft industry. At present monitoring techniques are primarily based on pessimistic prediction and periodic inspection. Flight parameters and a range of independent, nondestructive techniques are employed in practice.

1.2 SMART SOLUTION IN AIRCRAFT MONITORING

Structures which are able to sense and respond/adapt to changes in their environment are often referred to as *smart*. The design philosophy of smart structures is associated with the integration of sensors, actuators, controllers and signal processors. Smart solutions to structural health and usage monitoring relates to systems including sensors for damage detection combined with advanced signal processing and presentation. The sensitivity to damage and the reliability of performance are the major requirements with regard to smart technologies. In comparison to conventional solutions, smart sensors have to provide greater sensitivity, provided they are properly installed. This option is clearly related to a monitoring strategy being related to specific inspections at precisely known generally poorly accessible critical locations. On the other hand, smart sensor systems with advanced data processing may also be relevant for inspecting larger areas for a variety of defects, specifically in the sense of widespread and multi-site fatigue damage. If such smart systems virtually function continuously, the time between inspections effectively tends towards zero and then a moderate sensitivity might suffice, when compared to conventional inspection intervals.

Section 1.1 has identified a safety issue of structural health monitoring. Certainly in high performance transportation such as aerospace, high-speed trains and also automobiles, where structural failures may lead to fatal accidents, safety of operation is a prime consideration. Continuous research in the areas of fatigue and corrosion of metallic aircraft structures including inspection techniques (sometimes spurred and accelerated by dramatic accidents or incidents) has helped to achieve a very high level of structural reliability. Design for damage tolerance is now widely applied. It relies on a very profound understanding of material behaviour, on a very accurate description of the loading environment (both external and internal) all of this in combination with advanced manufacturing techniques and, of course, proven and reliable inspection and maintenance procedures. And in situations where brittle material behaviour or poor accessibility with regard to inspection are in the way of a damage tolerant design approach, detailed numerical analysis supported by advanced testing has produced the understanding of slow crack growth and allowed for the design of safe life structures. Interest for automated integrated inspection systems could thus result from a need for greater reliability of inspection. The damage tolerance chain is only as strong as its weakest link, which probably is inspection. Only in special situations an integrated sensor system may provide greater reliability than current methods. However, if in view of the rapidly growing air transport volume, expressed in billions of passenger miles flown, a significant reduction in structural failure rates is needed, smart solutions may become more relevant as a safety issue. The reality is that operators as well as technology providers have not sufficiently assessed the business case.

Another more important factor stimulating the development of smart systems, however, is the cost of inspection. There are very little published data on the potential for cost reductions but the inspection efforts applied in current aircraft maintenance procedures are very considerable and moreover inspector training and motivation require continuous attention. It must be mentioned here that significant improvements have been achieved in traditional inspection equipment with regard to inspector friendliness and quantitative data presentation. A recent study on inspection requirements for a modem fighter aircraft (featuring both metal and composite structure) revealed that an estimated 40 % plus could be saved on inspection time by utilising smart monitoring systems. The situation at hand is illustrated in Table 1.1. Another estimate derived for a fully automated impact sensing system for a composite structure, based on the use of integrated distributed piezoceramic sensors in combination with advanced signal processing software arrives at a 50 % saving on regular inspection time again for a fighter aircraft. Admittedly, these estimates are based on data derived from laboratory demonstrators. They provide a drive, however, for the development of full-scale demonstrators of smart structural health monitoring systems. As long as inspections can only be performed at discrete intervals, new sensing

Table 1.1 Inspection time effort for a modern fighter aircraft

Inspection type	Current inspection time (% of total)	Estimated potential for smart systems	Time saved (% of total)
Flight line	16	.40	6.4
Scheduled	31	.45	14.0
Unscheduled	16	.10	1.6
Service instructions	37	.60	22.2
	100		44.0

techniques will bring cost benefits. Continuous, in-service health and usage monitoring offers the potential to reduce the number of scheduled and unscheduled maintenance actions and the downtime. Monitoring of damage and loads will enable the assessment of the complete condition of structural components from cradle-to-grave. In fact major research programmes in this area assume that up to 20 % of current maintenance/inspection cost can be saved in civil and military transportation by the use of integrated on-line damage monitoring systems. This clearly suggests that the case for smart solutions to aircraft structural health monitoring requirements derives from cost considerations.

The development of structural health monitoring systems relies on different research disciplines and in addition it affects design and manufacture as well as operation and maintenance. As primary flight systems such as the airframe, landing gear or engines are considered, the airworthiness authorities will have to be involved. Novelty of the structural health monitoring systems is considerable and thus a broad acceptance among all parties involved is necessary to achieve implementation.

These considerations have led to a number of research programmes aimed at setting up collaborative research and development projects. Although, no aircraft operators currently use health and usage monitoring systems of the type envisaged in this book, such systems have been identified as key elements within United States, Europe and Japan. Not only countries that have significant aerospace programmes but also smaller nations with advanced system component expertise are involved in projects that are important for both ageing and future generation aircraft.

1.3 END-USER REQUIREMENTS

The development of structural health and usage monitoring systems depends largely on the available technology. However, the benefits derived from actual research and application programmes must be recognised and reflected by the potential end-users. Structural health and usage monitoring systems must satisfy realistic performance requirements to meet end-user needs. The main driver for automated inspection is reliability and cost reduction. Aircraft operators and maintenance providers will not buy expensive equipment otherwise. This is specifically true within the context of ageing aircraft and true as well with regard to any other ageing infrastructure.

Aircraft operators require in general: safe aircraft structures, low life-cycle costs, airframe operational life extension, high rate of operational reliability and potential for enhanced aircraft performance. The last requirement is particularly relevant to military aircraft operators. This section describes various requirement aspects of damage detection and load history monitoring. This includes: inspection types, global vs local monitoring, dynamic vs static techniques and special aspects of system performance.

1.3.1 Damage Detection

There are two possible options for automated inspection systems. These are: (a) retrofit in existing structures; (b) integration in the structural design of new aircraft. It seems very sensible to integrate inspection devices into the basic structural design to ease inter-service effort. Also, automated health and usage monitoring systems need to cover various inspection types. This includes not only levels of inspection but also areas, locations, types of damage and/or inspection techniques. All these elements can be referred to operational availability, accessibility and improved reliability, as summarised in Table 1.2. The nature and type of the damage mechanisms being of interests are discussed in more detail in Chapter 2.

Potential health and usage monitoring systems can offer either global or local inspection capability. Global techniques may be used to inspect relatively large areas with the aim of locating suspect positions that may then be covered in detail by a special inspection technique. There is a considerable interest among aircraft operators in automated global inspection techniques for:

Table 1.2 Inspection aspects nominated for coverage by integrated automated inspection techniques where underlying reasons are: a) operational availability; b) accessibility; c) improved reliability

	Underlying reasons		
	a	b	c
for cracks and corrosion in flight safety relevant areas at the squadron level	(√)	√	
for cracks and corrosion in larger areas at the depot level	√	√	√
crack detection and surveillance of crack growth	√	√	√
corrosion detection in internal wing skins			√
all NDT of known problem areas of ageing aircraft (e.g. lap joints and bonded joints)		√	
all scheduled inspection, where possible (reasons are important, not inspection types)	√	√	√
inspection after accidents or incidents (e.g. lightning strike)		√	√
water absorption in composite materials			√
ultrasonic inspection (cracks, disbonds, corrosion)	√	√	√
eddy current inspection (at large areas in ageing aircraft)			√
X-ray inspection	√	√	√
bond tester inspection	√		√
borescope inspection	√	√	√
Acoustic Emission techniques		√	
inspection technique based on fibre optics		√	
quantitative debris monitoring in helicopter gear boxes			√

- fatigue cracking, particularly in joints at countersunk hole edges,
- corrosion, particularly inside joints and closed compartments,
- paint damage as an impact event signal,
- disbond, possibly due to corrosion in joints and full depth honeycomb slats and flaps,
- impact damage in composites,
- manufacturing damage in composites,
- disbond in stiffened composite panels.

It is clear that aircraft operators will request at least the same performance as currently available systems and possibly even better to not compromise the overall performance of the aircraft. As a guidance, new systems will be required to detect: 1–2 mm cracks in aluminium sheet (at the base of a countersink), 5 mm cracks in a metallic frame, 100 mm cracks in large areas, 10 % of sheet thickness in corrosion or 15 × 15 mm debonds. Often the sensitivity of damage detection systems is motivated by the costs of such systems. The continuous automated monitoring will effectively reduce the inspection period and this should imply that considerably less strict requirements on sensitivity should be accepted.

Local inspection techniques are aimed at a specific damage of a known appearance and are by nature more focused than global inspection methods. Typical locations nominated by aircraft operators for dedicated inspection systems are:

- undercarriage areas, especially brackets and fittings, for cracking,
- joints connecting major subassemblies with low accessibility, for cracking, loose bolts and corrosion,
- engine blades,
- bulkheads, frames,
- fittings and brackets in general,
- stress concentration areas, such as bores in steel lugs, holes and cut-outs in composites,
- impact damage prone areas in composites,
- composite structures with high interlaminar stresses.

The sensitivity of local inspection techniques has to be specified by aircraft operators (typically 1 to 2.5 mm for cracks) but what really counts is the reliability of detection. Generally, the 90 % detectability requirement is acceptable, assuming that a 95 % confidence level is offered. The defect size that is associated with these levels will vary with defect type and inspection method. Clearly, the smaller the defect, the less the amount for repair and – in the case when a crack may be allowed to grow for reasons of ease in maintenance – the less frequent the re-inspections. But assuming continuous inspection the re-inspection issue becomes irrelevant and allowable defect sizes will then determine the required sensitivity.

Both global and local monitoring inspection can utilise dynamic and static techniques. Dynamic techniques monitor signals emitted during the process of damage propagation such as crack or delamination propagation both resulting from either impact events or in-service load cycles whereas static techniques measure the state of damage at the incident of the recording only. Dynamic techniques must therefore be active continuously in order not to miss any damage propagation events.

Aircraft operators will only consider health monitoring systems that do not increase technical workload and that will be communicating in the 'language' of maintenance personnel. It is estimated that by 2010 the aerospace industry alone will require about

20 % increase in the number of trained maintenance staff. The monitoring strategy, registration/handling of data and interpretation of results that such systems will generate, are very important. It is clear that any structural health monitoring system would have to be also acceptable from a health and safety aspect and environmentally friendly.

1.3.2 Load History Monitoring

Load history monitoring in support of fatigue life management is well established among a large number of military aircraft operators. Modern fighters designed nowadays are provided with a system upon delivery. In the civil aviation environment similar activities have been ongoing and are currently still emerging. However, the quantification of any benefits is still insufficient and thus the major supporting arguments have yet to be formulated. The exceptions are operators of commuter aircraft although these aircraft are not designed for damage tolerance. Some operators express their interest only in recording of irregular events, for example over-speed or heavy landings.

Local stresses, the more relevant inputs for fatigue life assessment, can either be measured directly (by local strain recording) or be derived from flight parameters of a more global character such as airspeed, altitude, mass, pitch and roll rates, and acceleration values. It is important to mention that this is a very unreliable method because one needs to convert from parameters to loads and then still perform fatigue life calculations with imprecise damage accumulation rules. Even then, aircraft life is expressed in logarithmic scales. Traditionally, parameter recording was the only option but strain sensors are increasingly used, albeit usually on a sample basis. Not surprisingly, for future systems the general interest is a combination of the two methods, i.e. applying strain gauges in a limited number of control points and using global flight parameters to determine stress histories in other (critical) locations of interest. Both systems as single entities have their advantages and disadvantages, which is why the one or the other is used. The requirement for number of control points is of direct importance for the definition of the load path monitoring systems to be developed and demonstrated. Most operators would like to see between five and 20 control points. However, more control points could be useful, provided this would not increase complexity of the aircraft.

It is generally observed that aircraft operators are currently satisfied with the performance of electrical strain gauges. Although it is believed that electrical strain gauges drift in service and occasionally are affected by corrosion or disbonding, the problems do not seem extensive or insurmountable. There is an interest, however, in optical fibre strain sensors and this is likely to become more pronounced when such sensors and the integrated data collection and processing systems have proven their advantages. Expected advantages of optical fibre systems include: improved reliability; immunity to electromagnetic fields; lower sensor system weight, specifically with regard to the significant reduction in wiring; accuracy; reduced fire hazards; and other benefits. Cost and ease of installation are also very important aspects of such systems and which are currently significantly improving with regard to the customer requirements. Many aircraft operators point out that efficient load/usage monitoring is not just a matter of having reliable sensors but rather depends on data handling and interpretation. There is a concern, first of all, about the controllable handling of the enormous amount of data a load history system can produce. Over long periods of time data drops and incorrect data are considered risk factors. It will further be necessary to keep track of exchange of components in the aircraft. Furthermore, for safe life aircraft a retrofit installation introduces uncertainties with regard to the remaining life.

Nevertheless, it is widely appreciated that load history data can be extremely useful for:

- risk assessment and other safety of flight issues;
- general insight in components failure records;
- determining long-term maintenance policies.

A number of other questions will have to be answered with regard to system performance. Is it necessary to install systems on all aircraft? Is sample tracking of loads on selected aircraft sufficient? Should airborne systems be merely data collection systems supported by a ground station or should some or all of the processing of data be carried out in the air? Should aircraft engines be monitored?

1.4 ASSESSMENT OF MONITORING TECHNOLOGIES

The major question driving the aircraft operator regarding structural health monitoring is: *How is structural health monitoring able to reduce the life-cycle cost of my aircraft*? A first approach to determine the answer to a similar type of question in other fields has been through the application of a procedure called Quality Function Deployment (QFD). QFD is based on establishing *a house of quality* being schematically shown in Figure 1.1. This scheme relates the customer requirements, and thus criteria, to the features and capabilities of the structural health monitoring techniques, and thus systems being considered. A rating system allows one to determine which of the techniques/systems and even features are best suited to meet the specific customer needs. In order to perform a ranking in a reasonable time, the assessment procedure can be split into the following parts:

- determination of possible technical options for structural health monitoring;
- assessment of the structural health monitoring techniques considered.

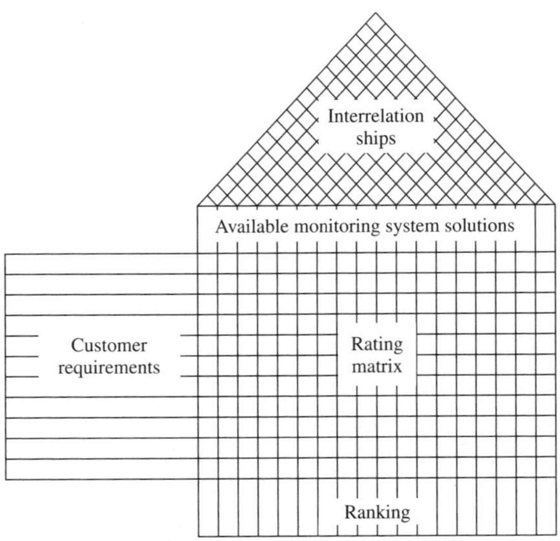

Figure 1.1 Schematic house of quality for damage monitoring

Within *the house of quality* shown in Figure 1.1 two blocks are essential as inputs. These are: the technical options; and the customer requirements. The technical options can be considered as the structural health monitoring systems available. The intention of a comparative study of different damage monitoring techniques is, however, also to determine which of the specifics used in one technique might even be beneficial in another technique. Possibilities of interrelationships, and thus technology transfer between the different structural health monitoring systems available as well as further options to emerge, can therefore be well determined using the interrelationship matrix shown as *the roof of the house of quality*. Structural health monitoring systems can thus be split into various elements in order to better structure the interrelationship matrix. These elements are for example:

- monitoring principle (physical parameter such as strain, light, vibration, sound);
- source of signal generation (from damage itself or external);
- sensor type (e.g. piezoelectric, optical fibres);
- hardware for signal processing (converters, amplifiers, data acquisition systems);
- software for data processing (algorithms, feature extraction procedures, etc.);
- visualisation (numbers, graphs, plots, colours);
- level of damage detection (location, severity of damage, type of damage);
- timescales for completion of the structural health monitoring technology in years.

An example of such an analysis is presented in Table 1.3. The technology validation process must be performed using criteria relevant to structural health monitoring. The aim of the assessment criteria is to provide a consistent framework for the assessment of the structural health monitoring technologies considered. The criteria need to meet as many as possible of the requirements having been set. The requirements need to be the basis for assessing the benefits that will be obtained by operators, if the structural health monitoring systems were deployed on production aircraft and other engineering systems. These factors relate to the regulatory and maintenance environment in which aircraft operate. The assessment should be based mainly on performance of monitoring technologies. The criteria need to be relatively simple and readily understandable.

The following three categories of criteria can be considered:

- **Individual Criteria Selection**

  The result of each structural health monitoring technique is compared, independent of the number of criteria selected for each monitoring technique.

- **Minimum Criteria Selection**

  The evaluation is redone including only the criteria, which have been selected for each technique. The criteria are therefore equal for all techniques.

- **Structural Health Monitoring Technique Related Criteria Selection**

  The evaluation is redone a few times, each time using the criteria that were selected for each specific structural health monitoring technique being considered as a reference, while omitting all criteria for the other structural health monitoring techniques ‘which were not selected for the reference technique.

Table 1.3 Example of interrelationship matrix for damage monitoring system

Mon. Principle	Sensing Type	Signal/Data Processing	Visualisation	Features/ Detection of
AS	Piezo.	Converter (e.g. AD/DA)	Waveform gen.	Damage Location
Strain	Opt. Fibre	Amplifier	Numbers	Damage Event
Light	Painting	Interface	Graph	Damage Size
Vibrat	Antenna	Photodetect/	Light	Cracking
Sound	MEMS	Photo-diode	Sound	Stress Corr. Cracking
Dam. induc.		Coupler/Divider		Corrosion
Non dam. ind.		Interferometer		Disbonding
		Oscilloscope		Impact damage
		Algorithms		Surface scores/Wear
		Neural Networks		Leaks
		Disc		Loose Bolts
		Computer		Delaminations (growth)
		Microprocessor		Humidity
		Filter/Polarizer		Time to completion
		Power supply		
		Transm./Receiv.		
		Switching Circuitry		
		Multiplexer		

The entire procedure is usually built upon a hierarchical system and always leads to a balanced quotation independent of the criteria being selected. The process involves the requirements of the aircraft manufacturers, operators and maintenance providers. However, the assessment results should also be valuable to the design authority and the regulatory authority. As shown in Section 1.3, the operator is principally concerned with three major factors: (a) is the aircraft safe to fly? (b) are defects growing that will need repair and require amendment to the maintenance schedule? (c) does the system provide the operator with any cost benefits? Unfortunately, these are not questions that may be answered with any certainty at the completion of the laboratory evaluation. Thus, often the initial assessment is based on a set of factors that relate to the performance of the systems to detect, locate and characterise any damage.

A significant difficulty for the assessment is that most, if not all, of the monitoring technologies are likely to be sensitive to a limited range of defect types. The requirements are for a wide range of defect types to be detected. Some defects are of more importance than others, and this can be reflected in the score weights. It may be that ultimately a system will need to incorporate a number of techniques. The assessment scoring system therefore needs to be a compromise between the various aspects indicated in this section and in Section 1.3. An example of assessment criteria and scoring system is shown in Table 1.4.

Table 1.4 Example of assessment evaluation table

Partner:
Technique:

Topic/Criteria	Score					Weight	Quote
	0	1	2	3	4		
System configuration							
– easily installed (yes = 4)	0	0	0	0	0	3	0
– easily maintained (yes = 4)	0	0	0	0	0	9	0
– weight (greater than 0.03 % of aircraft weight) (no = 4)	0	0	0	0	0	3	0
Monitoring requirements							
– continuous (Has to be active at all times?) (yes = 0)	0	0	0	0	0	3	0
– requires load (e.g. flight loads) (yes = 0)	0	0	0	0	0	9	0
Interpretability							
– assessment understandable by airline personnel (yes = 0)	0	0	0	0	0	3	0
– assessment requires extensive off-line processing (yes = 0)	0	0	0	0	0	9	0
– provides damage location (yes = 4)	0	0	0	0	0	9	0
Damage types detected							
Composites							
– delaminations (greater than 20 mm diameter) (yes = 4)	0	0	0	0	0	9	0

(*continued overleaf*)

Table 1.4 *(continued)*

Partner:
Technique:

Topic/Criteria	Score					Weight	Quote
– cracking (greater than 10 mm) (yes = 4)	0	0	0	0	0	3	0
– debonds (greater than 15 × 15 mm) (yes = 4)	0	0	0	0	0	3	0
– surface scores/wear (no = 0)	0	0	0	0	0	1	0
Metallics							
– cracks (greater than 1 mm) (detected = 4)	0	0	0	0	0	9	0
– cracks (greater than 5 mm) (detected = 4)	0	0	0	0	0	3	0
– cracks (greater than 100 mm) (detected = 4)	0	0	0	0	0	1	0
– corrosion (greater than 10 % thickness) (detected = 4)	0	0	0	0	0	9	0
– debonds (greater than 15 × 15 mm) (detected = 4)	0	0	0	0	0	9	0
Reliability							
– damage over threshold not detected (all damage over threshold always detected = 4)	0	0	0	0	0	9	0
– false calls (no false calls = 4)	0	0	0	0	0	9	0
– affected by aircraft environment (no = 4)	0	0	0	0	0	9	0
Total							0

1.5 BACKGROUND OF TECHNOLOGY QUALIFICATION PROCESS

Structural health monitoring systems based on smart sensors and materials form a collective set of various technologies, including actuation, sensing, processing and integration. A full qualification of these technologies is always required with regard to their suitability for measuring loads and detecting damage, on the types of structure envisaged and the environmental conditions that can be expected. Operational Load Monitoring (OLM) and automated damage detection systems are often referred to by the term *Sensory Structures*. This section briefly presents the background to the qualification process for technologies, which could form enhanced OLM and automated damage detection. It will identify the activities contained within the technology qualification phase and describes the flight vehicle certification work required for sensory structures. The qualification of structures with integrated actuation response will not be covered.

It is important to mention that the formal performance and operating requirements of damage detection and OLM systems must be studied in detail before any qualification tests are performed. This section describes a possible qualification route assuming the operator requirements from Section 1.3

The qualification route for any new technology is divided into two distinct phases. These are technology qualification and flight vehicle certification, as shown in Figure 1.2. The objective of the technology qualification phase is to assess the qualification evidence and from this to prove that the sensory structure system has comprehensively achieved

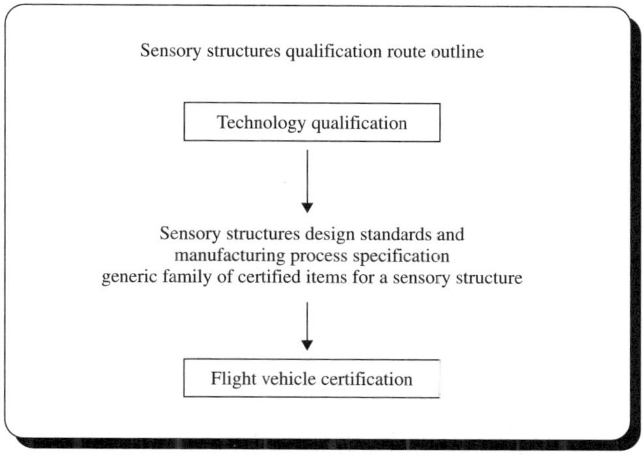

Figure 1.2 Sensory structure qualification route outline

its operating and performance requirements. It will authorise the Technology Design Standards and Manufacturing Processing Specification and certify a generic family of items to be used within sensory structures. Since a significant portion of sensory structures items are likely to be integrated within the structure and be difficult/expensive to subsequently add to or replace, sensory structures technology must be fully mature and the design must be mainly frozen with some technology upgrade compatibility given as an option. Technology development will cease once the technology qualification is achieved unless more stringent sensory structure requirements appear. The qualification of sensory structures is complicated by the novelty and multidisciplined nature of the constituent technologies.

Flight vehicle certification occurs for every new aircraft and for every significant aircraft modification. The flight vehicle certification phase for sensory structures will be achieved by demonstrating design and built compliance with the Sensory Structures Design Standards and Manufacturing Process Specification, supported by appropriate safety case analyses.

The qualification procedures, hazard analysis, safety classification and certification procedures are described in detail in various documents issued by appropriate ministries of defence and aviation authorities (see References). This section shows an example of qualification and certification procedures for military aircraft. The design certification procedure is shown in summary form in Figure 1.3. Within the accepted airworthiness clearance route shown in this figure, the technical work is broken down into various activities. Each activity is controlled by the single technical discipline having responsibility for producing the Statement of Design for that activity. The gathering of evidence to support certification is undertaken for each activity independently. Figure 1.4 gives an alternative presentation of the accepted airworthiness clearance route, showing how each technical discipline can gather its certification evidence in independence from the others. The qualification route for sensory structures will minimise qualification costs via:

- significant reduction in qualification expenditure through a substantial reduction in testing, enabled by improved modelling and analysis processes and through greater use of analogy;

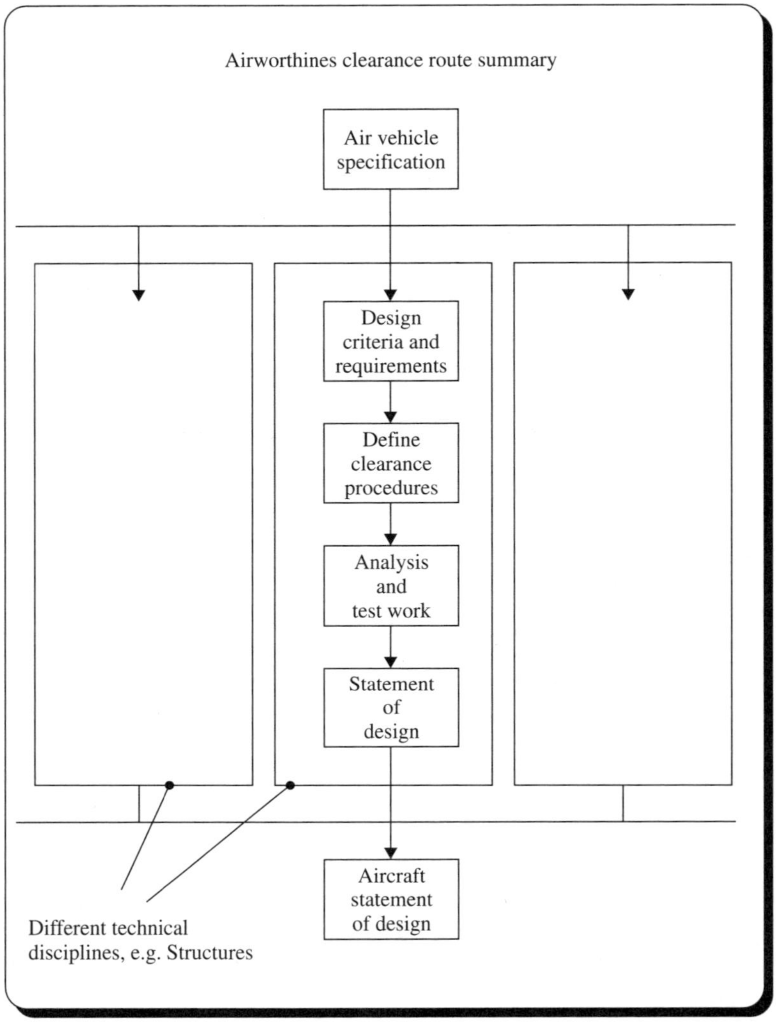

Figure 1.3 Airworthiness clearance route summary

- increased scheduled adherence by having detailed qualification programme plans available from the start, identifying qualification philosophy and from where the balance of qualification evidence should emanate.

Procedures for qualifying new technologies or new designs have generally been produced for single-discipline technologies. Sensory structures are in large part of multidiscipline nature and the existing procedures will therefore need to be applied in an integrated manner. Amongst the technical directorate, the disciplines of structures, nondestructive testing (NDT) and avionics are the main contributors to sensory structures.

The structural Airworthiness Clearance Procedure for the Structures is shown in an outline form in Figure 1.5. The Qualification Programme Plan defines the work and testing required for obtaining airworthiness clearance for each item of structure. The Statement

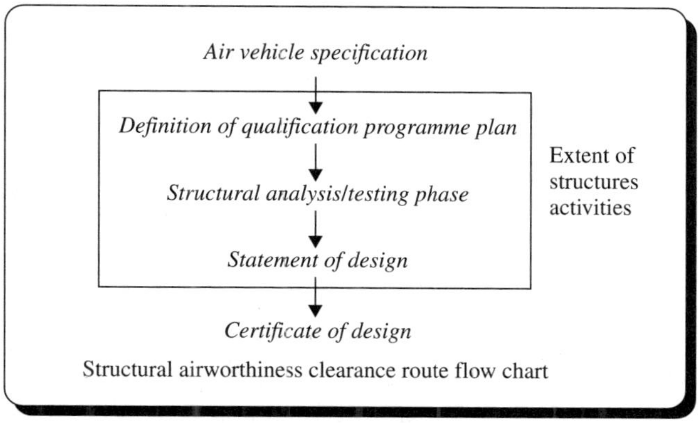

Figure 1.4 Alternative presentation of airworthiness clearance route summary for various single-discipline technologies

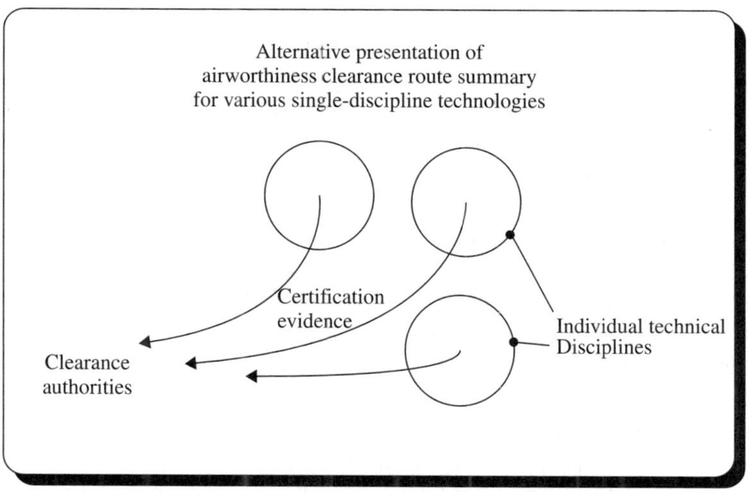

Figure 1.5 Structural Airworthiness Clearance Route Flow Chart

of Design provides the definitive statement of structural integrity. The availability of reliable strength and stiffness data is fundamental for the accurate prediction of structural performance. Within each project, the senior structures engineer has responsibility for identifying whether the strength and stiffness data are deficient and for requesting the necessary material qualification. Material qualification is most likely to be required for new materials or new hybrid materials. It is especially likely to be required for carbon fibre reinforced composites (CFC) whose mechanical properties vary according to lay-up. As an example, the material qualification process for the CFC structure in *Eurofighter Typhoon* military aircraft started with the down-selection of three material types from a group of at least ten. These three were then subjected to a Priority Programme of coupon testing and manufacturing trials to select a single material. This material was then subjected to four extensive levels of testing: coupon, element, subcomponents and component, as shown in

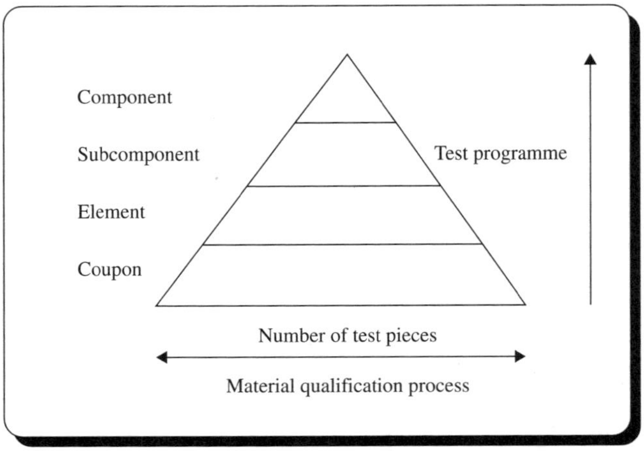

Figure 1.6 Material Qualification Process

Figure 1.6. Coupon testing for *Eurofighter Typhoon* included notch tension/compression, bearing strength and compression after impact, predominantly for flat panel coupons. The form taken by element, subcomponent and component testing depended upon the desired application. Element testing included the testing of the bond strength between the wing substructure and skin through shear, tension and combined shear/tension. Subcomponent testing included sections of the wing spar (i.e. three-point bending test) and the wing attachments. Component testing involved actual aircraft components. An example of the component testing undertaken for *Eurofighter Typhoon* was the testing of a component representing the wing undercarriage cut-out area with fittings and sufficient skin to represent the actual wing. The qualification of improved CFC materials for use on the next generation aircraft are likely to be more modest than for *Eurofighter Typhoon,* as a result of the lessons learnt on this earlier programme.

The formal procedure exists for the qualification of a new NDT method. However, formal procedures are in place for governing how established NDT methods are applied to structures, such as for parts manufactured from CFC, covering all stages after cure, including adhesive bonding and assembly operations. It allows the inspection frequency to be reduced if repeatability of good quality in production parts is demonstrated.

When an established NDT method is applied to a part, a structurally representative reference specimen containing artificial defects is used to confirm that the NDT method can reliably detect the minimum significant defect size. The use of a reference specimen formally calibrates the NDT method and guarantees reliability. Where a number of similar parts require NDT inspection, a Technical Sheet is written to describe the process and the appropriate reference specimens.

If a new NDT method became available and the type(s) of defect(s) to be measured were already subject to existing qualified NDT methods, then the relevant existing reference specimens could be used in its qualification. However, if the type(s) of defect(s) to be measured were new, then new reference specimens containing the new type of defect would be required. The primary technical objective of sensory structures technology is the development of structurally integrated facilities for damage detection/monitoring and for enhanced OLM. These technologies must be demonstrated at increasing levels of maturity.

1.6 TECHNOLOGY QUALIFICATION

1.6.1 Philosophy

The objective of the technology qualification process for sensory structures technology is to confirm that the detailed performance and operating requirements have been comprehensively achieved. Qualification evidence is gathered from the technology development programme and used to authorise Technical Standards and the Manufacturing Process Specifications and to certify a generic family of items to be used within sensory structures.

The technology development and technology qualification processes are closely interrelated, as shown in Figure 1.7. The performance and operating requirements for sensory structures must meet the operator's requirements for damage detection and OLM, while reflecting the expected capability of the developed technology. This process has been already described in Section 1.3. Requirements for qualification evidence are directly linked to the detailed performance and operating requirements. The specific performance

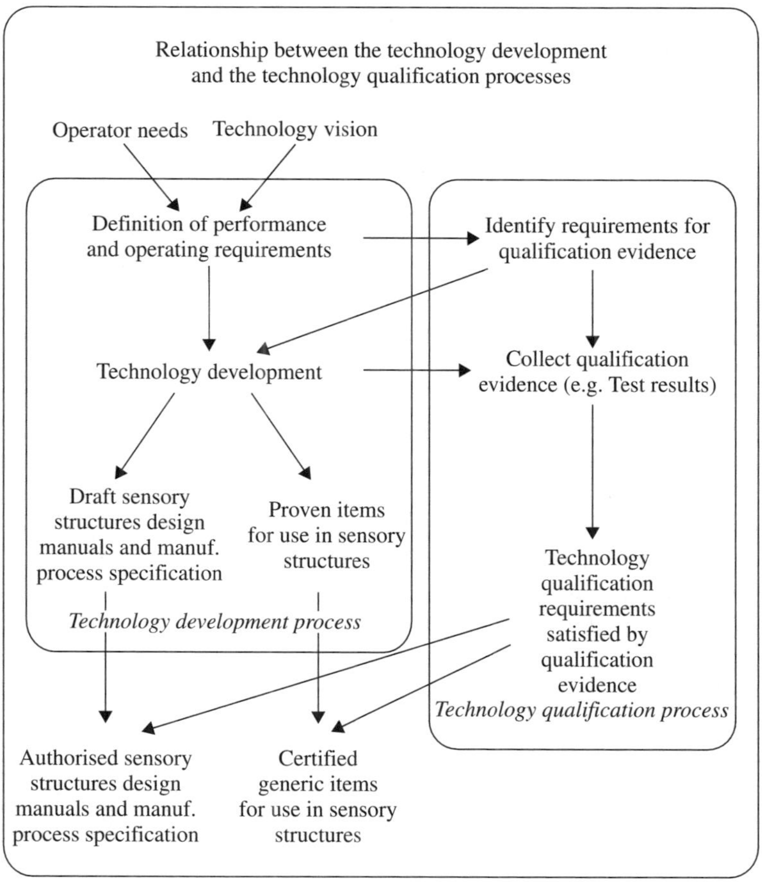

Figure 1.7 Relationship between the technology development and the technology qualification process

and operating requirements are product specific, however the qualification route needs to be identified to underwrite the feasibility of the technology. This section therefore presents a categorised overview of the likely performance and operating requirements and indicates the qualification evidence required to confirm that these likely requirements have been satisfied. Where suitable examples of requirements exist for current aircraft, these will be referenced. It will also identify methods for obtaining the required qualification evidence. In general the technology development programme should produce the following:

- qualification evidence;
- draft technical design manuals for sensory structures;
- draft manufacturing process specification for sensory structures;
- items proved within the sensory structures test environment.

The collection of satisfactory qualification evidence will permit the authorisation of the Sensory Structures Technical Manuals and the sensory structures manufacturing Process Specification plus certification of a range of generic items for use in sensory structures.

Qualification evidence is normally obtained by a combination of calculation, computer modelling, testing or by analogy. Typically, analysis or computer modelling methods have themselves been qualified by comparison with actual test results. Since the technology of sensory structures is immature when compared to the established methods, a sufficient quantity of tests results for qualifying useful analysis or modelling methods has not yet been accumulated. For this reason, the provision of qualification evidence for sensory structures must rely upon appropriate testing undertaken as part of the technology development programme. At the completion of the test programme, certain analysis and modelling methods for designing sensory structures might be qualified for use by the evidence gathered by the test programme and these methods would be incorporated into the sensory structure Design Manual. Maximum use of analogy will be used during the test programme to limit the amount of testing undertaken. Because of the nature of Sensory Structures technology, flight testing will be inappropriate for key aspects such as detecting the occurrence of structural damage. Ground testing will be used as much as possible.

To support the authorisation of qualification evidence by the relevant parent technical discipline, qualification evidence will be divided into individual elements each of which will be authorised by a single discipline. A single test involving several disciplines could therefore have several qualification evidence deliverables. The overall qualification of sensory structures would be achieved by confirming that each of the qualification deliverables have been accepted and authorised by its appropriate discipline.

Most tests will be multidisciplined in purpose and deliver several elements of qualification evidence. For each multidisciplined test, a lead technical discipline will be determined according to the significance of the various qualification evidence deliverables The major technical disciplines involved in the technology qualification process will be structures, materials engineering (for NDT) and systems engineering. More modest input will be required from design and flight test. Manufacturing departments will be closely involved with formulating and agreeing the Manufacturing Process Specification.

Each discipline will define the test requirements for delivering its own qualification evidence deliverables according to that discipline's local procedure(s). The lead test discipline will then be responsible for combining each of these test requirements into an overall test requirement and for managing the actual test. By using each discipline's local procedures to define the test requirements and to present the results, the validity of test results, as qualification evidence will not be questioned, even if the test is multidisciplined in nature.

The process of dividing the qualification evidence requirements into deliverables that are each to be authorised by a single discipline, enables existing procedures written for single disciplines to be applied to multidisciplined activities. For a multidisciplined test activity, the qualification evidence will originate from the separate disciplines as shown in Figure 1.8. The qualification evidence from the lead discipline is shown emboldened, since it is considered to be the more significant.

Procedures are currently either function or project based. It is expected that function based procedures will be used to provide the procedural framework for technology qualification.

Procedures describing test practice might be prepared to describe new fields of measuring or calibration which are developed during sensory structure testing, e.g. for governing damage detection testing. However, it is expected that almost all procedural requirements will be met by existing procedures.

Although the performance and operating requirements document for sensory structures will contain specific requirements it should also be of general applicability, permitting the methods and the certified items to be applied to other aircraft types with minimal requirements for additional technology qualification.

The individual technology qualification evidence deliverables will be authorised by the company specialists in the following disciplines: structures, materials engineering (for NDT), the relevant systems engineering departments, design and manufacturing. At all stages the qualification authorities will satisfy themselves that the qualification evidence is derived from techniques which are fully described in the Sensory Structures Design Manuals and Manufacturing Process Specifications.

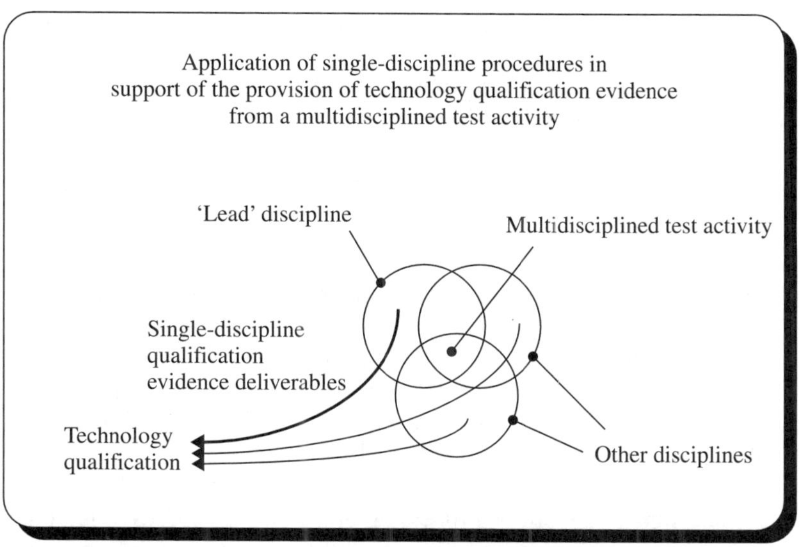

Figure 1.8 Application of single-disciplined procedures in support of the provision of technology qualification evidence from a multidisciplined test activity

1.6.2 Performance and Operating Requirements

The qualification evidence deliverables should fully demonstrate that the performance and operating requirements have been achieved. Performance requirements should address damage detection and operational load monitoring issues. This includes:

- effectiveness at identifying damage extent for a given minimum significant defect size;
- effectiveness at identifying damage location;
- structural configurations required;
- sensor accuracy.

Operating requirements should address reliability, maintainability and environmental issues. The reliability level for the OLM and damage detection elements of the sensory structure could be to achieve *occasional failure probability*, i.e. likely to occur once during the operational life of the aircraft. The reliability level required for the built-in, self-diagnostic system for flagging OLM and damage detection sensor failure could be to achieve *improbable failure probability*, i.e. very unlikely to occur during the operational life of the aircraft.

Operating requirements need to consider various environmental aspects related to the sensory structures. This includes:

- EMC/EMI/EMH protection, lightning protection, electrostatic discharge, solar radiation, X-ray radiation emission limits, fire resistance/proofness, contamination resistance and particularly associated with operation and servicing applicable to aircraft type;
- temperature, cooling and heating extremes, ambient pressure and rates of change, humidity, sand and dust, salt mist/moisture, ozone, rain, icing, freezing rain, water drip, weather erosion;
- vibration, acoustic noise (e.g. engine plume), mechanical shocks, explosive decompression, explosion damage, acceleration, other ground and flight loads, birdstrikes;

Normal design practice is to produce a document(s) for every aircraft type, which gives the environmental conditions divided into zones.

It is also important that the maintenance cost associated with OLM and damage detection system do not significantly reduce the costs benefits derived by the reduced inspections.

1.6.3 Qualification Evidence – Requirements and Provision

Qualification evidence deliverables are required only for areas of new technology or where technologies are integrated in a new manner. Perceived deliverables required for providing suitable qualification, and the responsible disciplines are outlined in Table 1.5.

The manufacturing test programme will establish qualified manufacturing techniques and the associated design practice for manufacturing embedded and surface items, i.e. sensors, data links, signal converters, processing units, power sources, remote interrogator electronics and antennae, diagnostic systems, recording systems, man – machine interface or surface connectors. Embedded items will be integrated with coupons manufactured from appropriate materials.

Table 1.5 Technology qualification evidence deliverables

Qualification evidence deliverable	Qualification authorisation discipline
Embedded Items Manufacturing Techniques	Manufacturing
Surface Mounted Items Manufacturing Techniques	Manufacturing
Embedded Items Design Practice	Design
Surface Mounted Items Design Practice	Design
Embedded Items Design Limitations	Structures
SubSystem Reliability	Systems
SubSystem Environmental Hardness	Systems
SubSystem Maintainability	Systems
Integrated System For OLM (*Test Phase A*)	Structures
Integrated System For Damage Detection (*Test Phase A*)	Structures
Swamp Processing For OLM (*Test Phase B*)	Systems
Swamp Processing For Damage Detection (*Test Phase B*)	Systems
OLM Sensor Positioning Design Practice	Structures
Damage Detection Sensor Positioning Design Practice	Structures

The lead discipline for manufacturing testing will be Manufacturing. Where items are to be manufactured by external suppliers, whether risk or nonrisk sharing, the external supplier will be responsible for manufacturing process testing and for providing the qualification evidence. The qualification of non-embedded, standard electronic items may be covered by the use of existing, qualified design and manufacturing practices.

Qualification evidence deliverables are shown in Table 1.6. These have not been subdivided into detailed subsystem components since the practical extent of item embedding has not yet been determined.

Interim qualification of the manufacturing techniques and associated design practice described in the draft Manufacturing Process Specification and the draft design manuals, will be achieved during manufacturing process testing. Final qualification will be achieved once any improvements found necessary during other testing are incorporated.

The material test programme will involve the testing of CFC coupons containing embedded items. It will ascertain the degradation on overall material properties beyond the existing *knock-down* incorporated into data sheets to allow for manufacturing defects and imperfections. It is assumed that degradation will be at an acceptable and predictable level such that the more demanding testing of elements, components and subcomponents with embedded items will not be necessary.

Table 1.6 Manufacturing process testing qualification evidence deliverables

Qualification evidence deliverable	Qualification authorisation discipline
Embedded Items Manufacturing Techniques	Manufacturing
Surface Mounted Items Manufacturing Techniques	Manufacturing
Embedded Items Design Practice	Design
Surface Mounted Items Design Practice	Design

Table 1.7 Material coupon testing with embedded sensor

Qualification evidence deliverable	Qualification authorisation discipline
Embedded Items Design Limitations	Structures

The lead discipline for material testing will be Structures. Qualification evidence deliverables are shown in Table 1.7.

Flight testing will have to be undertaken to identify the sensory structure airborne environment, e.g. acoustic noise from local fretting, bearing wear, pressure waves, stone attack, etc. in terms of its physical effects and the resulting electrical noise at the sensors. Measurements will be required from suitable aircraft locations.

From these environmental 'background' effects, noise signals will be developed to represent the airborne environment during ground testing. The different sensor types will be subjected to acoustic noise representative of the airborne environment to confirm that their function is not impaired. The airborne environment electrical noise signal will be superposed with the ground test sensor array signal(s) to confirm the function of the integrated signal processing, diagnostic, recording, and MMI systems.

Obtaining the airborne environment noise signature is part of the technology development process and is not itself qualification evidence. It is mentioned since its provision is essential to qualify sensory structures for flight, solely through ground testing. Otherwise much more expensive flight-testing would be required.

Subsystem components will be rig tested to demonstrate continued performance for the full environmental operating requirements and for periods representing up to the full aircraft life to confirm reliability and maintainability targets. The test period may be reduced if a replacement maintenance policy applies to the component; otherwise the test will extend to full aircraft life. To reduce risk, subsystem components should be tested separately since problems can then be resolved before overall system integration. However, it may be impractical to test each of the components separately due to difficulty in generating the inputs from connecting subsystems not included in the test or where the connecting subsystems are also embedded into the structure. The test groups used will be developed accordingly.

The lead discipline for environmental operating requirements testing will be Systems. Rig test vibration levels will represent the airborne environment measured during Airborne Environment Sampling. Qualification evidence deliverables are shown in Table 1.8. These

Table 1.8 Reliability/environmental/maintainability operating requirements subsystem rig testing

Qualification evidence deliverable	Qualification authorisation discipline
SubSystem Reliability	Systems
SubSystem Environmental Hardness	Systems
SubSystem Maintainability	Systems

have not been subdivided into grouped subsystem component tests since the detailed technical knowledge of the interrelationships between the subsystem components is not yet clearly defined.

Metrics will be collected from these tests to indicate the reliability level and maintenance man-hours to be expected for the fully integrated sensory structure system.

The fully integrated sensory structures system will later be cleared against its environmental operating requirements by read across from these tests. Testing of the fully integrated sensory structures system against environmental operating requirements will not therefore be necessary.

Two test phases will be undertaken to confirm that the fully integrated sensory structure meets the performance requirements. These are: Test Phase A – Integrated Sensory Structure System; and Test Phase B – Swamp Processing.

In Test Phase A, the test article will be a simple piece of structure containing a few sensors and equipped for OLM and damage detection. The OLM sensors will be calibrated against known loads. Damage will be progressively introduced to check the performance of the damage detection sensors. The electrical noise signature measured during Airborne Environment Sampling will be superimposed over the sensor response throughout the test.

The purpose of the Test Phase B test will be to prove that the system items used within sensory structures are able to handle the very large numbers of sensors thought typical of in-service applications. A structural test article will not be prepared and the signal processing, diagnostic, recording, and MMI systems will be integrated without any actual sensors. Instead the signature from a very large number of sensors will be artificially generated as signals containing damage and strain information plus the electrical noise associated with the airborne environment. These tests will not seek to prove requirements for reliability over aircraft lifetime since Operating Requirements Rig Testing will have already achieved this. The lead discipline will be Systems. Qualification evidence deliverables are shown in Table 1.9.

The performance requirements document will define which structural configurations will use sensory structure technology. This test phase will test the ability of sensors to measure load or detect damage for subcomponents and components representative of these structural configurations. Electrical noise representing the airborne environment will be applied throughout these tests. For test articles equipped with damage detection sensors, a phased introduction of damage will be introduced at the end of test.

Best practice for sensor arrangement and quantity will be identified. Therefore, the test articles will not necessarily need to use the fully integrated sensory structures system for processing.

Table 1.9 Integrated sensory structure rig testing

Qualification evidence deliverable	Qualification authorisation discipline
Integrated System For OLM *(Test Phase A)*	Structures
Integrated System For Damage Detection *(Test Phase A)*	Structures
Swamp Processing For OLM *(Test Phase B)*	Systems
Swamp Processing For Damage Detection *(Test Phase B)*	Systems

Table 1.10 Realistic structural configuration rig testing

Qualification evidence deliverable	Qualification authorisation discipline
OLM Sensor Positioning Design Practice	Structures
Damage Detection Sensor Positioning Design Practice	Structures

The lead discipline will be Structures with qualification evidence deliverables which are shown in Table 1.10. A more detailed breakdown of the qualification evidence deliverables will be possible once the structural configurations are defined.

1.6.4 Risks

The technology qualification route as described above, depends upon a certain level of confidence in the achievement of the technology development programme.

If technology development proves more difficult than anticipated, the provision of technology qualification evidence might be delayed even though the actual evidence requirements are unchanged. If technology development shows that certain technical assumptions are incorrect then the amount of qualification evidence required might increase markedly.

Potential risks within the technology development programme, which could adversely affect the technology qualification programme, have therefore been identified below.

All risk elements are contributing factors to the reliability of the system. This is the key performance requirement and hence the key risk. Overcoming manufacturing problems may delay the technology development programme.

It has been assumed that manufacturing processes can be developed to ensure repeatability of sensor performance and installation such that sensor calibration requirements can be minimised. However, if such repeatability cannot be achieved, significant additional cost will be incurred in supporting a more comprehensive sensor calibration programme during manufacture and function testing and through the extra sensors required to provide redundancy.

The technology qualification test programme assumed that any degradation in material properties is within acceptable limits already allowed for in data sheet values. If this assumption is proved to be correct, then the scale of the material test programme for a new aircraft will be largely unaffected by the introduction of sensory structures. The probable impact of the technology qualification process on the next new aircraft material test programme will be to add coupon specimens with embedded sensors (to confirm any degradation in material properties within limits) and to install sensors in certain areas of the subcomponents and components to prove sensor performance for realistic structural configurations. The testing of coupons with embedded sensors will be an essential part of every new aircraft material test programme. Sensor performance testing for realistic structural configurations need not be repeated unless a significantly different configuration is intended to be used.

However, if the embedding of sensors in CFC were found to produce a significant degradation in material properties beyond acceptable limits, then the material test programme for a new aircraft would become much larger. All the embedded elements of the smart sensors would need to be incorporated into the element, subcomponent and component material test articles as intended in the final design. If the embedded elements needed to be repositioned later in the design process, a repeat test series with new test articles would be required.

As explained above, any significant degradation of material properties by embedded sensors would significantly increase the cost of a new aircraft material test programme. Instead of extending the scope of the material test programme to allow for significant material degradation, a *blanket* increase in section thickness and area could be made in the vicinity of embedded sensors to maintain the structural strength and stiffness. With the emphasis upon minimum cost and mass, this is likely to be an unacceptable solution.

Confidence in the element, subcomponent and component testing requires that the items tested are similar or preferably identical to actual final design items. Any embedded sensors and associated local structural modifications should be incorporated in the test items reinforcing the need to consider sensory structures requirements very early in the design process.

The effectiveness of damage detection sensors has not been proven in high-energy acoustic environments. Using the vibration levels and sensor electrical noise signatures measured during Airborne Environmental Sampling has reduced this risk.

1.7 FLIGHT VEHICLE CERTIFICATION

A detailed description of the general Flight Vehicle Certification Process is shown in Figure 1.9. This process has been modified in Figure 1.10 for specific application to the flight vehicle certification of sensory structure technology.

For sensory structures, the code of general performance and operating requirements for sensory structures will be the performance and operating requirements document. If the performance or operating requirements for a particular project are not contained within this general code to which technology qualification was achieved then flight vehicle certification cannot proceed until additional technology qualification is undertaken. The design is proven solely by demonstrating compliance with the Design Manual. If compliance with the Design Manual cannot be demonstrated, then the project will need to revise its specification or the design; certification of a flight vehicle with derogation from the Design Manual is not to be allowed. Aircraft build will follow the Manufacturing Process Specification and use the generic items certified during technology qualification.

The Certificate of Design for Sensory Structures will be authorised by demonstrated compliance with the Sensory Structures Design Manual. The Certificate of Build for Sensory Structures will be authorised by demonstrated compliance with the Manufacturing Process Specification and by the use of sensors, connectors, avionics equipment, etc., which were individually certified during technology qualification. Flight vehicle certification for sensory structures is permitted by the functional analogy that exists between the flight vehicle and the system developed during technology qualification. The build application will be generic with that qualified during technology qualification. Build testing will involve checking that the damage detection sensor suite can detect known acoustic emissions from individual sensors plus the calibration of the OLM system through placing the flight vehicle in a loads rig.

If traditional damage detection maintenance is applied to the flight vehicle then sensory structure sensor function is not necessary, e.g. not required for first flight. Airworthiness

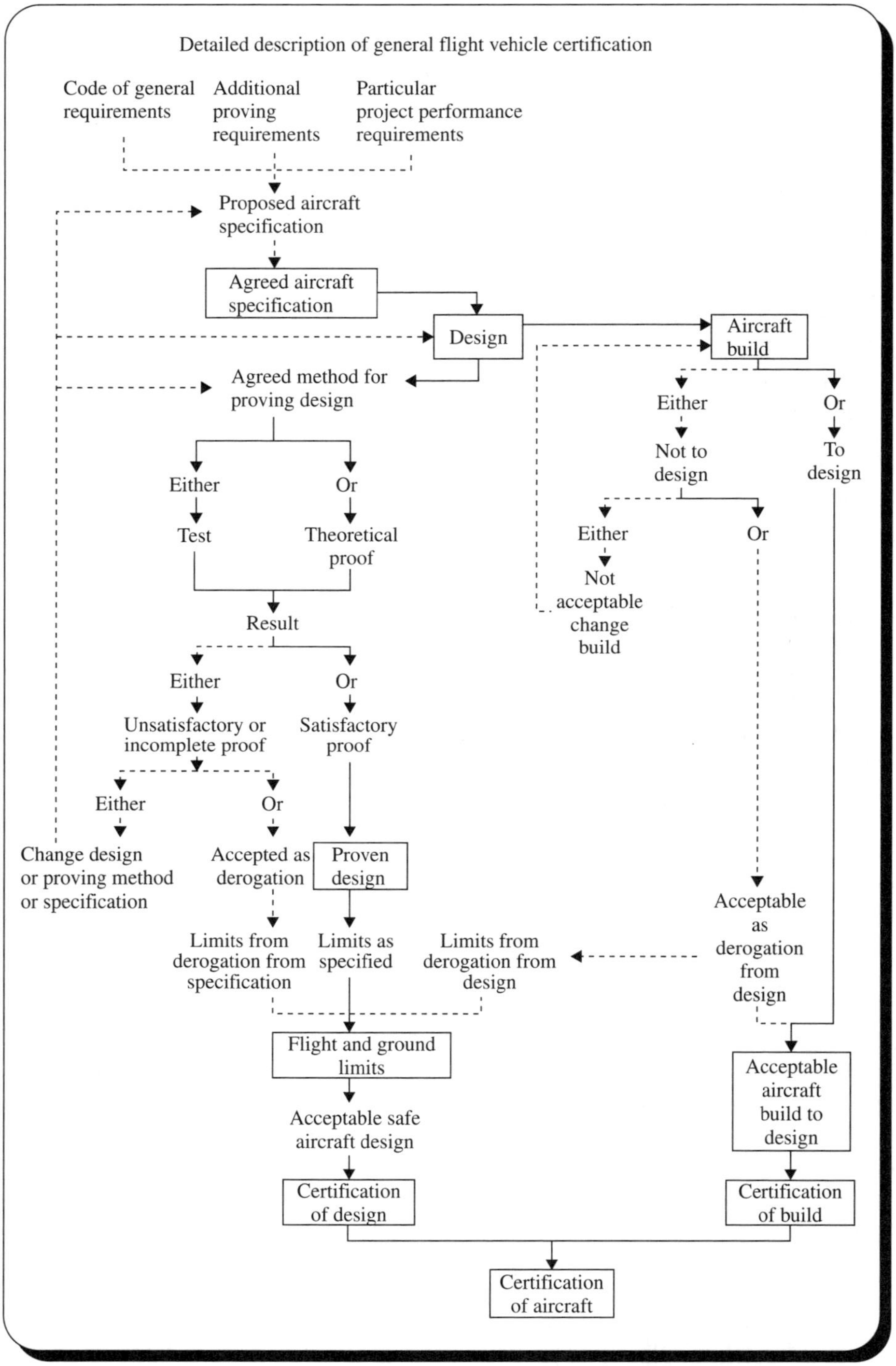

Figure 1.9 Detailed description of general flight vehicle certification

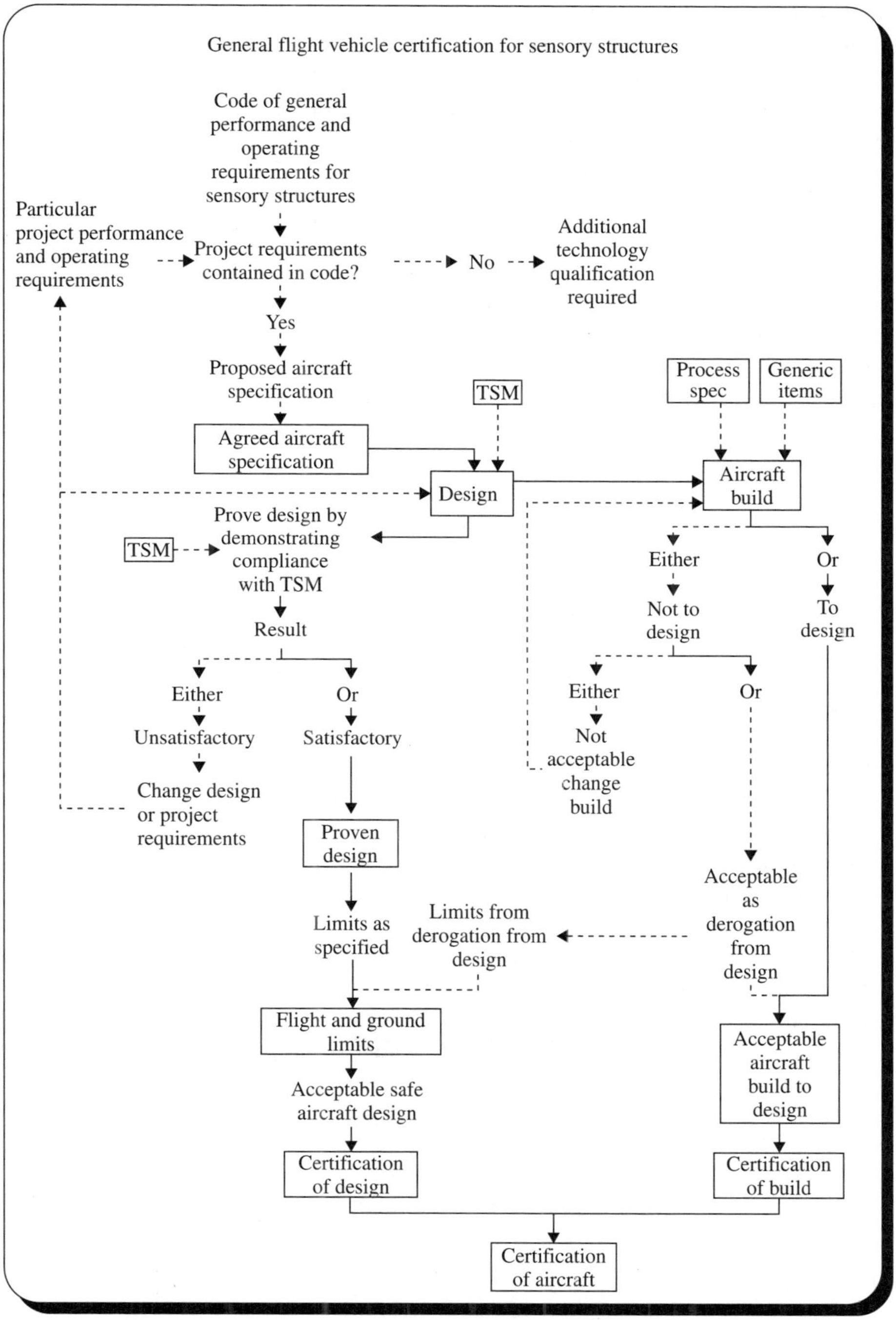

Figure 1.10 General flight vehicle certification for sensory structures

authorities would need to prepare a safety case for the built-in self-diagnostic system, which is considered to be flight safety critical. However, since the system applied will be generic, the safety cases for each project should be similar.

Technology qualification and certification information will be made open to operator representatives, to promote operator acceptance and revised maintenance practices.

1.8 SUMMARY

Automated structural health and usage monitoring system based on smart technologies will not solve all service problems of aircraft operators and maintenance providers. Detailed scheduled inspections of modern civil and military airframe structures is prescribed only after a long inspection-free initial life (up to 15 to 20 years of operation for some components depending on the usage of the aircraft) and it is done then only at rather large intervals. However, an increasing number of ageing aircraft beyond the age of 15 years becomes a significant problem for inspection.

There is very little precise quantitative information publicly available on the man-hours involved but general feeling is that any reduction in inspection time and cost will have to justify the installation of integrated and automated systems. The total effort spent on regular (daily, weekly or monthly) scheduled global inspection is significant but it concerns observations by visual means of obvious failures, missing or loose parts and corrosion. There is no doubt that unscheduled local inspections after unexpected service failure are often time consuming and at least disruptive with regard to operations. However, the major difficulty is that their occurrence and the structural parts affected are unpredictable by nature.

Nevertheless, inspection of structural zones that are extremely difficult to access or that require special measures in the design stage already, provide a very strong case for automated integrated inspection systems. Another case for automation but not for integration is the tedious inspection of extensive riveted structural joints. Thus the prospect of cost reduction arouses almost unconditional support to the new automated health and usage monitoring technologies, the only major boundary condition being its reliability.

Most of the conclusions regarding end-user requirements apply to damage in metallic aircraft structures. There is very little experience with structural health deterioration of composite structures, associated no doubt with conservative design practice. The situation might change in the future since future design effort will continue to aim at low cost structures.

An example of qualification route of sensory structures has also been outlined in this chapter. It is clear that more details will need to be added as the technology performance becomes better defined. In addition as applications/operating requirements are better understood so specific requirements should be identified.

REFERENCES

DEF STAN 00-56 (PART1)/2. Safety Management Requirements for Defence Systems Containing Programmable Electronics Part I: Requirements, Draft.

DEF STAN 00-56 (PART2)/2. Safety Management Requirements for Defence Systems Containing Programmable Electronics Part II: General Applications Guidance, Draft.

DEF STAN 00-970. Design and Airworthiness Requirement for Service Aircraft.

DEF STAN 05-123. Technical Procedures for the Procurement of Aircraft, Weapon and Electronic Systems.

2

Aircraft Structural Health and Usage Monitoring

C. Boller and W.J. Staszewski

Department of Mechanical Engineering, Sheffield University, Sheffield, UK

2.1 INTRODUCTION

Aircraft are highly complex systems composed of various structural, hydraulic, propulsion, electronic and avionic elements. Such complex systems require extensive maintenance. A major portion of the maintenance effort of aircraft structures is related to health and usage monitoring. The other significant portion of maintenance involves repair or replacement. All of this evolves from the safety criticality aspects that have to be set either prior to design or during in-service as a result from a changing operational environment. The health and usage monitoring for propulsion systems is highly advanced. Different in-flight Engine Condition Monitoring (ECM) systems have been gradually developed and are still further improved. The ECM systems have been approved by the Federal Aviation Authority (FAA) and are used by many aircraft operators (Haberding 1985; Spragg *et al.* 1989). Various parameters, including vibration, temperature, pressure, fuel usage and revolutions per minute, are utilised to monitor engine condition and detect/locate possible malfunctions a sufficient time before possible in-flight failures. Also, the control and avionic aircraft systems use built-in test equipment for monitoring and corrective action in the event of failure. Health and usage monitoring has been successfully introduced into helicopters, mainly to monitor vibrations on gears and specifically gear shafts, which once a crack has emerged have a relatively short crack propagation life.

It appears that monitoring of hydraulic subsystems and airframes is limited to on-ground inspection using various nondestructive testing (NDT) techniques. Determination of incidents where inspection becomes essential is either established by a prescribed inspection sequence and procedure and/or by monitoring the actual load sequence the

Health Monitoring of Aerospace Structures – Smart Sensor Technologies and Signal Processing.
Edited by W.J. Staszewski, C. Boller and G.R. Tomlinson © 2004 John Wiley & Sons, Ltd ISBN: 0-470-84340-3

aircraft has gone through. Current inspection techniques include visual observations, eddy current and ultrasonics. In-situ structural monitoring is particularly important for airframes. Lightweight and stiff aircraft structures are designed to withstand severe operational conditions and possible structural damage. This would not be possible without appropriate design concepts, advanced materials and maintenance effort. All these elements are briefly discussed in this chapter.

2.2 AIRCRAFT STRUCTURAL DAMAGE

Current aircraft structures are mostly built using metallic and polymer composite materials with metal-matrix composites in very exceptional cases. The mechanism of damage in these materials depends on the ductility and homogeneity. Materials, such as aluminium, develop cracks. Microscopic cracks generating in the grains of a metal propagate in the respective component during the life-time under variable loading conditions. Severe and unexpected loads or impacts can lead to significant plastic deformation on component surfaces. Polymer based composite materials do generate other mechanisms of damage where barely visible impact damage (BVID) is the mechanism of major concern.

The research project MONITOR (Staszewski 2000) funded under the EU Framework Programme IV performed an end user survey regarding the most common and/or important damage forms in aircraft structures. A summary of this survey is given in Figure 2.1, which shows that fatigue cracking is considered to be most important in metals, whereas impact damage is major for composites. This is then followed by corrosion and bonding/debonding, respectively.

The significance of fatigue failure has led to a differentiation in fatigue failure types, such as reported in (Suresh 1998). A sequence of variable loads and thus stresses and strains leads to mechanical fatigue or cracking. Cycling loads in conjunction with high temperatures results in creep-fatigue. The presence of a chemically aggressive environment causes corrosion. The variety of corrosion types includes pitting, galvanic, intergranular and exfoliation corrosion. Cycling loads with sliding and rolling contact lead to sliding and rolling contact fatigue, respectively. Fretting fatigue is another form of failure due to cyclic stresses and oscillatory frictional motion between components. Cracking and corrosion are the most common mechanisms of fatigue structural failure in aerospace

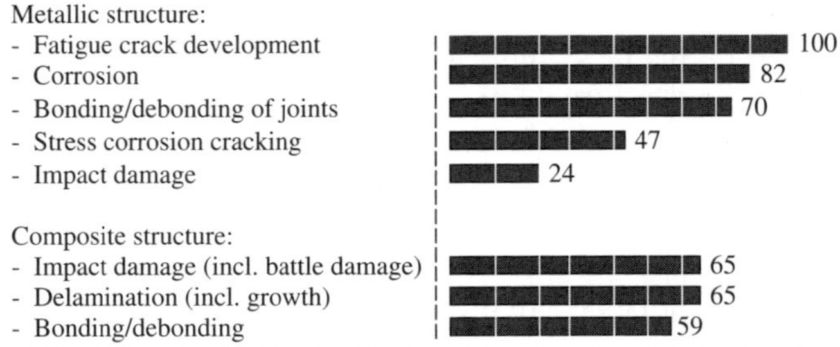

Figure 2.1 Damage statistics in metallic and composite structures taken from the MONITOR end-user surveys (reprinted with permission from Proc. of the 1st Internet. Workshop on Struct. Health Monitoring, 1997, Technomic Publ. Co. Inc., Lancaster PA, USA, pp. 293-300. Copyright CRC Press, Boca Raton, Florida)

engineering. All metallic components exhibit different stages of fatigue damage. These can be classified as (Boller 2001):

- substructural and microstructural changes,
- microscopic cracks,
- formation of dominant cracks,
- stable propagation of dominant cracks,
- structural instability and/or complete fracture.

One of the most critical tasks in fatigue analysis is to reliably identify the initial crack defined. Further areas to be considered include fracture mechanics with regard to short and long cracks as well as their phenomena of propagation under static and cyclic loading, cyclic deformation, fatigue-based design concepts and environmental interactions.

Fatigue research can be attributed to many researchers over the last 170 years. The work of Wöhler from the 1850s on the strength of steel railway axels subjected to cyclic load was one of the major developments in the nineteenth century. His work resulted in the characterisation of fatigue behaviour in terms of stress amplitude–life (S–N) curves. The aerospace industry started to fully appreciate the role of fatigue in structural integrity of airframe components only in 1940s which then resulted in the first damage-tolerant designs such as the first commercial jet aircraft *Comet*, manufactured by the *de Havilland Aircraft Company*. Significant drawbacks however resulted from serious accidents with these aircraft, which were attributed to fatigue cracking developed by stress elevation at rivet holes near passenger window openings due to a change of cabin pressurisation upon take-off and landing. A number of research studies, which followed these accidents, significantly contributed to design and safety of aerospace structures. The aerospace industries were taught the second major lesson of fatigue behaviour in another accident involving the *Boeing 737-200* aircraft operated by *Aloha Airlines* on Hawaii in 1988. The multisite fatigue cracking in the upper crown skin of the fuselage which resulted from the aircraft's age (80 029 flights) combined with operation mainly on runways next to the sea and thus corrosive atmosphere, led to the explosive cabin decompression. Mysterious piloting skills allowed the aircraft still to be landed and the short flight time with passengers having all seat belts on luckily reduced the number of fatalities to one. It has been this accident which initiated the major ageing aircraft initiative on the civil aviation side and which is a still ongoing initiative ever since. Military aircraft have also suffered a number of fatigue-caused accidents. Examples include crashes of *Vulcan, Harrier, Gnat, Buccaneer* aircraft in the UK and *B-47, F-111, F-15* aircraft in USA, as reported in literature. Due to the generally higher age of military aircraft and their frequent operation in corrosive environment (e.g. navy) the ageing aircraft discussion already started in the USA during the mid-1970s through implementation of the Aircraft Structural Integrity Programme (ASIP).

Statistical analysis of various types of structural damage can identify and improve fatigue-critical areas in structures. A highly important source of information in aerospace engineering is the so-called Major Airframe Fatigue Test (MAFT). As explained in more detail in Section 2.5, this test is required to be performed when a new type of aircraft is designed and realised in hardware for a first flight followed by a tear-down analysis after completion of the test. In some cases a MAFT is even performed along a mid-life update, where a used aircraft is put into a rig to allow for determining the residual life as well as damage critical locations, which may specifically result from a change in the operational

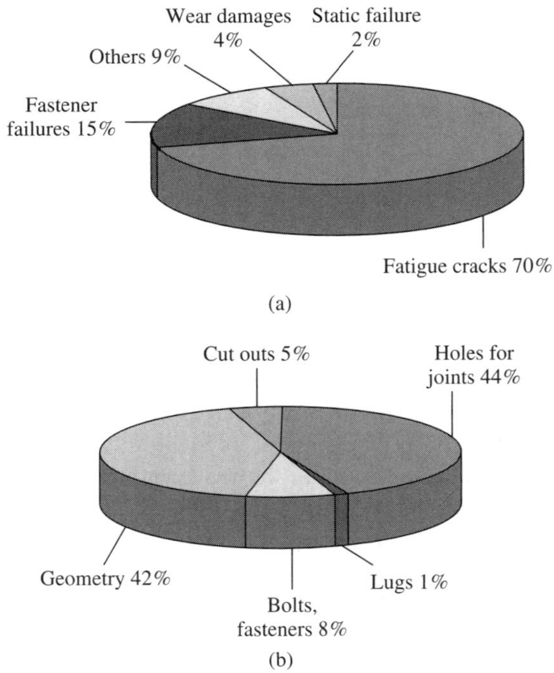

Figure 2.2 Structural damage after MAFT for *TORNADO* aircraft: (a) types of structural damage; (b) types of fatigue cracks (Boller 2001)

loads sequence. An example of damage statistics related to the *TORNADO* aircraft fighter along the MAFT in the design phase is given in Figure 2.2. Various types of damage found in the MAFT tests are classified in Figure 2.2a as: fatigue structural cracks, failures of fasteners (bolts, rivets, bonds, etc.), wear, static failures (as results of accompanying static tests) and other types of failure (e.g. failures of auxiliary structures as load introductions). The fatigue cracks reported in the MAFT are mostly due to geometry (e.g. notches), cut outs (e.g. open holes), holes for joints, bolts, fasteners and lugs, as shown in Figure 2.2b. Similar statistics can be observed for civil aircraft structures. Figure 2.3 gives an example of fatigue damage distribution and locations of fatigue cracks after in-service inspection where 714 cracks were identified in a fleet of 61 *Boeing 747* aircraft in Japan over a period of nearly three years (Asada *et al.* 1998). It appears that notch geometry and holes for joints/fasteners are the major source of fatigue cracking. However, this type of damage has been significantly reduced in absolute numbers over the past decades. The MAFT comparative results for the *Boeing 767* and *777* show a reduction of approximately 60 % (Goranson 1997).

Corrosion is another important type of aircraft structural damage. There are three major categories of corrosion in airframes. These are: time dependent, time independent and time related, as summarised in Table 2.1. Various factors, such as the environment, protective treatments and the inherent capacity of materials, contribute to the aircraft structural susceptibility to corrosion. The major problems are usually related to water intrusion into dry cavity areas or structural joints from exterior surfaces. This can result from: a breakdown of the sealant and interface layer that protects the mating surface, lack of

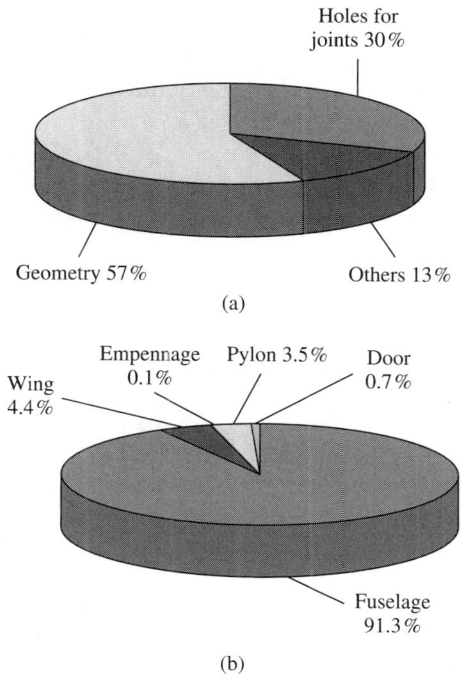

Figure 2.3 Structural damage after in-service inspection for civil aircraft (Boller 2001)

Table 2.1 Corrosion categories in ageing airframes. (Originally published in the corrosion of aging aircraft and its consequences, J. DeLucia, AIAA-91-0953-CP, Copyright © 1991 by the American Institute of Aeronautics and Astronautics, Inc. Reprinted with permission.)

Time dependent	Time related	Time independent
• General attack • Pitting • Exfoliation • Crevice corrosion • Filiform corrosion • Intergranular	• Corrosion fatigue	• Stress corrosion cracking • Environmental embrittlement • Hydrogen • Liquid metal

adequate draining/ventilation, inappropriate selection of protective coatings, contaminated fuels and dissimilar metal couples. Corrosion initiation points are very difficult to detect. Also, the severity of corrosion increases nonlinearly with the age of an aircraft.

The information gathered from MAFTs and tear-down inspections of aircraft is important for the risk analysis and prediction of the probability of fracture. The major instrumental initiatives, which compile this information, include the Aircraft Structural Integrity Programme (ASIP) and the Supplemental Structural Inspection Programme (SSIP). ASIP has been running for more than 25 years in the US Air Force. It summarises all the different activities performed with regard to experimentation, simulation and standardisation with regard to structural integrity. One of its outcomes has been a simulation environment for estimating the probability of fracture PROF (Barens *et al.* 1991) of which the schematic

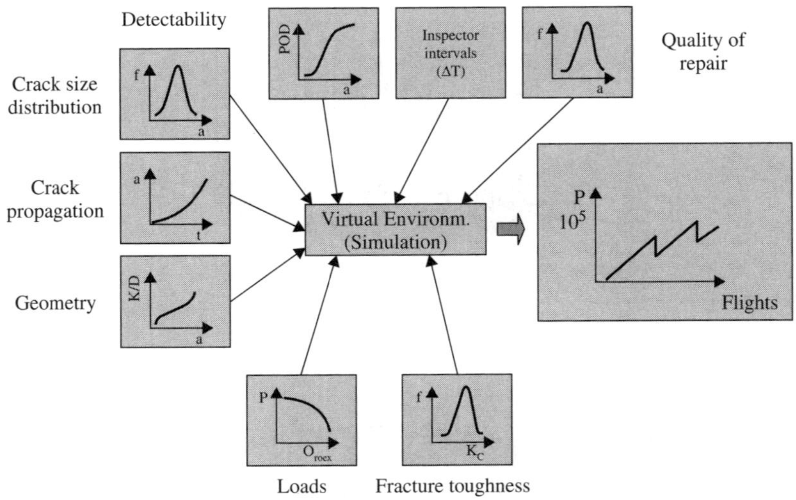

Figure 2.4 Probability of fracture – schematic diagram of the PROF code developed within ASIP (Barens *et al.* 1991)

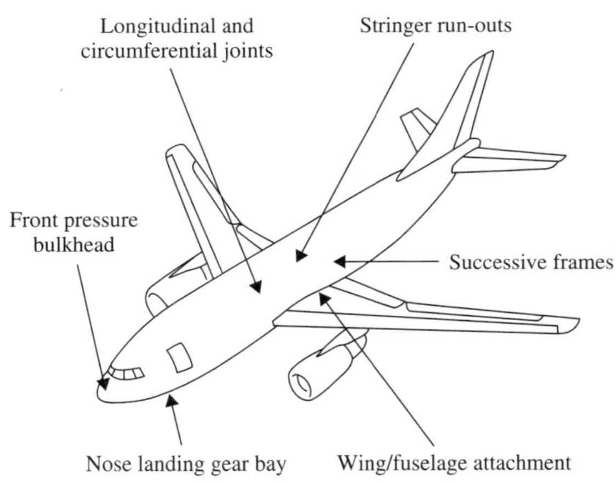

Figure 2.5 Widespread fatigue damage (WFD) areas identified for *AIRBUS A-300* aircraft (Brand and Boller)

is shown in Figure 2.4. Here various statistical parameters such as related to loads, crack detection (size, propagation, geometry), quality of repair, inspection intervals, probability of damage (POD), and fracture toughness are used to establish the overall probability of fracture. An example of the SSIP analysis is given in Figure 2.5, which shows structural areas being prone to fatigue damage in the early *Airbus A300*.

A significant amount of experience in the area of fatigue behaviour of aircraft structures has been gathered over the past years. This relates particularly to Widespread Fatigue Damage (WFD). As a result, aircraft have achieved the adequate levels of safety and reliability even in their old days. However, aircraft structures are designed for a specific period of lifetime and it is specifically the high amount of expensive maintenance required towards the end of the operational life which sets a natural end to all operational usage of the aircraft structure.

2.3 AGEING AIRCRAFT PROBLEM

It appears that a significant number of civil and military aircraft have exceeded their design lives. As mentioned before the resulting ageing aircraft problem has been specifically discussed since the *Aloha Boeing 737-200* accident. The statistical data given in Table 2.2 show that the number of aircraft older than 15 years is remarkable. It has increased from 4600 in 1997 to 4730 in 1999 for US and European built civil aircraft. Similarly, the number of civil aircraft older than 25 years has risen from 1900 in 1997 to 2130 in 1999. Nearly half of the entire *DC-8* fleet is still in operation. As illustrated in Table 2.3, the problem with military aircraft structures is even more serious. An increasing number of military aircraft (e.g. *F4, T-38, MiG-21*) exceed the age of 40 years. Ongoing mid-life updates of fighter airplanes show that the service life of 50 years and more is not exceptional. The $B-52$ aircraft, which needs to be retained for a few more decades, is one of the best examples. The *Boeing KC-135* in-flight tanker is intended to be kept in service until 2035 (Brand and Boller). In 1993, approximately 51 % of the aircraft in the US Air Force inventory were over 15 years and 44 % were over 20 years old. In 2000, over 75 % of US Air Force aircraft were more than 25 years old (Penney 2000).

The end of the cold war and the 'September 11' terrorist attack have contributed to many airplanes being retired. Almost 4600 military aircraft are held only in one place on the 6400 ha of desert in Arizona (Scott 2001). It is likely that many of these aircraft structures will be used in the future as many air forces shift away from new aircraft

Table 2.2 Ageing civil aircraft overview (Staszewski and Boller 2002)

Aircraft type	Total delivered	Fleet in service 09/01	Ageing aircraft in 1999		
			≥15 Years	≥20 Years	≥25 Years
A300	503	411 (82 %)	220 (46 %)	60 (12 %)	1 (0,2 %)
A310	255	218 (85 %)	54 (21 %)	–	–
707/720	1009	379 (37 %)	–	–	–
727	1831	1247 (68 %)	1381 (75 %)	1127 (62 %)	673 (38 %)
737–100/200	1144	901 (79 %)	853 (75 %)	442 (39 %)	222 (19 %)
737 CFMI	1988	1971 (99 %)	13 (0,7 %)	–	–
747–100/SP/200/300	724	562 (78 %)	490 (68 %)	317 (44 %)	154 (21 %)
757	968	943 (97 %)	51 (6 %)	–	–
767	840	820 (98 %)	109 (14 %)	–	–
DC-8	556	243 (44 %)	268 (48 %)	268 (48 %)	268 (48 %)
DC-9	976	727 (74 %)	776 (79 %)	739 (75 %)	588 (61 %)
DC-10	446	397 (89 %)	333 (75 %)	276 (62 %)	162 (36 %)
L-1011	249	155 (62 %)	185 (74 %)	113 (45 %)	60 (24 %)
Total	11489	8974	4733 (46 %)	3342 (33 %)	2128 (21 %)

Table 2.3 Ageing military aircraft overview (Staszewski and Boller 2002)

Aircraft Type	First built	Total in service 09/2001	Life extension until
BAE Hawk/Boeing T-45 Goshawk	1976	641	
Boeing F/A-18 Hornet/Super Hornet	1978	1762	2019
Boeing B-52H Stratofortress	mid 50ies	94	2045
Boeing 707/C-137/C-18/KE-3	60ies	97	2030
McDonnel Douglas F-4 Phantom	1958	889	
Dassault/Dornier Alpha Jet	1973	289	
Lockheed-Martin F-16	1974	3398	
Northrop T-38	1959	685	2040
Panavia Tornado	1974	746	2018
MiG-21 (incl. licences)	mid 50ies	3324	
L-39/L-39 Albatros/L-159	mid 70ies	2233	

and focus on through-life upgraded aircraft structures. Similar trends can be observed in converting older civil aircraft series such as the *Airbus A300* and *A*310 or the *Boeing 747, 757* and *767* to freighter aircraft. These upgrades and conversions are often also related to a change in the loading conditions of the aircraft structure, which by itself then also requires structural modifications. Upgraded fighters are possibly equipped with more powerful engines, new weapon loads, communication and electronic systems. These aircraft usually fly different envelopes and manoeuvres compared to those for which they were initially designed for. Additionally, all ageing aircraft suffer more from corrosion and cracking damage. In summary, the longer the operational life of an aircraft, the more likely the design load sequence may not meet the overall life requirements.

There are two major challenges associated with operating ageing aircraft: (1) keeping reliability standards of the aircraft structure along the extended operational life; and (2) controlling maintenance cost to the acceptable minimum. Means for extending the structural life of aircraft may include the introduction of compressive residual stresses that counteract to the stresses resulting from applied loading, or repair. Techniques for introducing residual stresses includes shot peening, laser shock peening and hole cold-working (Grandt 2000; I.Mech.E. 2000). Repair procedures can be performed using either component replacements or techniques utilising repair patches, welding and stop-hole-drilling (Grandt 2000; I.Mech.E. 2000). All these methods can significantly extend the fatigue life of structures. Nevertheless, the maintenance problem still exists and the ageing aircraft fleet leads to increased operation and maintenance cost that significantly contribute to the overall lifecycle cost of aircraft structures.

2.4 LIFECYCLE COST OF AEROSPACE STRUCTURES

The increasing number and age of aircraft and the desire of a high rate of military aircraft operational availability pose a big challenge to aircraft operators. One possible solution to this problem is to increase the effort for inspection and thus monitoring. However, this is inevitably associated with increased operational cost. The analysis of cost related to maintenance and new monitoring technologies has become essential for aircraft manufacturers and operators. This section shows how the cost associated with current and future inspection/monitoring technologies can be estimated. The cost benefit of new potential techniques will be also illustrated.

2.4.1 Background

A number of consecutive stages or phases can be considered in the life of all structures. This entire period of time is known as the life-time of structures. It includes the following stages (Blanchard 1978):

- *Research and Development (R&D)* – initial planning, market-analysis, feasibility studies, research, software development, documentation, project management, etc.
- *Production and Construction* – material acquisition, industrial engineering, manufacturing, process development, quality control, initial logistic support, deployment, etc.
- *Operation and Support (O&S)* – maintenance, repair, storage, transportation and handling, system modifications, etc.
- *Retirement and Disposal* – system retirement, disassembly, recycling, disposal of non-repairable elements, etc.

The total cost of these stages, or in other words the total costs of the structure's life, is known as the Lifecycle Costs (LCC). For example, the LCC of a typical *Boeing 747* aircraft is much higher than the purchase cost, which is up to 220 million US dollars. The LCC of defence systems is often called the Total Ownership Costs (TOC). The overall analysis of costs associated with the LCC is known as the Lifecycle Costs Analysis (LCCA).

The target of LCC analysis is the development of a cost profile that models the cost distribution over the complete lifecycle of a product as detailed as possible. The method for building the LCC models can be divided into six major steps (Blanchard 1978):

1. Identification of all activities contributing to cost within the structure's life cycle.
2. Assignment of activities identified under step 1 in a cost breakdown structure, which includes all possible variants of different structural prototypes, manufacturing processes and maintenance concepts.
3. Calculations of cost estimation relationships, which involves either self-developed procedures or parametric cost estimation models and tools, as presented in (Lockheed Martin).
4. Decision upon the reference date, which is either the current or future (e.g. the day of disposal) LCC value. The conversion for any given time can be obtained from

$$x(t_i) = \left(1 + \frac{y}{100}\right)^{t_j - t_i} x(t_j) \tag{2.1}$$

where x, y, t_i and t_j is the cost, percentage interests rate per annum and two different points of time, respectively.

5. Introduction of learning curves, to account for technological improvements over time.
6. Summary of different cost profiles within the cost breakdown structure.

The health and usage monitoring system is purely related to aircraft service and should be included in the O&S cost element. The O&S cost in the military aircraft environment can be determined following a methodology described in (MIL-PRF). This methodology requires information related to maintenance planning, repair analysis, support and test equipment, supply support and finally manpower, personnel and training. A number of software tools are available for the O&S cost assessment (OBS 1993). A simplified example procedure is illustrated in Figure 2.6. However, the analysis here does not give any details related to inspection cost.

2.4.2 Example

Following a case study presented in (Brand and Boller), an example showing how to estimate cost of a new damage inspection/monitoring procedure is given. This pragmatic analysis is based on a limited period of time (about ten years) and a small amount of maintenance data (average numbers only). Nevertheless, the information can be converted into a simple cost model.

The visual inspection effort of *TORNADO* airframe metallic components is considered in the first step. Figure 2.7 shows that this effort is related mainly to the aircraft fuselage and only very little is associated with checking for corrosion. Furthermore, the main part of the airframe inspection effort is due to visual inspection (61 %), followed by unplanned (31 %) and planned (8 %) nondestructive testing procedures, as illustrated in Figure 2.8.

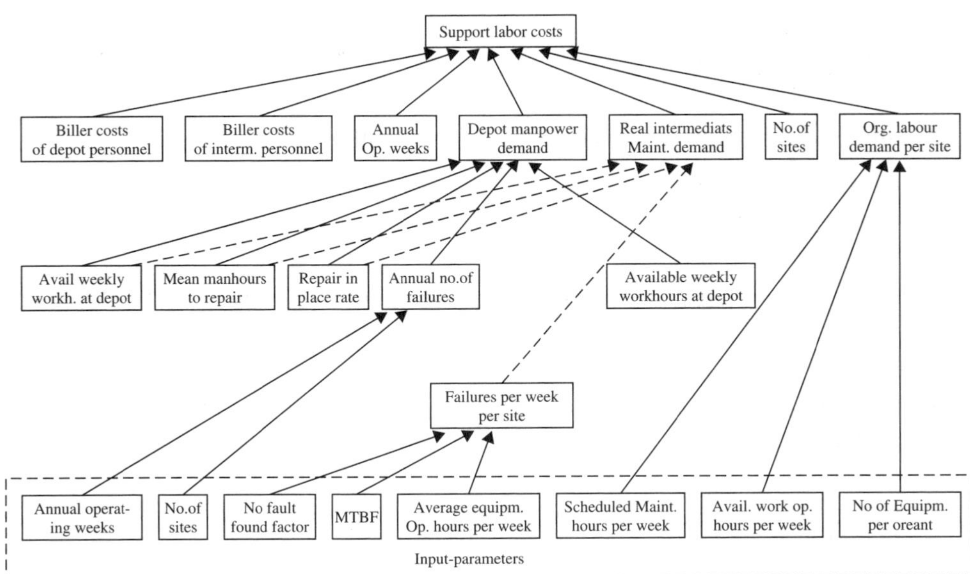

Figure 2.6 Simplified calculation procedure for the maintenance costs (Brand and Boller)

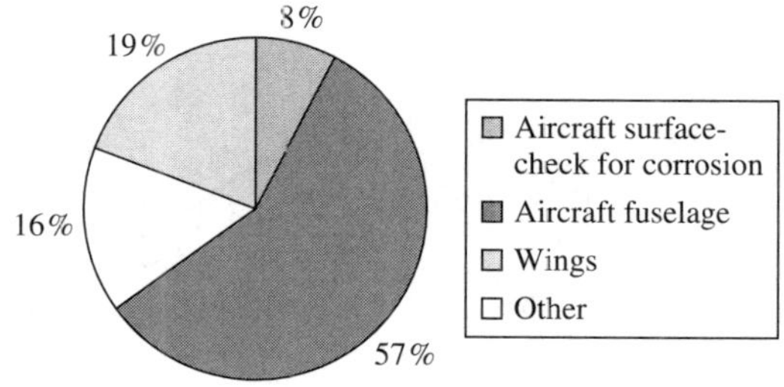

Figure 2.7 Statistical distribution of airframe visual inspection effort (Brand and Boller)

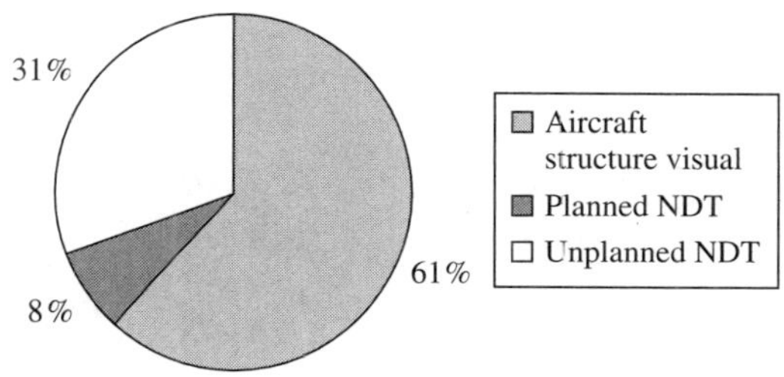

Figure 2.8 Statistical distribution of airframe NDT inspection effort (Brand and Boller)

To get a clearer picture regarding the effort required for the inspection of individual parts in an aircraft, six different components have been selected, which are shown in Figure 2.9. These include two types of fittings, two types of covers, a tail section skin and a taileron. The available inspection data for these components were: (a) average inspection time; (b) inspection frequency; (c) mean-time between failure (MTBF); (d) damage type; and (e) average repair effort. All inspections were defined as *minor, periodic* or *depot* and were performed before and after flights. From the available data given in Figure 2.10, a 50/50 percentage split of the total depot maintenance cost between inspection and repair can be roughly assumed as a first guideline. The cost estimation relationships were tried to be estimated for different inspection operations on the basis of the very limited data available, i.e. for dismantling/assembly, visual inspection against corrosion, visual inspection against loosening and nondestructive inspection against ruptures/corrosion respectively. More details about these calculations can be found in (Brand and Boller). Figure 2.11 gives two examples for two operations, namely assembly and visual inspection respectively. The analysis has finally led to the comparison between the predicted and actual inspection-related LCC costs given in Figure 2.12. With the exception of *cover 1* and *fitting 1*, where the difference of up to 80 % can be observed at, however, relatively low percentages for inspection in general, the predictions are fairly acceptable.

A similar analysis can be performed for composite components. However, the inspection and/or maintenance data in this area is much more limited. Although most of these components were designed to be maintenance free, inspections are still performed. This inspection effort however mainly results from the relative novelty of the material and its use in partially high performance areas. To obtain a feeling on what amount of effort

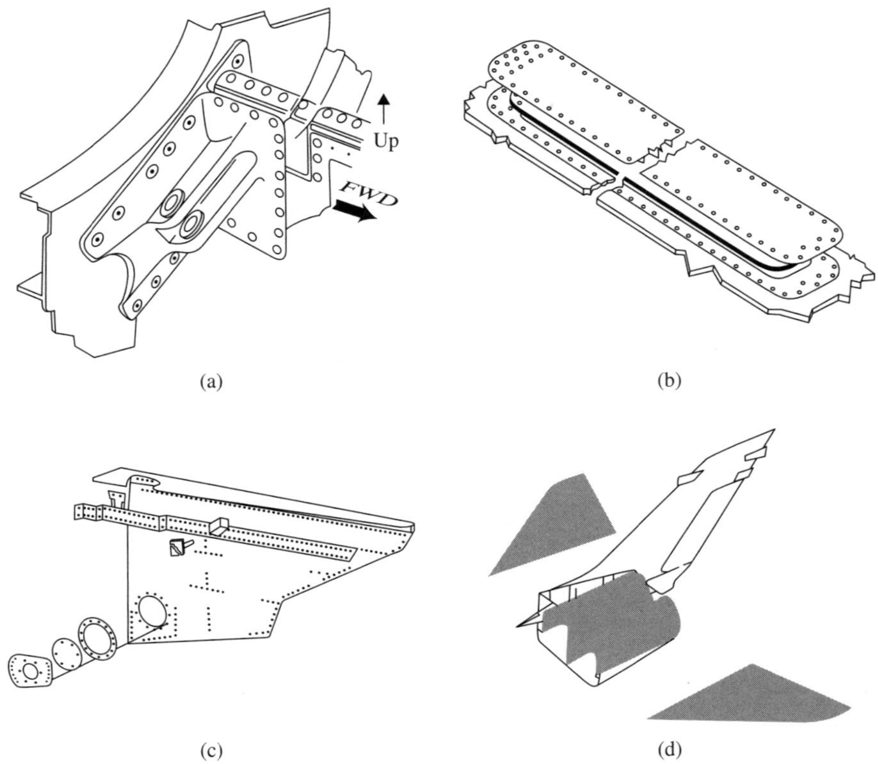

Figure 2.9 *TORNADO* airframe components used for cost analysis: (a) main landing gear fitting; (b) cover; (c) tail section skin; (d) taileron (Brand and Boller)

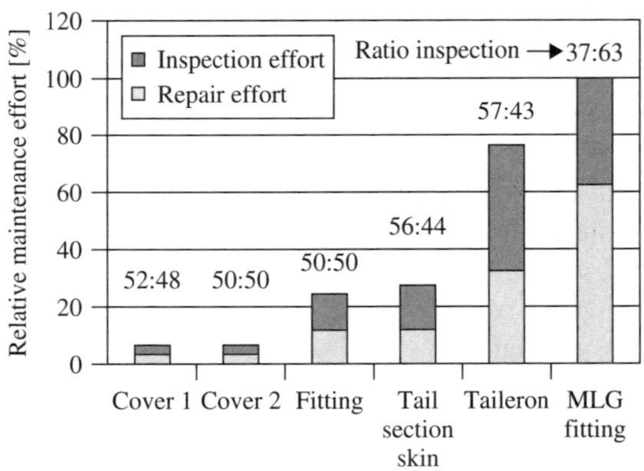

Figure 2.10 Statistical distribution of depot maintenance effort for metal parts (Brand and Boller)

may be required to inspect an aircraft which per se may not be inspected due to its safe-life design, data can be taken for safe-life built versions of the *Boeing 707*, which are given in Figure 2.13. This figure shows that, after 20 years, the inspection effort for these components has not vanished but has at least been reduced significantly following an

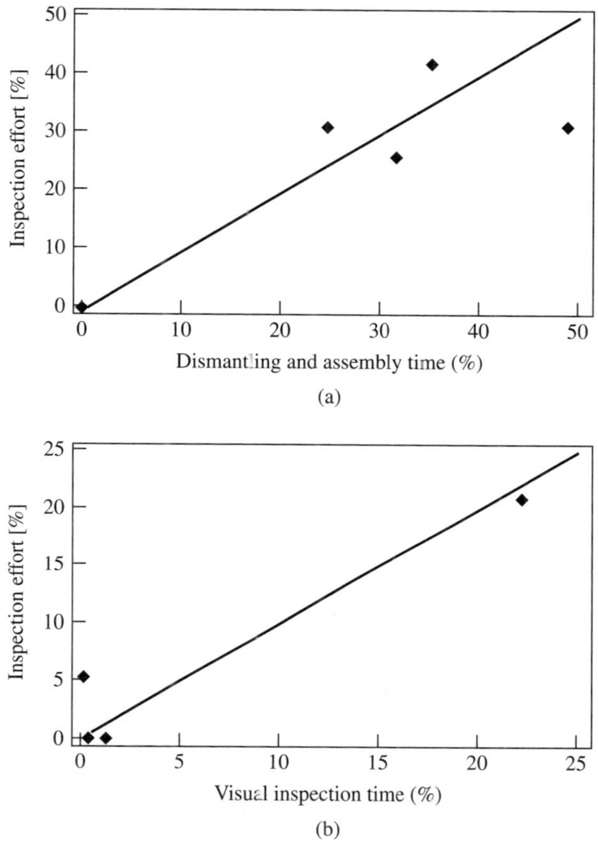

Figure 2.11 Cost estimation relationship (CER) for: (a) assembly effort; (b) visual inspection effort (Brand and Boller)

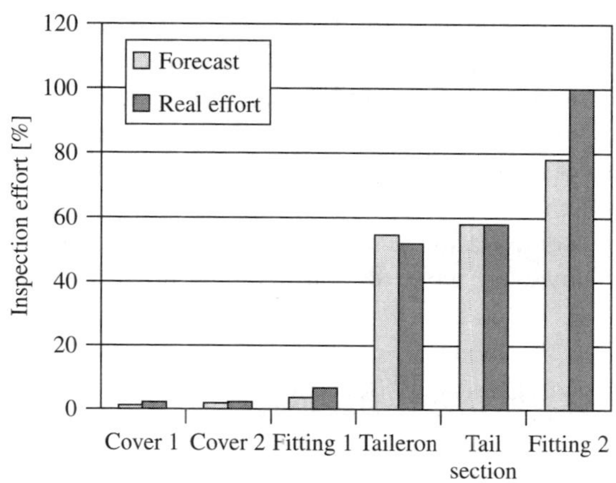

Figure 2.12 Validation of the cost estimation relationship model (Brand and Boller)

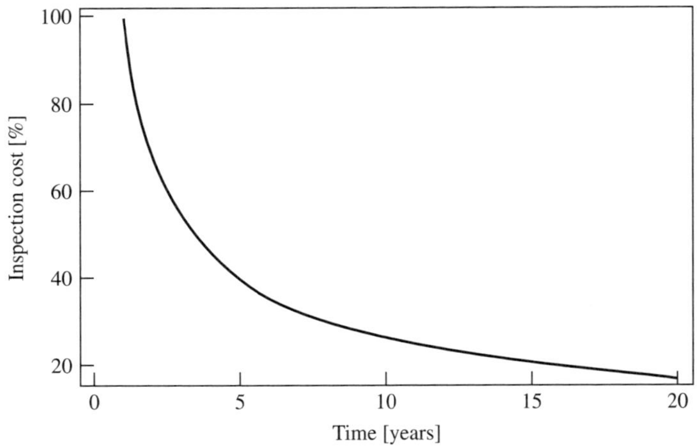

Figure 2.13 Inspection cost model for *Boeing 707* aircraft (Brand and Boller)

exponential function and thus a learning curve respectively. Clearly, similar experience can therefore be concluded for composite materials.

Estimation of LCC regarding new inspection/monitoring technologies is not an easy task. One of the reasons for this is that some of these monitoring technologies have not yet passed the R&D stage. Nevertheless, a similar analysis has been performed in (Brand and Boller) assuming a scenario that a serial monitoring system based on Lamb waves (see Chapter 4 for more details related to the method) and utilising the *Smart Layer®* technology (Chang 1998) is available. The results of this are presented in Figure 2.14, where the total length of each bar represents the inspection cost normalised on the inspection cost of the main landing gear fitting. Subtracting the LCC of the *Smart Layer®* from this inspection cost, which is the upper part of each of the columns in Figure 2.14, leads to the remaining possible gain in inspection cost. The composite element analysed here was the *TORNADO* main landing gear door. The results in Figure 2.14 show that it is specifically the highly loaded and difficult to access components which may benefit from a new inspection/monitoring technology. However, certain types of metallic components being easily accessible may not be too much suitable for an integrated structural health monitoring system. It is therefore generally advisable to perform a cost analysis beforehand, which allows identifying which of the components of the engineering system to be monitored are the ones with a true positive LCC impact when integrating a structural health monitoring system.

Simple examples presented in this section illustrate the benefits of new inspection/monitoring technologies. However, more data are still required to improve the real costing models.

2.5 AIRCRAFT STRUCTURAL DESIGN

2.5.1 Background

Different approaches to fatigue lead to different design concepts. Aircraft design procedures have significantly contributed to two major engineering design concepts: the safe-life

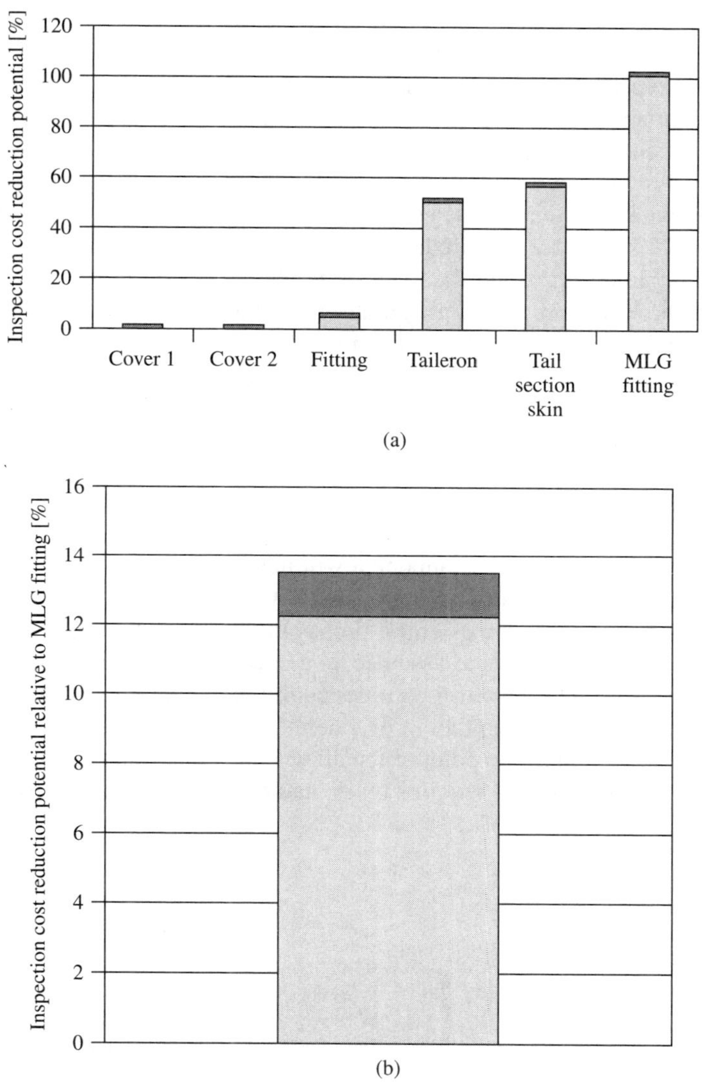

Figure 2.14 Potential life cycle costs savings through structural health monitoring systems for: (a) metallic components; (b) composite components (Brand and Boller)

and damage-tolerant design. Both concepts require the cyclic load spectra as an input. Generating the representative spectra needs a substantial amount of information, which in the case of an aircraft is related to: flight envelope including runways to be used, loads to be carried, aerodynamics, past experience, statistics and any sort of specific information to be provided by the operator. The load spectra are first determined and used to evaluate the service fatigue life of many critical structural components. Fatigue testing includes material or coupon tests, element, subcomponent and component tests and finally full-scale structural tests ending up with the Major Airframe Fatigue Test (MAFT) where the full-scale airframe structure is tested on the ground. The safe-life concept assumes that all components have no margin for a fatigue failure over a defined fatigue life, usually nominated in flight hours or flight cycles respectively. This safe life is therefore estimated on the basis of experimental results combined with imposed safety factors, which cover any type of uncertainties, such as resulting from scatter in loads, material condition, notch factors, surface conditions and much more. Fatigue design is, however, only a fraction of an aircraft structure's design process. Figure 2.15 shows the aircraft design process chain together with some specifically related questions and issues in a very condensed form, while Figure 2.16 shows the multitude of elements in an aircraft structure, that have an influence on fatigue design along the aircraft structure's design process chain.

The lack of knowledge in structural and thus material damaging behaviour has led safe-life designed components to be stressed at relatively low level and thus to be of higher weight. Further information on a material's and thus structure's damaging behaviour can therefore help to make the structure lighter weight designed. This situation has been provided by introducing the damage-tolerant design, which can be achieved by either including redundancy such that a component can fail without compromising the structure (fail-safe) or by taking advantage of fracture mechanics (slow crack growth). In the case of fail-safe, a crack may grow at any time up to a certain length where it will be stopped by a crack stopper or the component will have fractured and the loads transferred by that component will be transferred by some other component (multiple load path). In the case of slow crack growth crack propagation size and endurance is estimated on the basis of the load sequence assumed and fracture mechanics models as well as crack propagation data. This however assumes additionally that: (a) a location of potential damage is known; (b) damage initiation of a defined length can be reliably detected; and (c) damage propagation until the point of the allowable damage can be tolerated. Thus the damage-tolerant design concept requires by its nature already an inspection effort. A big

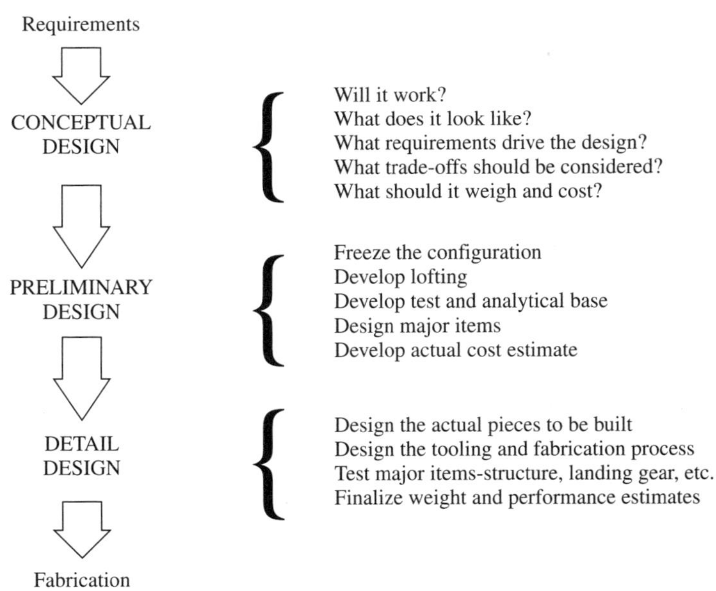

Figure 2.15 Aircraft design phases (Raymer 1992)

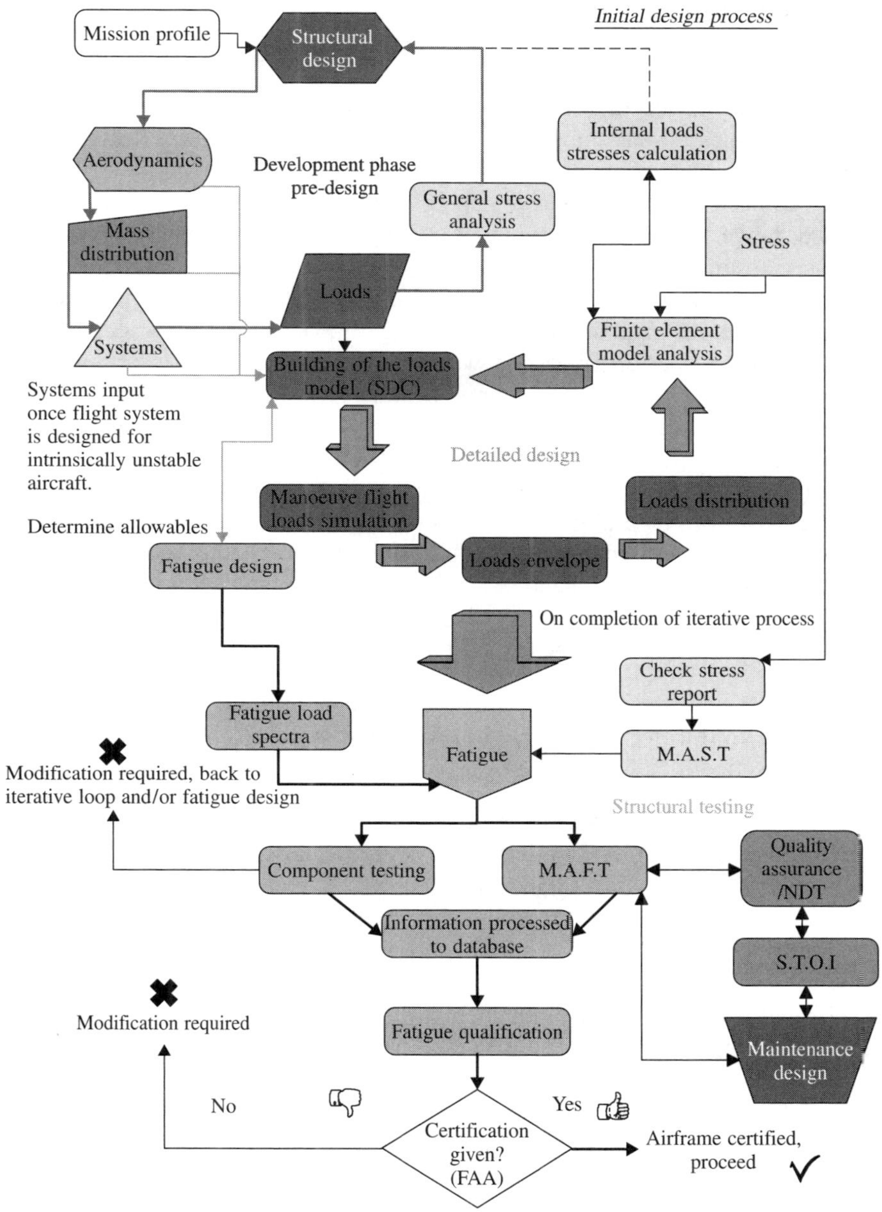

Figure 2.16 Initial aircraft design process (Banks 2000)

challenge in that context is therefore to use reliable damage detection techniques, which can not only fulfil the above requirements but also reduce related maintenance effort and cost. Finally, it has to be mentioned that inspection is required irrespective of the design principle once the design limits have been exceeded, which is the case with any sort of accidental damage and thus overloading. Principally it is however the damage-tolerant design which requires the much higher amount for inspection.

2.5.2 Aircraft Design Process

As already shown in Figure 2.15, the aircraft design process can be divided into three major phases being conceptual, preliminary and detail design respectively. This design process also includes all structural testing, which comprises material tests, fatigue tests of individual components and complete airframe tests. After knowing the task of the aircraft to be designed in general, a first step is to establish mission profiles (i.e. payload, performance, range, speed, manoeuvres, etc.). This is when the boundary conditions of structural aspects are discussed for the first time and where the aircraft is shaped according to aerodynamic requirements in a first approach. In a multidisciplinary process between aerodynamic, aeroelastic and strength performance the aerostructure is shaped and optimised with regard to performance and weight. The study takes into account different mass distributions and possible shape modifications. Loads and stresses are calculated and referred back to structural design. The loads are not limited to aerodynamic and mass loads resulting from flight manoeuvres and gusts only but also include environmental conditions (e.g. temperature, humidity, corrosion, etc.) hazardous conditions (e.g. bird strikes, impacts with foreign objects, lightning, etc.) and human error (overloads, impacts due to tool drops, ground collisions, etc.). Load sequences, amplitude levels, and frequency of occurrence can significantly vary in practice. Assumptions regarding scatter in loads can therefore be mainly based on utilising previous experience only. Further to this the following three measures can be taken to cope with unexpected detrimental overloads:

- reduce allowable stresses to a level where damage is not critical;
- compile as much fracture mechanics knowledge as possible to create design guidelines;
- increase inspection/monitoring effort.

The first measure is in fact nothing more than to apply the safe-life design concept. Increased structural weight, and perhaps also operational costs, is the price paid for the reduced allowable stresses. The knowledge gathered from coupon, component and major airframe fatigue testing, together with fracture mechanics development, being the second of the measures mentioned above, is nothing else than the damage-tolerant design concept. The third measure is, finally, structural health monitoring, which is the integration of sensing elements in the structure and which will be described in much more detail in the subsequent chapters.

Once the aircraft's weight is estimated, an appropriate propulsion system is initially selected. Further design phases incorporate more details, such as interior, following specifications from the system, and more detailed load and stress analysis. This is followed by a detailed design process, which is focused on loads. Initial load models are built, which allow simulating specific manoeuvres and are crucial for fatigue structural testing. The analysis leads to the loads envelope, which is further used to establish load distributions. This is the starting point for the Finite Element (FE) models. Revised FE models are produced subsequently using current FE models, aerodynamics data and mass distributions. The fatigue design process is performed simultaneously utilising various modifications. Updated load models are taken and allowable stresses are calculated from a defined design life. The detailed design process is carried out in an iterative loop, which allows obtaining more accurate allowable stresses. The FE analysis allows for internal stresses to be calculated. When an accurate conclusion is reached, the design process proceeds to structural testing.

After material testing, the first structural tests are carried out under static stress, which in the final stage is done in the Major Airframe Stress Test (MAST). Besides determining the ultimate strength and stability criteria, the results of these tests are also used to better design the fatigue tests, i.e. component testing and the MAFT. Fatigue tests are performed to specifically simulate the conditions that are expected to occur in real flights. The tests are usually carried beyond the design life (flying hours) which allows determining the safety factors (e.g. factor 2 in fatigue life) for military and civil structures, respectively. The MAFT results are used as reference information, should the aircraft encounter any structural problems. Finally, fatigue qualification is given to all components for a specific number of flight hours. The entire design process is completed by a certification from the respective authorities such as the Joint Aviation Authority (JAA) or the Federal Aviation Authority (FAA).

2.6 DAMAGE MONITORING SYSTEMS IN AIRCRAFT

Damage monitoring through a system inherent in an aircraft can be principally done in two ways. The one is already applied today, which is to monitor loads or better load sequences, which are then used to estimate the accumulated fatigue damage indirectly by means of analytical procedures. This is what is commonly also known as Operational Loads Monitoring (OLM). The other way is to integrate systems onto or into the structural component, which allows directly determining the occurrence, size and possibly even location of damage and which mainly works on an actuator–sensor basis but can also be reduced to a sensor basis only in the specific case of Acoustic Emission.

Various procedures developed and used in practice are briefly discussed in this section. The focus here is on general monitoring methodology with examples. Damage inspection technologies are then discussed in more detail in the next section.

2.6.1 Loads Monitoring

Loads can only be monitored through the parameters describing them. This may be either given globally by sensing accelerations or masses or any parameters again influencing accelerations or masses such as flap positions, speed, height, fuel flow or any other. It is what is also considered in the context of aircraft to be a flight parameters-based loads monitoring system. The other approach is more of a global nature through monitoring of strain sequences at discrete locations and then converting this to a more global load sequence of the structure considered. Both approaches have their justification and are thus described in the subsequent paragraphs.

2.6.1.1 Flight Parameters-Based Loads Monitoring

Application of this approach in aviation dates back to the 1950s, where the so-called *fatigue meter* was installed, which does nothing else than counting exceedances at different levels of vertical acceleration. The cumulative counts of the reached or exceeded values of vertical acceleration are recorded on-board and stored in a nonvolatile counter or more recently a memory/processor. Flight-parameters, such as speed, altitude, acceleration, flap positions, fuel content, etc., are parameters which have been added to this and may lead to loads monitoring for further enhancement. The use of sensors already built into the aircraft system and used for the OLM flight-parameter systems additionally is definitely advantageous because neither overall reliability nor complexity of the aircraft is negatively affected by additional sensors to be built in. Furthermore the existing sensors in the aircraft have already proven airworthiness and are thus widely accepted. However, the OLM systems based on flight-parameters are still very costly and their exactness does often not provide the precision required for analytical fatigue life evaluation, because loads need to be analytically and indirectly determined from the sensor information provided. Further reasons include a coarse differentiation between the different exceedance levels which do not allow differing between different loads in between, the differentiation between and superposition of different loading cases, the lack of transferring loads from the global to the local view or the missing information on the sequence of the loads applied. Statistically only 1 % of the flight data recorded is used for monitoring.

Figures 2.17 and 2.18 give an overview of the philosophy of the OLM system developed for the *Airbus A320* (Ladda and Meyer 1991) and the On-Board Life Monitoring System (OLMOS) used for the *Panavia TORNADO* (Bauer 1987; Krauss 1988) aircraft, respectively. Both systems are designed to perform on-board processing of special events (e.g. hard landing, exceedance) and flight-parameter data.

2.6.1.2 Strain Gauge Based Loads Monitoring

The partially unsatisfactory precision of loads determined through the flight parameters has led some aircraft manufacturers to using a loads monitoring system based on conventional strain gauges bonded at discrete well selected locations of the aircraft structure. Figure 2.19 gives examples for the *AMX* trainer aircraft where 10 to 20 strain gauges have been fixed. Strain sequences are monitored for the different strain gauges, stored on-board the aircraft in a data acquisition unit (DAU) and downloaded within specified intervals. Improvement in the strain gauge bonding process has increased confidence in this type of sensing. However, operational conditions for the strain gauges have to be checked beforehand to avoid facing operational difficulties.

2.6.2 Fatigue Monitoring

Fatigue monitoring is one of the consequences of loads monitoring when combining the latter with analytical fatigue life evaluation. Figure 2.20 gives an example philosophy of a military strain gauge based OLM system (Amabile and Giacobbe 1991). Here, strains from the sensors are converted to digital signals and stored in the DAU. The signals are then converted to stress histories. This information is used to obtain the load sequence for the given locations including load transfer functions where available and useful. The load sequences are analysed and synthesised using a rainflow cycle counting procedure. This evaluation procedure can either be performed in-flight or on-ground. The *Eurofighter Typhoon* aircraft is due to be equipped with the loads monitoring systems directly linked

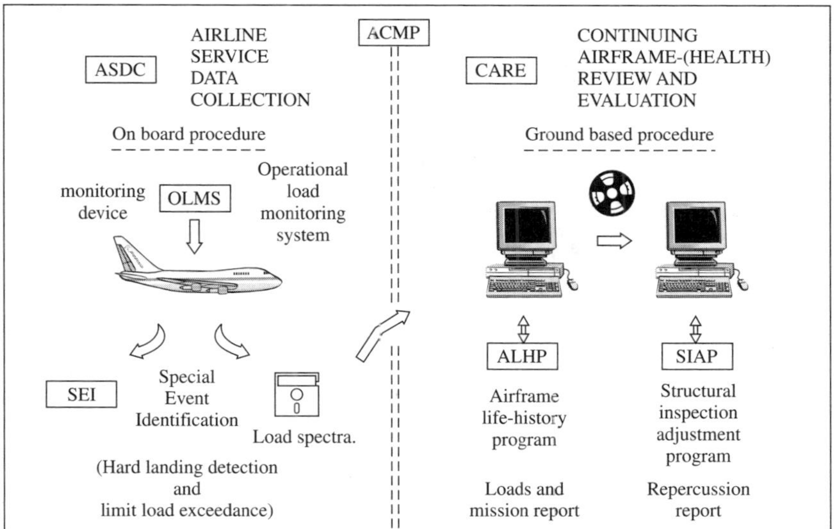

SOURCE	PARAMETER
ADIRU Air Data Inertia Reference Unit	ALT, MN, TAS, AOA, PTCR ROLR, LATG, VRTG
FMGC Flight Manag. Guld. Computer	GW, CG
PRESS. CONTR. Pressure Control System	PDC (cabin different pressure)
DMC Data Management Computer	RALT
FQIS Fuel Quantity Indication System	FUEL QUANTITIES WING (Inner and outer cell)
FCDC Flight Control Data Computer	STAB, AIL ELEV, RH SPL 1-6
SDAC System Data Analog Converter	RUDD
SFCC Slat Flap Control Computer	SLAT/FLAP
FWC Flight Warming Computer	FLIGHT PHASE
FDIU Flight Data Interface Unit	TIME, DATE, AC-TAIL NUMBER

Figure 2.17 On-board life monitoring system (OLMOS) using flight parameters for the *Airbus A-320* aircraft (Ladda and Meyer 1991. The original version of this material was first published by the Advisory Group for Aerospace Research and Development, North Atlantic Treaty Organisation (AGARD/NATO) in AGARD Conference Proceedings, CP506, "Fatigue Management" in December 1991 and later in Conference Proceedings, CP-531, "Smart Structures for Aircraft and Spacecraft" in April 1993)

to ground-based maintenance (Hunt and Hebden). In order to perform real-time fatigue calculations and estimations of the life consumed by the airframe, the system utilises the auxiliary data (e.g. flying log data, design/performance parameters), events monitoring (e.g. hard landing, exceedance) and loads monitoring (using either strain gauges or flight parameters). The entire philosophy is illustrated in Figure 2.21.

Using a fatigue – life (S–N) curve for the material and notch considered as well as a damage accumulation rule allows determining a Fatigue Index (FI), which is nothing more than the estimated accumulated damage. The extent of fatigue damage induced by m blocks of constant σ_i stress amplitudes can be estimated using the well-known Palmgren–Miner damage accumulation rule, which states that failure will occur when

$$\sum_{i=1}^{m} \frac{n_i}{N_{fi}} = 1 \tag{2.2}$$

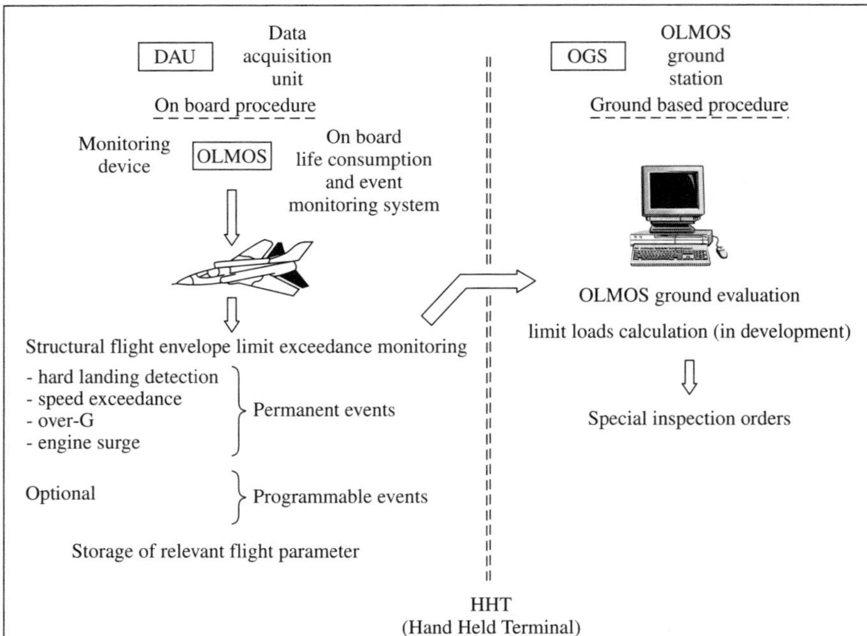

Summary of begleitparameter' permanent events
BANK, INCLINATION, HEADING,VX, VY, VZ, Ma, VCAS,
ALPHA, SPOILER, ROLLRATE, RUDDER, L/R, YAWRATE,
PITCHRATE, TAILERON L/R, FLAP/SALT, WINGSWEEP,
NZ, NY, MASS, EXTAST, HEADING RATE

Figure 2.18 On-board life monitoring system (OLMOS) using flight parameters for the *TORNADO Panavia* military aircraft

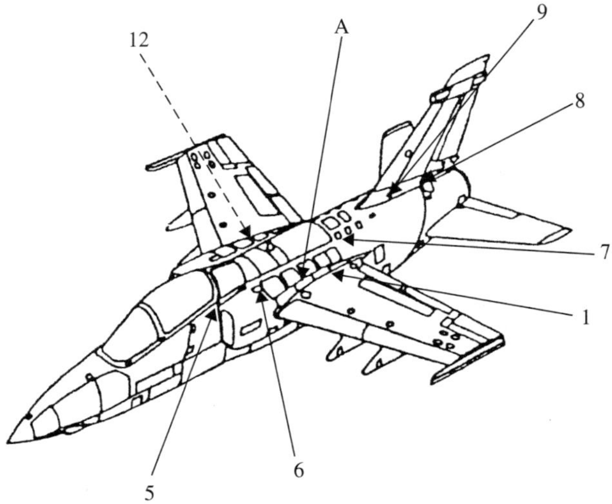

Figure 2.19 Strain gauge monitored locations on AMX trainer aircraft (Amabile and Giacobbe 1991)

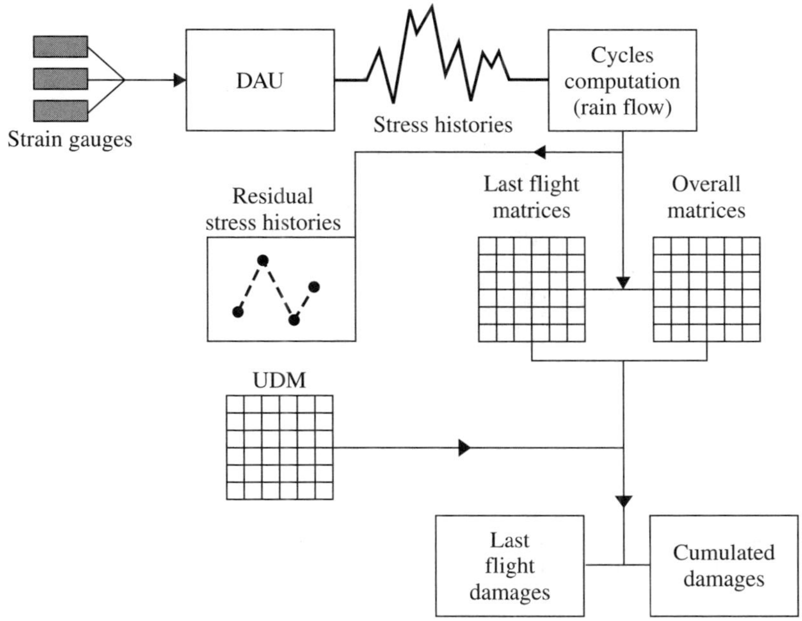

Figure 2.20 Loads monitoring system using strain gauges for *TORNADO* aircraft (Boller 1996)

where n_i is the number of fatigue cycles corresponding to each block of load and N_{fi} is the number of fatigue cycles to failure at the amplitude stress level σ_i. For each of the cycles at stress level σ_i, unit damage $1/N_{fi}$ is determined and then multiplied by n_i of the respective stress level σ_i and accumulated to obtain the FI. The fact that damage often does not accumulate linearly has raised a long-standing discussion over decades now and nothing seems to be more welcomed than additional information that describes the material's true nonlinear damage accumulation behaviour. This is therefore one of the areas where structural health monitoring using structure integrated devices comes into play.

2.6.3 Load Models

As to the descriptions given before, loads are not only essential to describe a structure's environmental condition but also to estimate its accumulated life as well as residual life. Load models, such as obtained through an extensive FE analysis, are therefore needed to design the component and structure considered in the way described also in more detail in Section 2.5. Load models could however also be used in practice in the context of the above-mentioned tasks. Based upon recordings of some strain-sequences at discrete locations on the structure, this strain information can be fed back into the loads model, which virtually allows estimating analytically stresses and strains as well as their time sequence at any location of the structure considered. However, these models are very expensive to obtain in the first place as well as their updates, specifically when structural modifications and/or mission changes have to be considered. This is why load models are still not often used directly for health monitoring so far. However, with the increasing

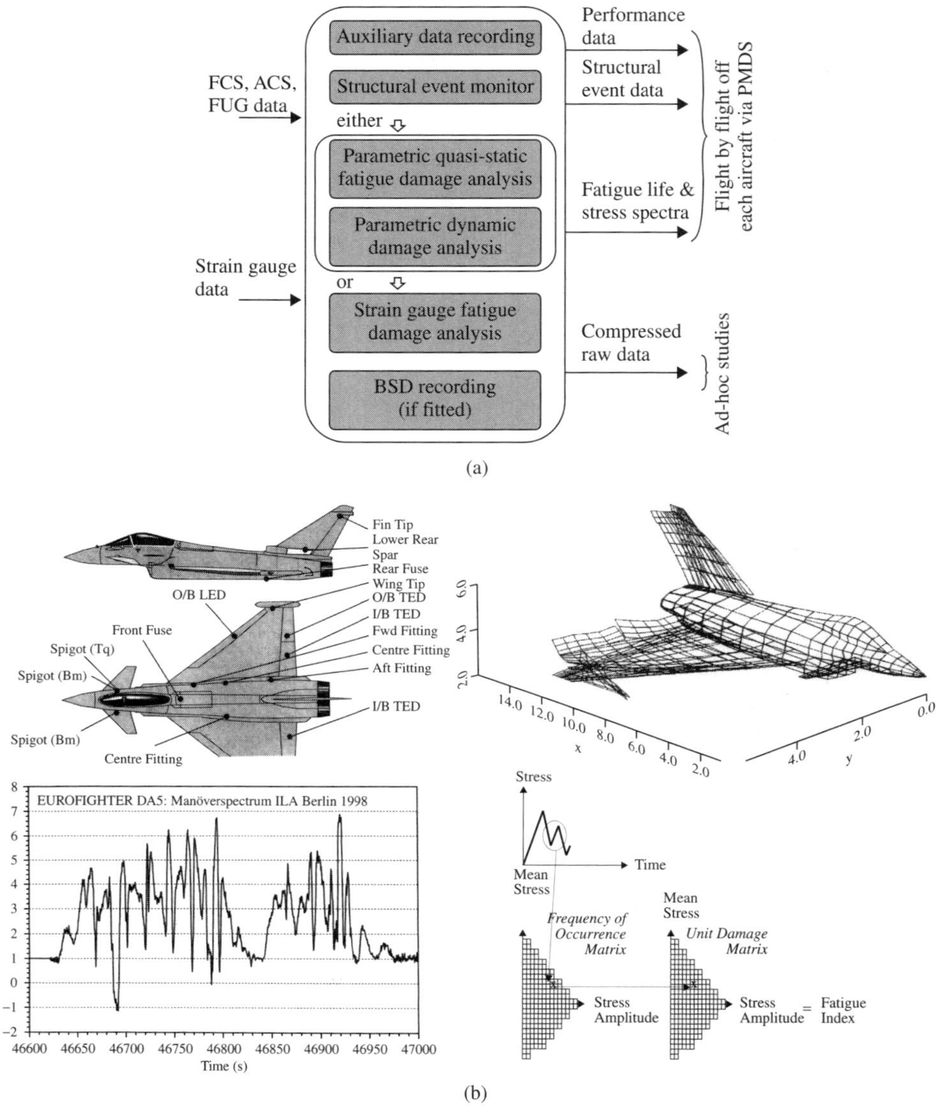

Figure 2.21 Health and usage monitoring system for the *Eurofighter* military aircraft: (a) schematic diagram (Hunt and Hebden); (b) loads monitoring logic (Boller 2001)

trend and ease of virtual simulation this approach is likely to become more appealing in the near future.

2.6.4 Disadvantages of Current Loads Monitoring Systems

Systems based on loads monitoring, commonly known as Operational Loads Monitoring (OLM) systems, are used in different military aircraft types so far only. Experience gathered in this field over the last 30 years leads to the conclusion that the vast amount of data collected needs a more efficient data management. This issue becomes specifically relevant when looking at the increasing average age of aircraft, which will significantly increase the amount of ageing aircraft obstacle data to be handled such that the fatigue life consumption and possible retirement time of individual aircraft can be better controlled. This procedure can result in significant cost savings.

OLM systems are not able to directly detect and monitor structural damage. The OLM system can only provide a major input for analytically determining when damage might occur. Being a well-known experience, the experimental and thus real fatigue life can often be two to three times higher than the analytically predicted (Boller 1996), which is mainly resulting from scatter in material properties but also due to some security factor which has to be included in the analytical calculation. This gap between predicted and real damage becomes even larger for composite materials where fatigue analysis and fracture mechanics of these materials is relatively less known when compared to metallic materials. Loads monitoring is only of limited use for detecting BVIDs as long as impact loads are not detected specifically. Loads monitoring does also not provide any information regarding environmental effects such as corrosion in metallic and humidity/temperature in composite materials as long as these environmental loads such as temperature, humidity or others are not recorded with appropriate sensing devices and the respective analytical procedures are not available. These aspects become increasingly important for the case of ageing aircraft and may be alleviated by advanced sensing such as it is currently proposed with the usage of micro-electro mechanical systems (MEMS) (Hautamaki *et al.* 1999; Matzkanin 2000).

2.6.5 Damage Monitoring and Inspections

It may have become obvious from the paragraphs above that OLM combined with analytical fatigue life evaluation is not sufficient to determine damage in a structure accordingly. Compared to reality there may still be a factor of two to three in fatigue life to be gained if damage could be monitored more adequately. Neither knowing a load sequence nor the status of damage will require the aircraft operator to inspect his aircraft every time he suspects a load exceeding the design spectrum may have occurred, specifically when his aircraft is designed safe-life. This unscheduled maintenance can become very costly in LCC terms, specifically when the aircraft has to be taken out of service and has to be dismantled for inspection to a remarkable extent. Aircraft damage inspection procedures recommend nondestructive techniques such as described in Section 2.7. This way of damage monitoring can become quite costly due to relatively short inspection times and labour intensive procedures. More than 70 million hours per year, equivalent to US$ 10.5 billion, is invested in civil aircraft maintenance. A typical *Boeing 747* aircraft is inspected every 12 to 17 months specifically for signs of fatigue damage. An air force or navy may often need about 6000 man-hours per aircraft and year for maintenance where a significant portion (easily around 50 %) may be devoted to inspection only. The overall cost of this inspection effort may go into the multi-billion US$ per year range for a single air force, such as the US Air Force or the Royal Air Force. The manpower effort required for inspecting an aircraft increases with the aircraft's life. An example from the military aircraft area shows that the number of man-hours per year required for inspecting and repairing one *EF-111A* fighter has increased from 2200 in 1985 to 8000 in 1996 (Sampath

1996). The weight of this effort also very much depends on the usage, complexity and size of the aircraft. Some comparative numbers show that the 1993 inspection cost for the $F-18$ fighter and $T-38$ trainer aircraft were 88.4 and 29 US$ per flight hour, respectively (Kudva *et al.* 1993).

2.7 NON-DESTRUCTIVE TESTING

Conventional aircraft damage inspection methods are based on either visual inspection or different nondestructive testing (NDT) methods. Most of the NDT techniques were developed in the early to mid-1960s. A number of theoretical models and simulation/analytical techniques were developed in 1970s. Recent developments in physics, electronics, computer technologies and signal processing have shown the application of a few new techniques and significant improvements in detection reliability of existing techniques. An excellent overview of various NDT techniques can be found in (Bar-Cohen 2000a, 2000b). The applicability of these techniques for aircraft damage detection is discussed in (Boller 1996). The NDT procedures used for aircraft damage monitoring are in fact one of the elements of the ASIP discussed in Section 2.2. It appears that eddy current and ultrasonic inspections are the most established techniques beside visual inspection within aircraft maintenance operators. In what follows, a summary of currently used and potentially applicable NDT techniques is given.

2.7.1 Visual Inspection

Visual inspection is the natural form of evaluating structural integrity of material components. The method is effective for detection of surface and near-surface damage. Visual inspection is the most common damage inspection technique applied in aircraft service. Several variants of this approach are used in practice. This includes various levels of sophistication from a simple examination by eye to the use of a static optical or scanning electron microscope. The eye alone can determine little detail about the damage mechanism or its severity. Visual inspection by the unassisted eye is limited in composite elements when damage occurs below the surface. While microscopy can provide detailed information on micro-cracks and crack initiation in metallic elements or delamination areas in composite elements, it can only be used in laboratory conditions since a section of the component considered must be removed from the aircraft structure. Recent development in the area of visual inspection utilises various illumination techniques that allow improving the inspection capability. This includes the use of retro-reflective screens and the scattered light from service deformations.

2.7.2 Ultrasonic Inspection

Ultrasonic inspection is based on various properties of ultrasonic waves propagating in monitored structures. Damage detection utilises wave attenuation, reflection, scattering, diffraction, harmonic generation, wave mode conversion and other physical phenomena. This includes the application of longitudinal, shear, surface waves and the so-called leaky

Lamb waves. Tests are conducted with either pulse-echo mode using one probe or pitch-catch mode using two probes. When one probe is used it acts as an actuator and sensor. When two probes are used, they are positioned or move in tandem on either one or two sides of the specimen's surface. One probe is then used as an actuator sending ultrasonic waves through the material whereas the other probe is used as a sensor collecting the transmitted acoustic waves on the opposite or possibly even the same side. Normal and angle ultrasonic beam inspections are possible for all probe configurations. Both types are defined by the angle of the induced ultrasonic waves.

The technique is often referred to as A-, B- and C-scans, where an A-scan refers to a single point measurement, a B-scan measures along a single line, and a C-scan is a collection of B-scans forming a surface contour plot. The C-scan has become common practice in industry specifically since the introduction of composite materials. Its results are widely understood and can be used to scan a large area of structure in a relatively short time period. Figure 2.22 gives an example of a typical A-scan plot. It describes a pulse going through the thickness of the component and the reflected pulses (peaks) recorded over time. Since the pulses are reflected every time they hit a border, a decaying sequence of pulses shows that the signal has been reflected different times. This is given as an example in Figure 2.22 for the case of a reflection due to a flaw inside and the backwall respectively.

Traditional ultrasonic inspection approaches utilise different types of gel couplants between ultrasonic probes and monitored specimens. Often probes and monitored specimens are immersed in water, which acts as a couplant. Recent development in this area includes noncontact techniques, which use air as a couplant. Various types of Electromagnetic Acoustic Transducers (EMATs), capacitance transducers and lasers are used in practice. Laser induced ultrasonic inspection is particularly attractive and allows for automated scanning of monitored specimens. Ultrasonic waves are induced to the material through the thermoelastic expansion caused by a series of short-time laser impulses. Often high-power Nd:YAG types of lasers are used for signal generation. Reflected signals are sensed using different types of laser interferometry (e.g. Fabry–Perot and Mach–Zehnder interferometers). The major difficulty with air-coupled probes is the acoustic impedance

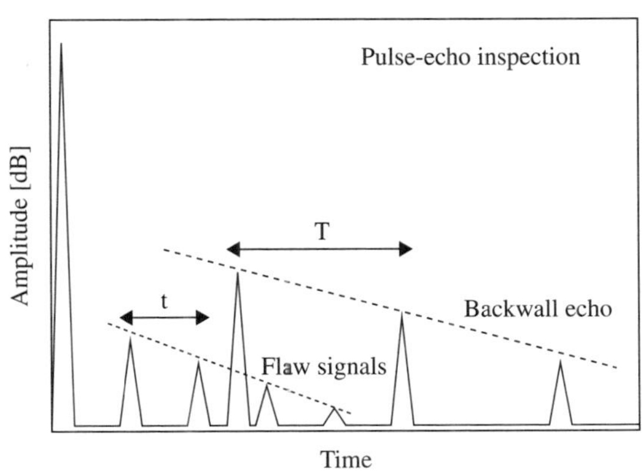

Figure 2.22 Ultrasonic inspection – schematic diagram of an A-scan

mismatch between air and the material of the specimen tested. Ultrasonic wave attenuation in air is also a significant problem. The sensitivity of the air-coupled ultrasonic inspection is therefore relatively poor when compared to traditional coupling techniques.

Altogether, ultrasonic inspection methods are highly sensitive to small surface and deep flaws in the material. Difficulties with coupling and scanning requirements are the major problems with the method. The scanning time of ultrasonic C-scan inspection is quite significant. Additionally, the geometric sensitivity of the method often requires repeated scans with different probe orientations. The size and cost of the ultrasonic inspection equipment is also one of the limiting factors. There is also the problem that access is required to both sides of the structure, so parts must often be disassembled for testing.

2.7.3 Eddy Current

The eddy current technique is another very valuable monitoring technology used in aerospace maintenance. This method is the third most commonly used for in-service aircraft inspection next to visual and ultrasonic inspections. Eddy current methods function by detecting changes in electromagnetic impedance due to strain in the material. A probe, which is in fact a coil, is excited with sinusoidal alternating current to induce closed loops of current in the material to be monitored. These closed loops, called eddy currents, are distorted due to material defects. Research work in this area, performed successfully in a number of academic and industrial laboratories, has materialised in various monitoring equipment (e.g. *MIT, JENTEK Sensors Inc.* and *General Dynamics*). Figure 2.23 explains graphically the principle of the eddy current method and gives an example of the typical eddy current aircraft inspection procedure and results. The method is suitable to sense strains and cracks in short specimens and around holes with conformable sensors. Riveted lap joints are often inspected in aircraft using the eddy current technique. Detection of corrosion and erosion with the eddy current method is also possible due to the method's ability to measure the thickness of the material specimens. However, the technique is not as mature for composite materials as it is for metals.

Recent development in this area includes pulse eddy current techniques, which utilise multiple frequency and sinusoidal excitation, and thus allows increasing sensitivity and reducing effects not related to damage. Eddy current C-scans (Lepine *et al.* 1998) are also possible but tend to be very time consuming.

Eddy current methods are often used because they are simple to implement and do not require expensive equipment. However, their disadvantage is that they require a large amount of power and that the data they produce are among the most complicated to interpret which finally makes detection of damage difficult. The technique requires extensive calibration before any characterisation of defects can be done. Despite major technological improvement in recent years, well-qualified and experienced technical staff are often still required to perform the tests.

2.7.4 Acoustic Emission

All solid materials have a certain level of elasticity and plasticity before they finally fracture. The application of external forces can exceed this level and finally result in fracture

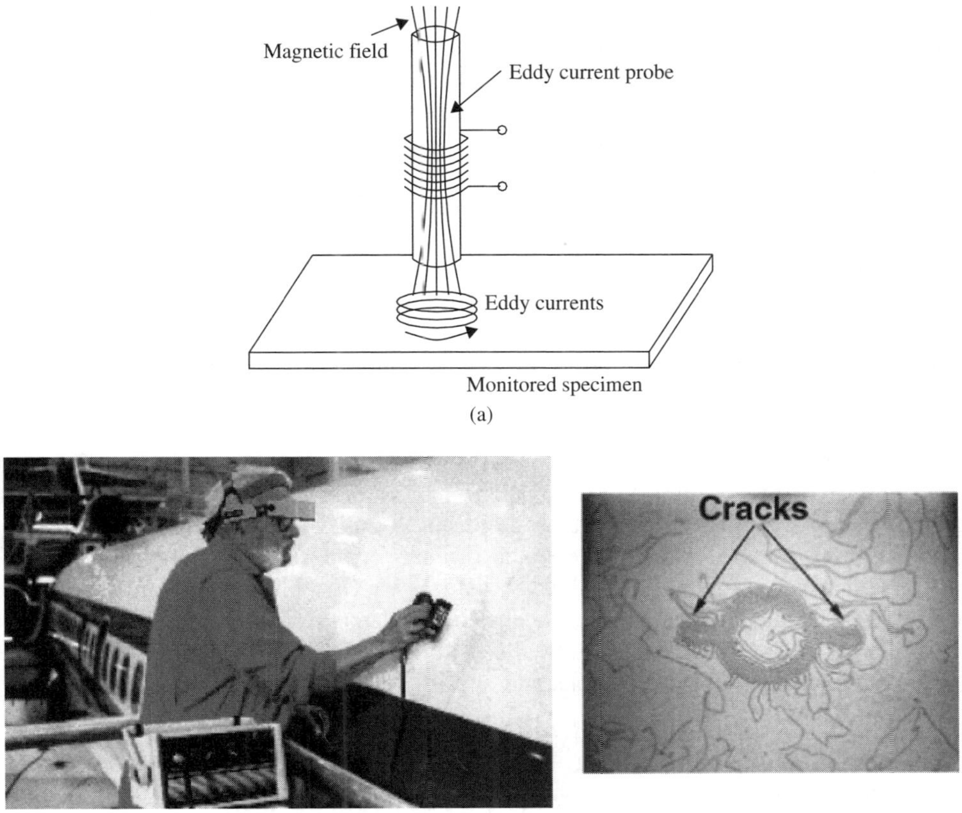

(b) (c)

Figure 2.23 Eddy current inspection: (a) physical principle; (b) civil aircraft inspection; (c) detected cracks around a rivet. Figures (b) and (c) are courtesy of *PRI*, Torrance, California

of the material. The rapid release of the elastic energy is known as the Acoustic Emission (AE). The AE energy may be released for example due to dislocation movements, microscopic deformation, friction, crack nucleation/propagation, fracture and corrosion in metals or matrix cracking, fibre fracture, fibre debonding and delamination in composite materials. There are many other possible sources of AE events in various materials. The energy emitted from the damage can be registered in forms of burst or continuous acoustic signals. A number of specific signal features are used for damage detection and location. These features include: signal duration, maximum amplitude, signal threshold level, signal energy, number of threshold crossings (counts), arrival time and signal energy. The frequency range of AE events is usually between 10 kHz and 1 MHz.

The release of AE can be registered using various types of sensors such as special accelerometers, piezoelectrics or microphones. An array of multiple sensors is capable to triangulate the location of damage by the signal time of flight. Recent advances in this field include the development of micro-electro-mechanical systems (MEMS) technology to manufacture extremely small, inexpensive, conformable and accurate AE sensors that can be either bonded onto or embedded into the structural component (Schoess 1999).

Data from these sensors hold much potentially useful information for the detection of damage but is complicated to interpret. An array of multiple sensors can be used to triangulate the location of damage by the signal time of flight.

AE has been successfully used in many engineering areas for monitoring discontinuities, fatigue failures, material flaws, welding flaws and stress corrosion cracking. Examples include specifically pressure vessels, rotating machinery, seismic applications or materials testing in general. Various studies have been performed over the past decades using AE for monitoring damage in aircraft structures (Bailey 1976; Scala 1986; Carlyle 1989; Fotos 1989; McBride *et al.* 1989). Most of these studies have, however, been limited to ground and laboratory tests. Examples to in-flight applications include (Scala 1986; McBride *et al.* 1989). Other studies have been performed by pressurising the cabin of a commercial aircraft fuselage for the AE detection of fatigue cracks, corrosion, cracked lap joints and cracks around rivets and in forgings and wing splices (Fotos 1989). The $F-111$ fighter/bomber aircraft has been tested in a chamber where the structure was periodically chilled to $-40\,^{\circ}\mathrm{C}$, stressed between $+7.3$ and $-3.0\,\mathrm{g}$ and monitored using an AE system to locate sources of structural failure. The handling, processing and interpretation of AE data have been significantly improved over recent years. AE being a passive monitoring technique however always requires external loading for the AE events to be released.

The major power of the AE technique is that it is well established and understood. The method offers damage detection and localisation in large structures and is not sensitive to geometry. The method has also limited sensitivity and is only arbitrarily reproducible. A good summary of AE fundamentals and application examples can be found in (Miller and McIntire 1987).

2.7.5 Radiography, Thermography and Shearography

A number of different methods utilising advanced development from physics has been implemented to provide images that are easy to interpret. This includes the radiography, thermography and shearography. The physical principles of these methods are illustrated in Figure 2.24. Radiographic techniques utilise various forms of gamma rays and X-rays for material scanning. Some energy of these rays is absorbed by the specimen to be monitored. The level absorption is measured by exposure of the rays to a photographic film. Thermography uses the thermal conductivity and emissivity of material defects. The surfaces monitored radiate energy at wavelengths corresponding to their temperatures. This radiated energy is transformed into thermal images. Shearography uses laser light to detect small surface deformations due to subsurface flaws. Shearographic images are created from the difference between the stressed (loaded) and unstressed (unloaded) surfaces. Often two beams of laser light are used leading to 3-D holographic images. One beam is reflected from the monitored specimen and the other – reference beam – is sent directly to the detector. The method measures the out-of-plane surface displacement in response to two different stressing levels.

The major advantage of these three techniques is their ability of rapidly inspecting relatively large surfaces in real time. Their major disadvantages are related to high cost and damage detection sensitivity. Despite major technological development in the last ten years, the methods are not as widely used in aerospace maintenance as other NDT techniques. Application examples include damage inspections in bond-line areas in composite as well as in honeycomb materials (Bar-Cohen 2000a; Davis 1996).

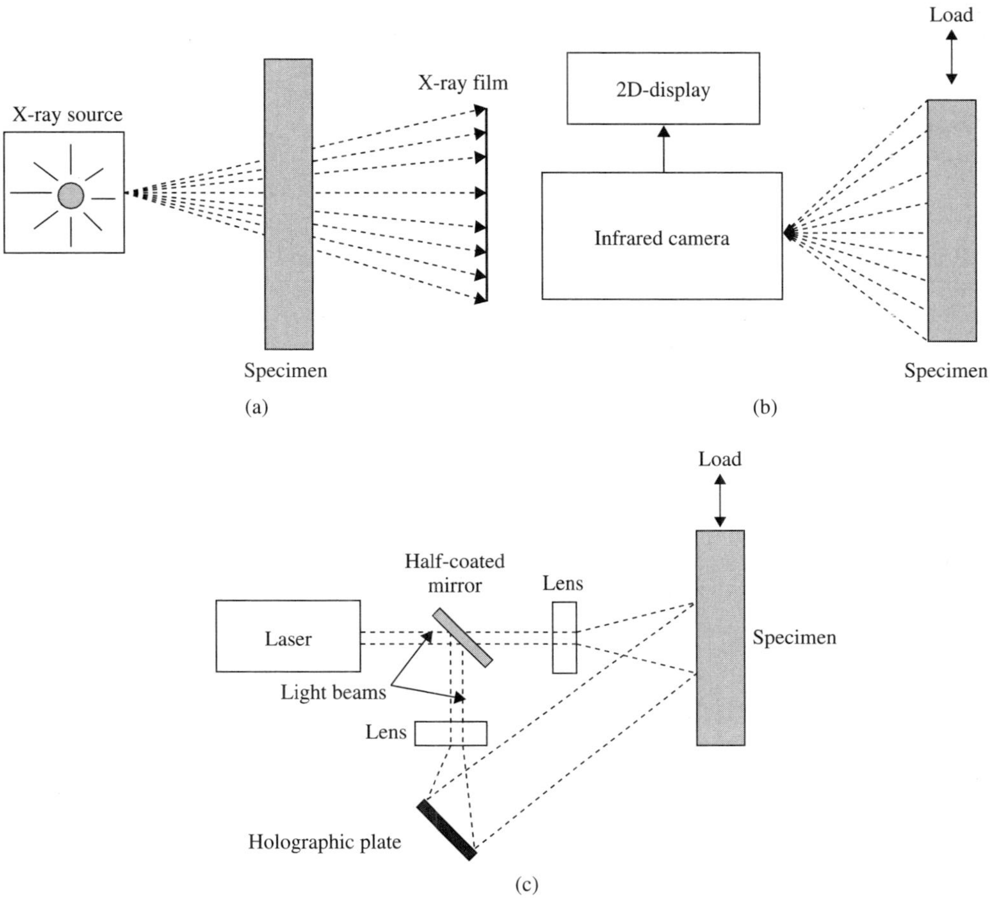

Figure 2.24 Physical principles of: (a) radiography; (b) thermography; (c) holographic shearography

2.7.6 Summary

It appears that most of the NDT techniques are nowadays well established in the aerospace engineering community. A summary of NDT techniques briefly described in this section is given in Table 2.4. The major criteria used for this simple comparative exercise are: the degree of development, cost, the applicability of in-flight (on-line) monitoring and types of damage to be monitored.

NDT aircraft inspections are generally performed manually. Various research activities are ongoing in order to automate this effort. An extensive description of the state-of-the-art in robot-assisted aircraft inspection is given in (Siegel *et al.* 1998). The major idea is to use robots for NDT scanning. Further activities being specifically related to the area of smart structures tend to let NDT become an integral part of a material or a component itself and will be described in more detail throughout the following chapters.

Table 2.4 Summary of NDT techniques used for aircraft damage monitoring and inspection

	Damage Type	Advantages	Limitations
Visual Inspection	• Fatigue cracks • Delamination	• Does not require sophisticated equipment • Relatively inexpensive	• Time consuming • Limited accuracy
Ultrasonics	• Fatigue cracks • BVID • Delamination • Corrosion (?)	• Well established and understood • Sensitive to small damage • Possible damage location • Good depth ranging • Relatively inexpensive • Possible in-flight monitoring	• Point monitoring (requires scanning) • Requires coupling • Often does not detect closed crack • Sensitive to geometry
Eddy Current	• Fatigue cracks	• Detection of small crack • Possible noncontact testing • Does not require coupling • Possible data storage • Relatively inexpensive	• Used mostly for crack detection • Point monitoring (requires scanning) • Requires specific monitoring skills • Requires calibration • Poor penetration (max. monitored thickness - 6 mm)
Acoustic Emission	• Fatigue cracks • BVID • Delamination • corrosion	• Well established and understood • Large structures can be monitored • Not sensitive to geometry • Possible in-flight monitoring	• Requires load (passive technique) • Not reproducible • Not sensitive to small damage
X-Ray Radiography	• Fatigue cracks • BVID • Delamination • Corrosion	• Fast monitoring • Good penetration • Relatively inexpensive	• Not sensitive to small damage • Not possible for large structures
Thermography	• Fatigue cracks • BVID • Delamination	• Fast monitoring • Large structures can be monitored	• Expensive • Not sensitive to small damage • Poor penetration
Shearography	• Fatigue cracks (?) • BVID (?) • Delamination (?) • Corrosion (?)	• Fast monitoring • Large structures can be monitored	• Very Expensive • Not well developed • Requires load (passive technique)

2.8 STRUCTURAL HEALTH MONITORING

A number of different definitions can be proposed to describe *damage, health* and *monitoring* of structures. Intuitively, *health* is the ability to function/perform and maintain the structural integrity throughout the entire lifetime of the structure, *monitoring* is the process of diagnosis and prognosis, and *damage* is a material, structural or functional failure. In this context, damage detection/monitoring and structural health monitoring have the same meaning. One of the major problems in this area is to establish relevant parameters used for monitoring damage as well as subsequent damage accumulation. All aircraft structural health and usage monitoring approaches consider stress as a symptom used for damage monitoring. Levels of stress can be estimated relatively easy from either load models, strain and/or flight parameters. Depending on criticality and access of the damage NDT techniques used for aircraft inspection/monitoring are based on direct visual observations and/or various physical phenomena. These techniques are often limited to single-point measurements but allow surface scanning if a complete structure is considered to be analysed. There are numbers of other approaches available which indirectly relate various parameters or symptoms to possible structural conditions. These techniques monitor for damage globally and do not require single-point measurements. Examples include the use of structural vibration for damage detection in civil engineering or the use of impact energy for impact damage detection in composite materials. Both approaches utilise the relationship between the symptom of damage and the actual damage condition. Different physical models and/or system identification procedures are used to establish this functional relationship. More recently, pattern recognition methods have been used to solve the problem. In this context, the group of methods based on the symptom – damage relationship is commonly known as *Structural Health Monitoring*. However, *Structural Health Monitoring, Damage Detection/Monitoring* and *NDT* are very often used synonymously in many areas of engineering.

2.8.1 Vibration and Modal Analysis

Damage can be often considered as a modification of physical parameters such as mass, stiffness or damping. A number of vibration-based parameters have been used for structural health monitoring. The application of modal analysis is one of the most popular approaches since the classical work on the use of natural frequencies for damage detection in structures (Cawley and Adams 1979). Previous studies show that modal shapes and damping can also be used to detect damage. Other applications in this area involve modal energy, curvatures and transfer functions. Vibration-based data have been employed with some success to detect aircraft structural damage (Hickman *et al.* 1991; Bristow 1992; Robinson *et al.* 1996; Manson *et al.* 2002, 2003a, 2003b). However, the major problem in this area is related to damage sensitivity. Modal and/or vibration based techniques are in fact global methods. A number of studies have been performed on beams and plates where cracks originated from the specimen's surface perpendicular to the applied normal stress (Pandey *et al.* 1991; Campanile 1993). However, very long cracks or delaminations are required to affect the structural physical and/or modal parameters in the case these cracks and delaminations are parallel to the loading direction Despite different reports on successful crack detection, the ability of vibration/modal techniques for damage inspection/monitoring in aerospace structures becomes somewhat questionable and

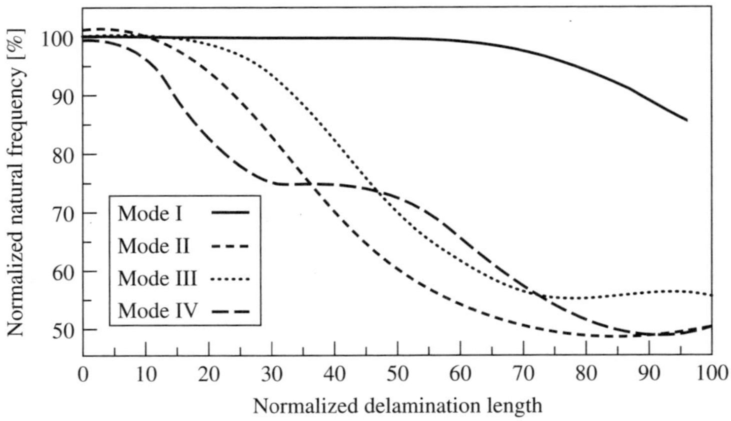

Figure 2.25 Influence of delamination size on natural frequency (Boller 1996)

leads the ongoing discussion on global and local monitoring. Experimental results show that the size of damage (e.g. delamination in composite materials) must be at least 10 % of the area monitored to be reliably detectable (Balis Crema *et al.* 1985; Lee *et al.* 1987; Tracey and Pardeon 1989). This is illustrated in Figure 2.25, where normalised natural frequencies of various vibration modes have been used for damage detection on a delaminated beam. The study involved a classical fourth-order differential equation beam model with simply supported conditions. This theoretical analysis supports previous experimental results. Clearly vibration/modal based damage detection methods are useful for global monitoring.

2.8.2 Impact Damage Detection

A passive approach, in which the energy of impact is related to damage severity, can be used for impact damage detection. A simple example relating sensor parameters to the impact energy is given here following the studies presented in (Boller 1996).

Figure 2.26 shows a simple model of a beam structure under impact excitation. The equilibrium impulse equation for the analysed system can be given as

$$m_I u_I + m_B u_B = m_I v_I + m_B v_B \tag{2.3}$$

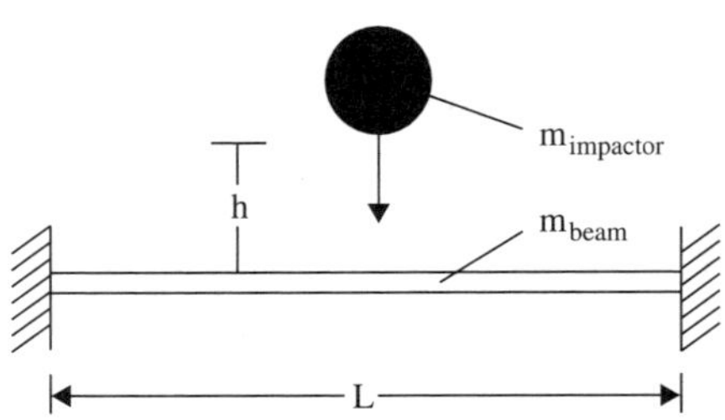

Figure 2.26 Impact model (Boller 1996)

where m is the mass, u, v are the velocities and I, B denote the impactor and structure, respectively. The equilibrium of energy can be described using the following equation

$$\frac{m_I u_I^2}{2} + \frac{m_B u_B^2}{2} = \frac{m_I v_I^2}{2} + \frac{m_B v_B^2}{2} \tag{2.4}$$

Equations (2.3–2.4) can be solved assuming the beam and impactor vibration as

$$w_B(t) = \sum_i A_i \sin \omega_i t \tag{2.5}$$

$$w_I(t) = \frac{gt^2}{2} + v_0 t + w_0 \tag{2.6}$$

respectively. Here, A_i are the vibration amplitudes, v_0 is the initial velocity and w_0 is the initial displacement. Two possible conditions can be considered in this simple study: (a) the impactor's mass is less than or equal to the mass of the beam; and (b) the impactor's mass is larger than the mass of the beam. Figure 2.27 shows examples of experimental vibration signals for both conditions. The data have been acquired using piezoelectric sensors. Two types of vibration behaviour can be observed in the data. The first stage is the impact interaction between the impacting mass and the structure, whereas the second stage is the free vibration of the structure. The study in Figure 2.28 shows that the impact contact time increases with the mass of the impactor. Although the experimental data show some scatter, a linear relationship in a logarithmic scale can be observed and modelled as

$$t = Bm^s \tag{2.7}$$

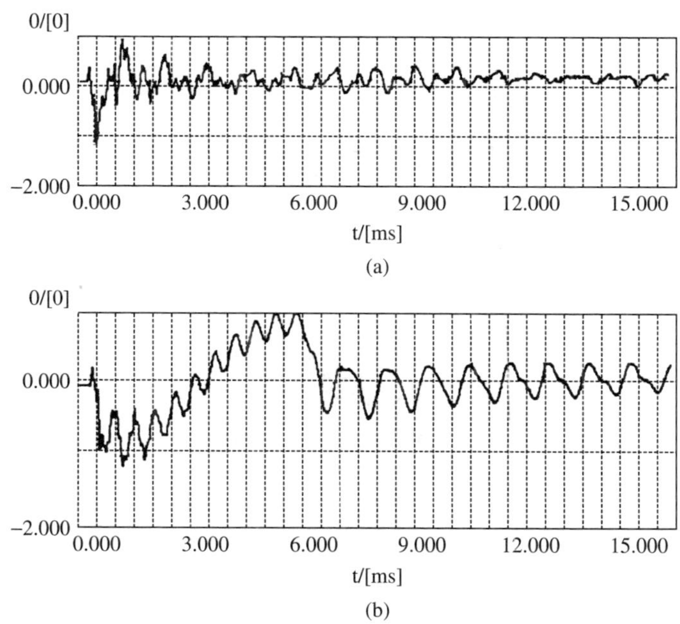

Figure 2.27 Examples of impact signals: (a) impactor's mass equal to structural mass; (b) impactor's mass much greater than structural mass (Boller 1996)

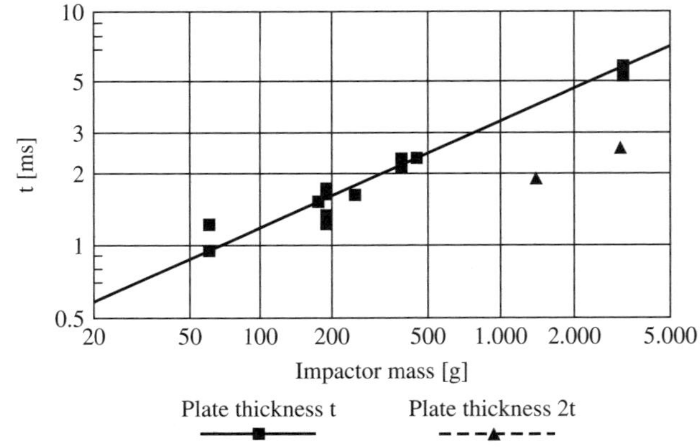

Figure 2.28 Impact contact time vs impactor's mass (Boller 1996)

Similar studies have been performed for the impactor velocity. The experimental results for two different thicknesses of the beam shown in Figure 2.29 are almost a linear relationship. This can be modelled as

$$U = Cv \tag{2.8}$$

where U is the voltage from the piezoceramic sensors, v is the impactor's velocity just before the impact and C is the slope coefficient. The impact energy can be estimated from Equations (2.7) and (2.8) as

$$E = \frac{mv^2}{2} = \frac{1}{2}\left(\frac{t}{B}\right)^{1/s}\left(\frac{U}{C}\right)^2 \tag{2.9}$$

The theoretical value of impact energy calculated using the mass m and velocity v can be plotted in Figure 2.30 against the energy value estimated from the right-hand side

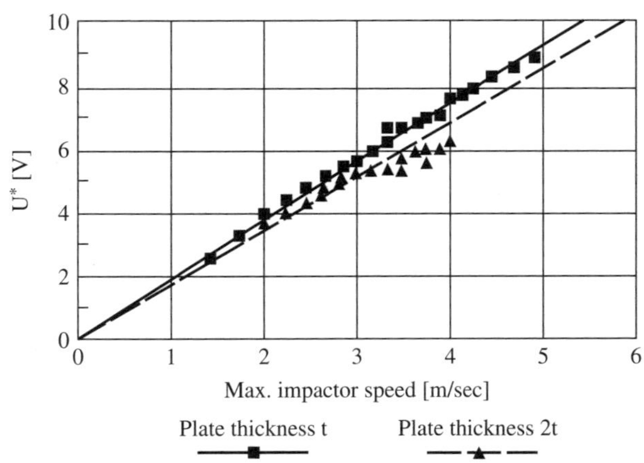

Figure 2.29 Maximum impact amplitude vs maximum impactor's velocity (Boller 1996)

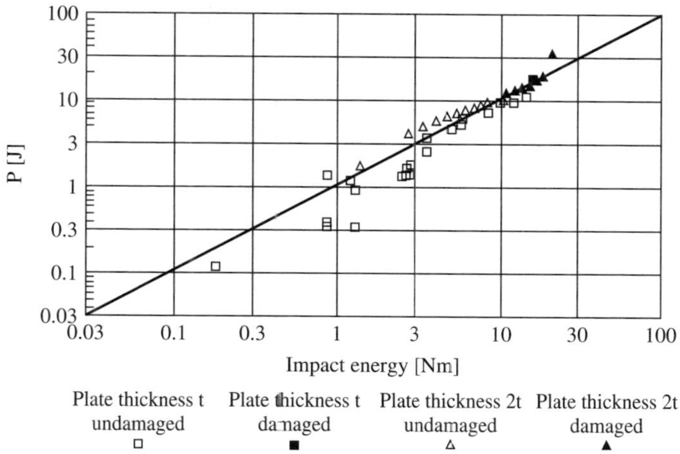

Figure 2.30 Experimental vs theoretical impact energy (Boller 1996)

of Equation (2.9). The results show very good correlation. Impact energy can be estimated from the impact strain data and damage can be detected when its level exceeds a certain threshold value. The severity of damage can be confirmed and quantified using conventional NDT techniques. Although conventional NDT is still required, the amount of monitoring is significantly reduced. More experimental results related to passive impact damage detection will be given in Chapter 4.

2.9 EMERGING MONITORING TECHNIQUES AND SENSOR TECHNOLOGIES

Recent developments in sensor technologies, signal processing and electronics have shown the potential for new monitoring techniques that could be used for aircraft damage detection. One of the key elements is the integration of health monitoring systems into something to be denoted as monitored structures. The number of academic and industrial publications in this area is enormous. Figure 2.31 gives statistics gathered on damage detection patents published by relevant industries. Some of these new developments are briefly described in this section.

2.9.1 Smart Structures and Materials

Materials and structures which are able to sense and perhaps respond/adapt to a change in their environment are commonly known as *smart*. Smart structures and materials have opened new opportunities for damage monitoring. In general damage monitoring systems which utilise smart structures and materials technologies are concerned with a design philosophy directed to the integration of actuators, sensors and signal processors. The attractive potential of such technologies arise from the added value in terms of more reliable damage monitoring systems, reduced inspection monitoring cost and improved

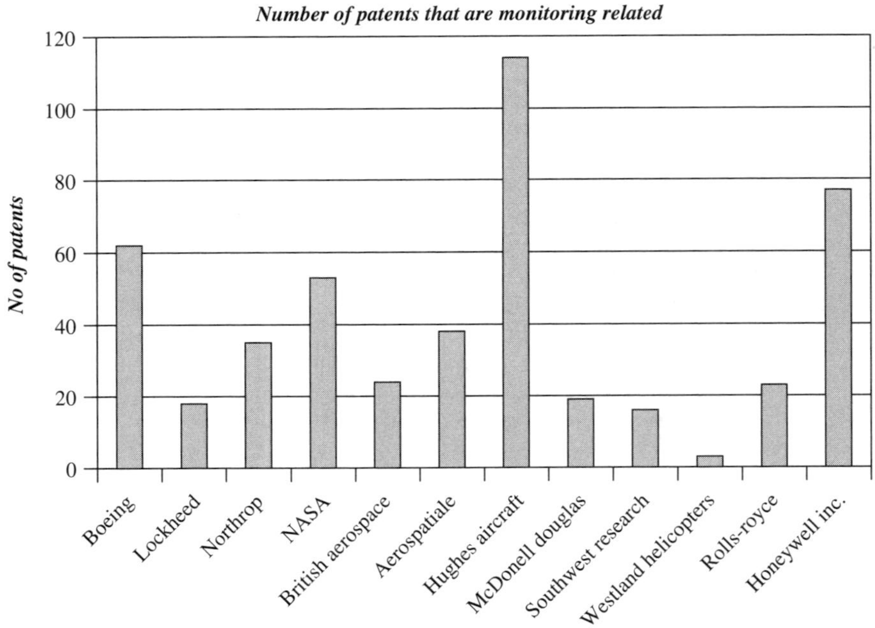

Figure 2.31 Statistical distribution of damage detection patents (Banks 2000)

safety. The last ten years have seen an enormous amount of research in this area. This includes new materials (piezopolymers, piezoceramics), sensors and actuators (MEMS, Micro-Surface Acoustic Waves – MSAW devices) and intelligent data processing (pattern recognition, data fusion, neural networks, combinatorial optimisation based biological and physical systems, and much more). Most of these developments have been reported in the literature (Culshaw 1996; Srinivasan and McFarland 2001; Bishop 1995; Goldberg 1989). Current research publications appear in new journals such as *Smart Materials & Structures, Intelligent Materials Systems & Structures, Microtechnology* and *Nanotechnology*. Applications are being driven towards multifunctional damage monitoring systems integrating actuators, sensors, processing together with self-validation, reliability, redundancy and autonomy aspects.

2.9.2 Damage Detection Techniques

In principle all the NDT techniques described before can be considered as implemented onto or into a component to be monitored, which in the end is already some initial type of a smart structure. With regard to the simplicity and availability of sensing and possibly also actuation elements, piezoelectric elements have turned out to be one of the types being highly viable. The acousto-ultrasonic technique therefore looks to be one of the very promising techniques to start with. It is based on stress waves introduced to a structure by a probe at one point and sensed by another probe at a different position (Vary and Lark 1979; Hillger and Block 1986). The frequency of these waves can go up to MHz. Various types of signals are used as input excitation including impulse, sine burst, sine sweep and Gaussian white noise signals (Staszewski *et al.* 1999). Damage in a structure can be identified by a change of the output signal. Often attenuation is sufficient to detect defects.

Lamb wave inspection is based on the theory of guided waves propagating in plates (Viktorov 1967; Rose 1999). In general, the principles of acousto-ultrasonic and Lamb wave inspections are similar; Chapter 4 gives more details about both techniques. Also, signal processing used for damage detection is similar and is often based on wave attenuation and/or wave dispersion. The factors, which determine the Lamb wave inspection, are related to properties of the structure under inspection and transducer schemes, as reported in (Wilcox *et al.* 1999). Other important elements which form the monitoring strategy include various aspects related to transducer coupling methods, types of excitation signals, optimal sensor location, sensor validation and intelligent signal processing (Staszewski and Boller 2002). Figure 2.32 gives a comparison between classical NDT (ultrasonics and eddy current) and acousto-ultrasonics for Lamb wave monitoring in a multi-riveted metallic panel structure using different methods for processing the sensor data (Boller

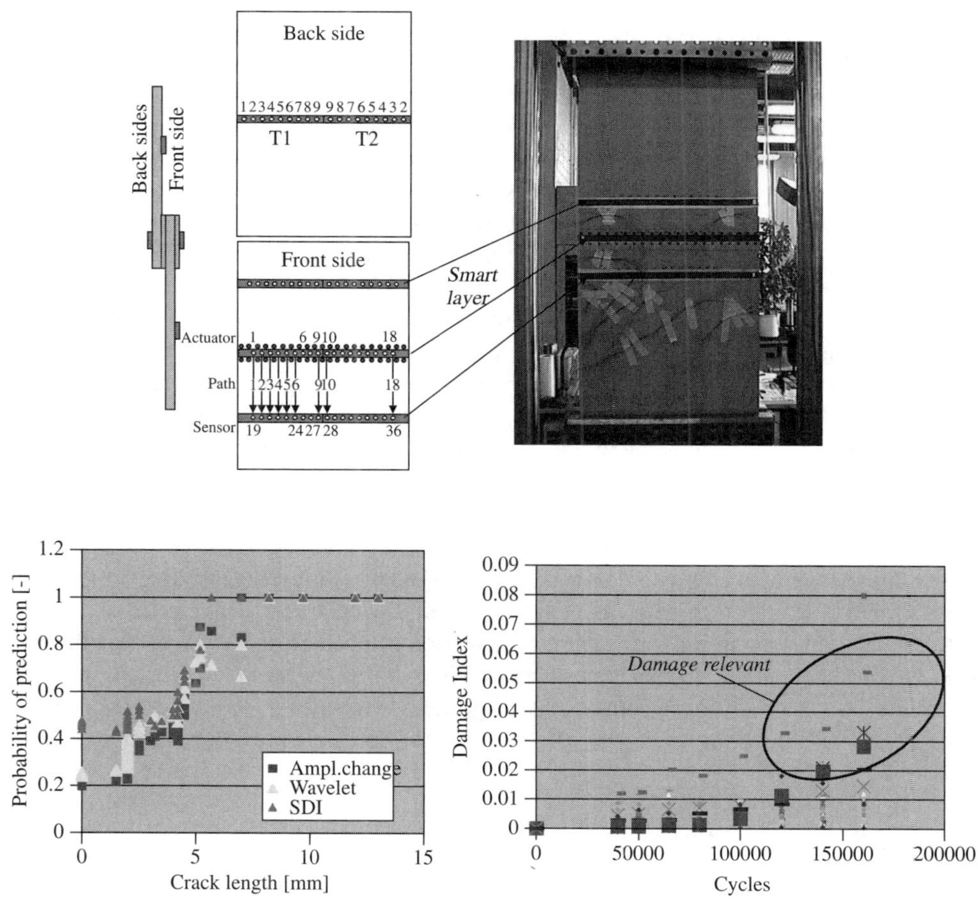

Figure 2.32 Lamb wave based damage detection results using *Smart Layer*™ sensors vs eddy current inspection results (Boller *et al.* 2001)

et al. 2001). The bottom left-hand diagram shows a summary curve for all crack events monitored while in the right-hand diagram the acousto-ultrasonic results are referenced to the results obtained by classical NDT. It can be specifically recognised from the latter diagram that cracks above 5 mm in length could be reliably detected by acousto-ultrasonics when compared to classical NDT for the component considered here.

2.9.3 Sensor Technologies

Various sensor technologies are currently available which can be either adapted onto or integrated into the structure to be monitored. These include optical fibre sensors, piezoelectric sensors and Micro-Electro-Mechanical Systems (MEMS), to just name the ones mostly discussed. Further options exist with eddy current layers bonded around fatigue or corrosion critical areas. A very simple monitoring method exists when taking advantage of the electrical conductivity of carbon fibres in a composite material, where the conductivity changes when the fibres break or delamination occurs. A major challenge with this technique is related to adequately contacting the carbon fibre layers.

A lot of development work has been done in the area of optical fibre sensors (Glossop *et al.* 1990; SPIE; Udd 1992; Pearson 1992; Measures 2001). The major advantage of these sensors is their immunity to electromagnetic fields and their compatibility with data transmission systems. However, more work needs to be done in this area regarding material/structures integration and repairability procedures. Optical fibre sensors have been used for monitoring the curing process and/or damage induced by impact and overloads in composite materials. Optical fibre sensors are also increasingly used for strain and temperature measurements. Recent development in this area shows applications of Bragg – Grating sensors for acousto-ultrasound monitoring (Betz *et al.* 2002). It is quite feasible that multifunctional optical fibre sensors will be soon available for both strain and damage monitoring.

Piezoelectric materials have been used for years for actuating and sensing stress waves. However, only recently these materials have become available in the form of ceramic elements that can also become an integral part of a structure to be monitored. Piezoceramic sensors are also available on Kapton layers in the form of so called *SMART Layers*® (Chang 1998), which can be embedded or bonded on structural components and here specifically in areas prone to damage such as notches. A variety of sizes and shapes for these sensor layers can be made available and basically tailored according to customer needs where a selection of patterns is shown in Figure 2.33.

Actuating and sensing for active damage detection can be accomplished using other new technologies such as interdigital transducers (Wilcox *et al.* 1997), phase array transducers (Blanquet *et al.* 1996), piezoelectric paints (Egusa and Iwasawa 1993) and MEMS (Khuri-Yakub *et al.* 2000). Some examples of recent sensor concepts are given in Figure 2.34.

2.9.4 Intelligent Signal Processing

Intelligent signal processing is the key element, which builds the bridge between the sensor signal and the structural integrity interpretation (Worden *et al.* 1997). Various methods have been developed in recent years. This includes: data pre-processing techniques (e.g.

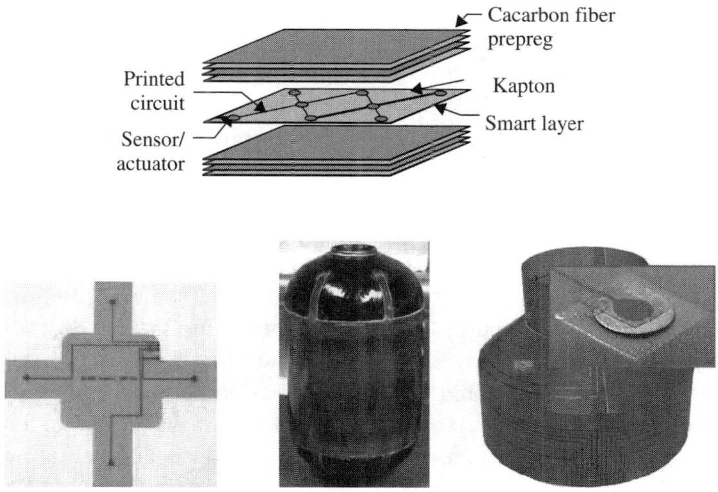

Figure 2.33 *Smart Layer*™ sensors. Courtesy of *Acellent Technologies Ltd*, California

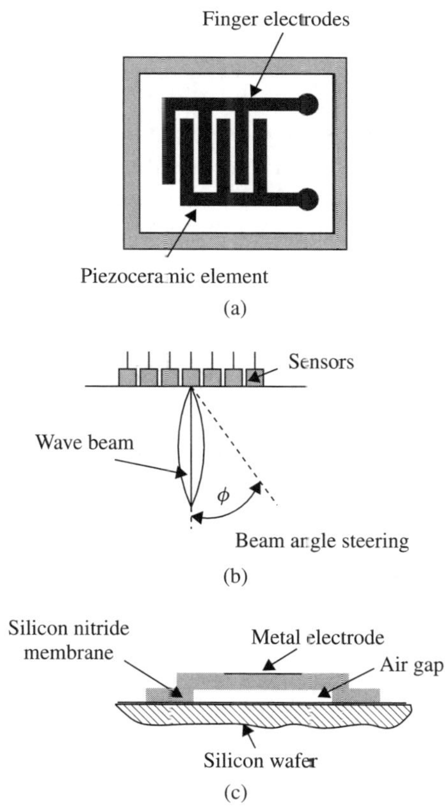

Figure 2.34 Examples of recent sensor design concepts: (a) interdigital sensor; (b) phased-array sensor (c) MEMS capacitance transducer

trend/outlier analysis, signal denoising), feature extraction/selection (e.g. signature analysis, time–frequency analysis, wavelets) and pattern recognition (e.g. neural networks, novelty detection). All these methods lead to signal features, which are sensitive to damage but insensitive to boundary, load or environmental conditions.

Sensors are usually deployed in arrays. Multi-sensor architectures not only improve the signal-to-noise ratio but also offer better robustness, reliability and confidence in the results. Sensor data and information (e.g. flight parameters) can be combined using various fusion techniques such as physical models, parametric methods, information techniques and cognitive-based models.

The fewer sensors need to meet the requirements set for structural health monitoring, the better the overall reliability, signal processing effort and thus smaller cost for the damage monitoring system. The optimal sensor number and their locations can be established using various combinatorial techniques and mutual information approach. Sensor architectures also require validation procedures, which are important to detect sensor failures. There exist various methods in this area based on statistical analysis and neural networks.

Recent developments in signal processing for damage detection are discussed in more details in Chapter 5.

2.10 CONCLUSIONS

Damage detection in aircraft has become a major issue in aerospace engineering. Aircraft structures are safe and reliable but designed for a specific period of time/flight cycles only. This time has become specifically long or better some designs have even received a substantial extension of their operational life after having been fully revisited and upgraded. The longer, however, the life of an aircraft becomes, the more likely changes in operational conditions have to be expected, which also includes the payloads transported, the maneuvers it is requested to fly or the environment it may have to fly in. Even a safe-life designed aircraft, which per se has been designed inspection free, may suddenly have to be inspected. This scenario has specifically become relevant in military aviation where *B52, F4, C130* or *MiG21* aircraft can be looking forward to quite long operational lives with *F*16, *F*18 and *TORNADO* aircraft to follow into similar conditions already. Similar trends can be also observed in civil aviation where an increasing number of older *Airbus A300* and *A*310 as well as *Boeing* 747, 757 and *767* aircraft are converted to freighters. The emerging development in advanced sensor technology combined with sophisticated signal processing and computation hardware will help to keep these aerostructures operational for a longer period, enhancing their performance without compromising reliability and safety and thus helping to conserve natural resources with regard to aerospace structural materials. Enhanced monitoring technology will, however, also help to reduce maintenance cost in general, which is important with regard to the large number of damage-tolerant aircraft flying around. It may even allow switching from safe-life to damage-tolerant design in case this may allow and help to extend an airframe's performance.

Whatever the scenario is, reliable and automatic damage detection systems will be essential for future developments. There are a number of emerging monitoring techniques, sensor technologies and intelligent signal processing methods available for damage detection systems which could be integrated with aircraft structures and play an important role in aircraft maintenance.

REFERENCES

Amabile, P. and Giacobbe, T. 1991. Proposal for the new fatigue management system for the AMX, *AGARD Lecture Notes (AGARD-CP-506)*, paper 6.

Asada, H., Sotozaki, T., Endoh, S. and Tomita, T. 1998. Practical evaluation of crack detection capability for visual inspection in Japan, *NATO RTO Meeting Proc. 10: Airframe Inspection Reliability Under Field/Depot Conditions*, paper 5.

Bailey, C.D. 1976. Acoustic emission for in-flight monitoring on aircraft structures, *Materials Evaluation*, Vol. 34, No. 8, pp. 165–171.

Balis Crema, B., Castellani, A. and Peroni, I. 1985. Modal test on composite structures application in damage detection, *Proceeding of the 3rd International Modal Analysis Conference, Orlando, Florida*, pp. 708–713.

Banks, D. 2000. *A Look Into the World of Structural Health Monitoring*, Industrial Placement Report, EADS/Sheffield University.

Bar-Cohen, Y. 2000a. Emerging NDE technologies and challenges at the beginning of the 3rd millennium – Part I, *Materials Evaluation*, Vol. 58, No. 1, pp. 17–30.

Bar-Cohen, Y. 2000b. Emerging NDE technologies and challenges at the beginning of the 3rd millennium – Part II, *Materials Evaluation*, Vol. 58, No. 2, pp. 141–150.

Barens, A.P., Howey, P.W. and Skinn, D.A. 1991. *Risk Analysis for Aging Aircraft Fleets*, Vol. 1 WL-TR-91-30066, Wright Laboratory, Air Force Systems Command, Wright–Patterson Air Force Base, OH 45433-6553, USA.

Bauer, W. 1987. Event monitoring functions introduced by the onboard life monitoring system (OLMOS) into German aircraft, *DFVLR-Mitt.* 88 – 04, pp. 131–154.

Betz, D., Thursby, G., Culshaw, B. and Staszewski, W.J. 2002. Acousto-ultrasonic sensing using fiber Bragg gratings, *Smart Materials and Structures*, Vol. 12, No. 1, pp. 122–128.

Bishop, C.M. 1995. *Neural Networks for Pattern Recognition*, Clarendon Press, Oxford.

Blanchard, B. 1978. *Design and Manage to Life Cycle Costs*, M/A Press.

Blanquet, P., Demol, T. and Delebarre, C. 1996. Application of array transducers to health monitoring of aeronautic structures, 14th World Congress on Nondestructive Testing (New Delhi, India) 8-13 December Vol. 4, pp. 2057–2060.

Boller, C. 1996. Fundamentals on damage monitoring, *AGARD Lecture Series 205 (AGARD-LS-205)*.

Boller, C. 2001. Ways and options for aircraft structural health management, *Smart Materials and Structures*, Vol. 10, pp. 432–439.

Boller, C., Ihn, J.B., Staszewski, W.J. and Speckmann, H. 2001. Design principles and inspection techniques for long life endurance of aircraft structures, *Proceedings of the 3rd International Workshop on Structural Health Monitoring, Stanford, California, 12–14 September*, pp. 275–83.

Brand, C. and Boller, C. Identification of life cycle cost reduction in structures with self-diagnostic devices, *NATO RTO Symposium on Design for Low Cost Operation and Support*, Ottawa, Canada, 21–22 October.

Bristow, J.W. 1992. Integrated health and usage monitoring system (IHUMS), *Aircraft Engineering*, February, pp. 12–13.

Campanile, L.F. 1993. A modal method for structural health monitoring, *Proceedings of the 4th International Conference of Adaptive Structures (ICAS)*.

Carlyle, J.M. 1989. Acoustic emission testing the F-111, *NDT-International*, Vol. 22, No. 2, pp. 67–73.

Cawley, P. and Adams, R.D. 1979. The location of defeats in structures from measurements of natural frequencies, *Journal of Strain Analysis*, Vol. 4, No. 2, pp. 49–57.

Chang, F.K. 1998. Smart layer: built-in diagnostics for composite structures, *Proceedings of the 4th European Conference on Smart Structures and Materials and the 2nd International Conference on Micromechanics, Intelligent Materials and Robotics, Harrogate, UK*, 6–8 *July*, pp. 777–781.

Culshaw, B. 1996. *Smart Structures and Materials*, Artech House Books.

Davis, C.K. 1996. Shearographic and thermographic NDE of the Space Shuttle structure and thermal protection systems, *Proceedings of the Technology 2006, Anaheim, California, October, 29–31.*

DeLuccia, J. 1991. The corrosion of aging aircraft and its consequences, *Naval Air Development Center*, AIAA-91-0953-CP.

Egusa, S. and Iwasawa, N. 1993. Poling characteristics of PZT/epoxy piezoelectric paints, *Ferroelectrics*, Vol. 145, pp. 45–60.

Fotos, C.P. 1989. Acoustic emission technique tests aircraft integrity, *Aviation Week & Space Technology*, 28 August, p. 76.

Glossop, N.D., Dubois, W., Tsaw, W., LeBlanc, M., Lymer, J., Measures, R.M. and Tennyson, R.C. 1990. Optical fiber damage detection for an aircraft composite leading edge, *Composite*, Vol. 21, pp. 71–80.

Goldberg, D.E. 1989. *Genetic Algorithms in Search Optimization and Machine Learning*, Adison–Wesley Publishing Co., Reading, MA.

Goranson, U.G. 1997. Jet structures performance monitoring, *Proceedings of the 1st International Workshop on Structural Health Monitoring*, Stanford, California, 18–20 September, pp. 3–17.

Grandt, A. 2000. Analysis of aging aircraft scenarios and models for fracture mechanics research, *Notes of Open Lectures Presented at the 2000 USAF Aircraft Structural Integrity Program (ASIP) Conference*, San Antonio, Texas, 4 December.

Haberding, R. 1985. Schwingungsüberwachung an Turboflugtrieben am Biespiel des Airbus A310, *VDI-Berichte*, 568, pp. 167–181.

Hautamaki, C., Zurn, S., Mantell, S.C. and Polla, D.L. 1999. Experimental validation of MEMS strain sensors embedded in composites, *Journal of Microelectromechanical Systems*, Vol. 8, No. 3, pp. 272–279.

Hickman, G.A., Gerardi, J.J. and Feng, Y. 1991. Application of smart structures to aircraft health monitoring, *Journal of Intelligent Materials, Systems and Structures*, Vol. 2, pp. 411–430.

Hillger, W. and Block, J. 1986. Ultraschall- und Schallemissionsprüfttechnik zur Zerstörungsfreien Schadensanalyse an CFK-Proben, *Zeitschr. f. Flugwiss. u. Weltraumforschung*, Vol. 10, heft 4, pp. 273–280.

Hunt, S.R. and Hebden, I.G. Validation of the Eurofighter Typhoon structural health and usage monitoring system, *Smart Materials and Structures*, Vol. 10, pp. 497–503.

I.Mech. E. 2000. In-service structural integrity, NDE and repair, *Notes of Lectures Presented at I.Mech.E. Meeting on 30 May*, Bristol, UK.

Khuri-Yakub, B.T., Cheng, C.H., Degertekin, F.L., Ergun, S., Hansen, S., Jin, X.C. and Orlakran, O. 2000. Silicon micromachined ultrasonic transducer, *Jpn. J. Appl. Phys.*, Vol. 39, pp. 2882–2887.

Krauss, A. 1988. Betriebslastenermittlung für Flugzeugentwurf unt – entwicklung, 14, *Vortgasveranst*, DVM-AK Betriebsfestigkeit.

Kudva J.N. *et al.* 1993. Smart structures concepts for aircraft structural health monitoring, *SPIE Smart Structures Conference, Albuquerque, New Mexico.*

Ladda, V. and Meyer, H.J. 1991. The operational loads monitoring systems, *AGARD Lecture Notes*, AGARD-CP-506, paper 15.

Lee, B.T., Sun, C.T. and Liu, D. 1987. An assessment of damping measurement in the evaluation of integrity of composite beams, *Journal of Reinforced Plastics and Composites*, Vol. 6, pp. 114–125.

Lepine, B.A., Wallace, B.P., Forsyth, D.S. and Wyglinski, A. 1998. Pulsed eddy current method developments for hidden corrosion detection in aircraft structures, *Proceedings of the 1st Pan-American Conference for NDT, Toronto, Canada, September*, pp. 118–124.

Lockheed Martin. *PRICE Systems Executive Overview*.

Manson, G., Worden, K. and Allman, D.J. 2002. Experimental validation of a damage severity method, *Proceedings of the First European Workshop on Structural Health Monitoring, Paris, France, 10–12 July*, pp. 845–852.

Manson, G., Worden, K. and Allman, D.J. 2003a. Experimental validation of a structural health monitoring methodology, Part II: Novelty detection on a Gnat aircraft, *Journal of Sound and Vibration*, Vol. 259, No. 2, 345–363.

Manson, G., Worden, K. and Allman, D.J. 2003b. Experimental validation of a structural health monitoring methodology, Part III: Damage location on an aircraft wing, *Journal of Sound and Vibration*, Vol. 259, No. 2, 365–385.

Matzkanin, G.A. 2000. Technology assessment of MEMS for NDE, *Proceedings of the SPIE's 5th Annual Internation Symposium on Non-Destructive Evaluation and Health Monitoring of Ageing Infrastructure, 5–9 March 2000, Newport Beach, California*, paper No. 3994–30.

McBride, S., Viner, M., Pollard, M.D., MacPhail, Bouman, P.S. and Peters, D.T. 1989. Acoustic emission detection of crack presence and crack advance during flight, *AGARD Lecture Notes 462 (AGARD-CP-462)*, paper 16.

Measures, R.M. 2001. *Structural Monitoring with Fiber Optic Technology*, Academic Press, San Diego, California.

Miller, R.K. and McIntire, P. (eds). 1987. *Non-Destructive Testing Handbook*, 2nd edition, Vol. 5: Acoustic Emission Testing, American Society for Non-Destructive Testing.

MIL-PRF. *Performance Specification Logistics Management Information*, 49506.

OBS. 1993. EDCAS *User Manual*, OBS GmbH.

Pandey, A.K., Biswas, M. and Samman, M.M. 1991. Damage detection from changes in curvature mode shapes, *Journal of Sound and Vibration*, pp. 321–332.

Pearson, J.E. (ed.). 1992. Optical technologies for aerospace sensing, *Critical Reviews of Optical Science and Technology*, Vol. CR44, SPIE Optical Engineering Press, Bellingham, Washington, USA.

Penney, S. 2000. Geriatric ward, *Flight International*, 12–18 December.

Raymer, P.R. 1992. *Aircraft Design: A Conceptual Approach*, AIAA Education Series, Washington DC, USA.

Robinson, A., Peterson, L.D. and James, G.H. 1996. Health monitoring of aircraft structures using experimental flexibility matrices, *AIAA-96-1304-CP*.

Rose, J.R. 1999. *Ultrasonic Waves in Solid Media*, Cambridge University Press, Cambridge, UK.

Sampath, C.P. 1996. Aging combat aircraft fleets – long-term applications, *AGARD Lecture Series 206 (AGARD-LS-206)*.

Scala, C.M. 1986. A semi-adaptive approach to in-flight monitoring using acoustic emission, *Proceedings of the Review of Progress in Quantitative NDE, San Diego, California*, pp. 361–369.

Schoess, J.N. 1999. *Development and Application of Stress-Wave Acoustic Diagnostic for Roller-Bearings*, Honeywell Tech. Center report.

Scott, R. 2001. Home on the range, *Flight International*, 19 December–1 January.

Siegel, M., Guanatilake, P. and Podnar, G. 1998. Robotic assistance for aircraft inspectors, *IEEE Instrum. Meas. Mag.*, No. 3, pp. 16–30.

SPIE. *SPIE Proceedings – Fiber Optic Smart Structures and Skins*, Vol. 989 (1988), Vol. 1170 (1989), Vol. 1370 (1990), Vol. 1588 (1991), Vol. 1798 (1992).

Spragg, D., Ganguli, R., Thamburaj, R., Hillel, R. and Cue, R.W. 1989. The role of inflight engine condition monitoring on life cycle management of CF-18/F404 engine components, *AGARD R-770*, paper 4.

Srinivasan, A.V. and McFarland, D.M. 2001. *Smart Structures*, Cambridge University Press, Cambridge, UK.

Staszewski, W.J. 2000. Monitoring on-line integrated technologies for operational reliability – MONITOR, *Air & Space Europe*, Vol. 2, No. 4, pp. 67–72.

Staszewski, W.J. and Boller, C. 2002. Acoustic wave propagation phenomena modelling and damage mechanisms in ageing aircraft, *Aircraft Integrated Monitoring Systems (AIMS), Garmisch-Partenkirken, Germany, 27–30 May, CD-ROM Conference Proceedings*, pp. 169–184.

Staszewski, W.J., Biemans, C., Boller, C. and Tomlinson, G.R. 1999. Crack propagation monitoring in metallic structures, *Proceedings of the International Conference on Smart Materials, Structures and Systems*, Bangalore, *India, 7–10 July*, pp. 532–541.

Suresh, S. 1998. *Fatigue of Materials*, Cambridge University Press, Cambridge, UK.

Tracy, J.J. and Pardeon, G.C. 1989. Effect of delamination on the natural frequencies of composite laminates, *Journal of Composite Materials*, Vol. 23, pp. 1200–1215.

Udd, E. (ed.). 1992. Fiber optic sensor, *Critical Reviews of Optical Science and Technology*, Vol. CR44, SPIE Optical Engineering Press, Bellingham, Washington, USA.

Vary, A. and Lark, R.F. 1979. Correlations of fiber composite tensile strength with the ultrasonic stress wave factor, *Journal of Testing and Evaluation*, pp. 185–191.

Viktorov, I.A. 1967. *Rayleigh and Lamb Waves*, Plenum Press, New York.

Wilcox, P.D., Castaings, P., Monkhouse, R., Cawley, P. and Love, M.J.S. 1997. An example of the use of interdigital PVDF transducers to generate and receive a high order Lamb wave mode in a pipe, *Rev. Prog. Quantitative NDE*, Vol. 16, pp. 919–926.

Wilcox, P.D., Dalton, R.P., Lowe, M.J.S. and Cawley, P. 1999. Mode transducer selection for long range wave inspection, *Proceedings of the 3rd International Workshop on Damage Assessment Using Advanced Signal Processing Procedures (DAMAS), Dublin, Ireland, 28–30 June*, pp. 152–161.

Worden, K., Staszewski, W.J. and Tomlinson, G.R. 1997. Smart systems – the role of signal processing, *Proceedings of CEAS (Confederation of European Aerospace Society), International Forum on Aeroelasticity and Structural Dynamics, Rome, Italy, 17–20 July*.

3

Operational Load Monitoring Using Optical Fibre Sensors

P. Foote[1], M. Breidne[2], K. Levin[3], P. Papadopolous[4], I. Read[1], M. Signorazzi[5], L.K. Nilsson[2], R. Stubbe[2] and A. Claesson[2]

[1] *BAE SYSTEMS, Sowerby Research Centre, Filton, UK*
[2] *Institute of Optical Research (IOF), Stockholm, Sweden*
[3] *Aeronautical Research Institute of Sweden (FFA), Bromma, Sweden*
[4] *Association for Research, Technology and Training (ARTT), Heraklion, Greece*
[5] *Alenia Research Department, Rome, Italy*

3.1 INTRODUCTION

The first and most common application of optical fibres has been for data transmission in telecommunication. Recent years have yielded many applications for sensing physical parameters such as strain, vibration, temperature and pressure. Optical fibres can be easily integrated within structures that can then be monitored. Various smart structures utilising optical fibre sensors have been investigated over the last 15 years. Optical fibre sensors are also considered for possible damage detection in aerospace applications. This includes Operational Load Monitoring (OLM) and impact damage detection systems. OLM systems are used to estimate the pattern of structural fatigue life and provide valuable information about possible structural damage, as described in Chapter 2. Impact damage detection systems have the ability to obtain information about impact energy and locations. There is strong evidence that damage severity in composite structures can be correlated with impact energy (see Section 2.8.2), with no damage occurring below a certain energy

Health Monitoring of Aerospace Structures – Smart Sensor Technologies and Signal Processing.
Edited by W.J. Staszewski, C. Boller and G.R. Tomlinson ISBN: 0-470-84340-3

threshold. In contrast to strain gauges, optical fibre sensors can also be used to measure temperature, to monitor curing of composite structures and to acquire information about chemical states in structures (e.g. corrosion due to de-icing substances). Recent applications have also showed that optical fibre sensors can be used to sense acousto-ultrasonic signals. The multifunctional optical fibre sensors are particularly attractive for aerospace applications. Examples include monitoring curing processes and/or damage induced by impact in composite materials, strain monitoring in airframes, indication of icing, monitoring of temperature and vibration (Pearson 1992) monitoring of strain and ultrasound (Betz *et al.* 2003).

This chapter describes various aspects related to optical fibre load and damage detection systems considered for aircraft health and usage monitoring. The analysis includes a brief introduction related to optical fibres and Bragg grating fibre sensors. However, the reader is referred to (Culshaw 1988; Krohn 1988; Udd 1991, 1992; Claus 1991) for more details related to the theory of optical fibre sensors. The focus of the material presented is on various aspects related to the development of an optical fibre sensor based OLM system. This includes target performance specifications of monitoring systems, design features and manufacturing specifications. One impact detection application example is also provided.

3.2 FIBRE OPTICS

3.2.1 Optical Fibres

Optical fibres were originally developed for data transmission in telecommunications. They have been successfully used for this purpose for many years. Their low mass, small cross-section and immunity to electromagnetic interference also make them attractive for sensing applications. Although, in general, the theory of electromagnetic waves needs to be used to study the propagation of light in optical fibres, in practice simple optical laws can be used for explanation. Light is guided in optical fibres because of total internal reflection. Optical fibres have a high refractive index core surrounded by lower refractive index cladding. Light is reflected at the boundary between the fibre core and cladding. As the core is completely surrounded by the cladding light in the core once reflected is reflected again and again, and thus light is guided along the fibre. All the light is reflected only if the angle of incidence is greater than the critical angle. Light incident at more acute angles will quickly be lost from the fibre. As light travels a wave interference effect will occur. This means that light arriving at a point within the fibre core via one path will interfere with light arriving at the same point via a different route. Bands of constructive and destructive interference will occur. These interference effects mean that in practice only certain light pathways, called spatial modes, can actually travel along the fibre. Fibres can be manufactured to support many or just one spatial mode.

The core and cladding of a typical telecommunications optical fibre are manufactured from glass (silica). The core is doped to give it a slightly higher refractive index than the cladding. Tight quality control during manufacture of the optical fibre ensures its properties are constant throughout the entire length. Finally, a coating (typically acrylate or polyimide) is applied to fibres for protection against the environment. Optical fibres potentially have an enormous information bandwidth, which is measured in THz.

3.2.2 Optical Fibre Sensors

Optical fibres can also be used to for sensing applications. They have been used to measure temperature, pressure, strain, curvature, rotation, flow, refractive index, concentration, electric current and many other parameters. They have many advantages over electrical sensing devices. Their small size means that they can be embedded into a composite structure without significantly degrading its strength. They can also be used to make measurements in confined spaces. Their low mass means that they can respond quickly and make measurements on small samples. The dielectric nature of fibre optic sensors renders them immune to electromagnetic interference. Thus they can operate in places that would be unsuitable or impractical for electrical sensors. Fibre optic sensors also have the potential to be multiplexed, so that many sensors can be accessed via a single fibre optic cable (Kersey 1992). Multiplexing can greatly reduce cabling requirements and system complexity.

Optical fibre sensors can be either intrinsic or extrinsic. Intrinsic sensors use a length of the fibre itself as the sensing element. Extrinsic sensors use the fibre merely as a transmission medium to deliver light to and from a sensing head at the end of the fibre. The sensors can use a number of mechanisms.

Probably the simplest fibre optic sensors use light intensity to represent the property being measured. Here the light transmitted by or reflected from the sensor varies with the measured property. This type of sensor can be proned to errors due to unexpected or variable losses in the fibres that connect to the sensor and to variations in the optical power delivered by the light source. These problems can be alleviated if a reference is used that can account for losses and changes in the source.

The phase or optical path length is used in fibre optic sensors based on interferometry. Here the light is split between two (or more) paths. One path acts as a reference and the other is subject to change because of the parameter to be measured. Recombining light from these two paths generates optical interference. The measured property is thus related to the interferometric phase. These sensors have a limited range due to the periodic nature of the optical interference. Fringe counting or multiple wavelengths can be used to extend the range. Another variant uses a light source with a broad spectral content to extend the range. Here a receiving interferometer is used in conjunction with the sensing interferometer. Interference is only observed when the path differences of the sensing and receiving interferometers are matched.

A change of the polarisation state of light has also been used to make optical sensors. Here a well-defined polarisation state is launched into the sensor. The property to be measured causes a change in the state of polarisation in the sensor (for example, magnetic field via the Verdet constant of a glass rod). Then the light emerging from the sensor is analysed. The measured property is thus related to the change in polarisation state. Sensors using polarisation are likely to require connection via expensive polarisation maintaining fibres.

As light propagates through a fibre some of it is scattered causing loss. A small portion of the scattered light remains in the fibre but propagates in the opposite direction, which is known as back scattering. Optical sensors have been developed to take advantage of back scattering. A short light pulse is launched into a length of fibre. As the light pulse travels through the fibre some of the light is back scattered. The time that back scattered light takes to travel back to the end of the fibre is dependent on how far along the fibre the scattering occurred. Thus examination of the light returning to the end of the fibre at a given time after the launch of the light pulse provides information on a particular point on the fibre. This technique is known as optical time domain reflectivity (OTDR). Both linear and nonlinear scattering can be used to provide different kinds of information. This type of sensor can address long lengths (several kilometres) of fibre. The spatial resolution is limited to about a metre. This type of sensor is also quite slow as the scattering tends to be weak and averaging of result over many light pulses is required.

Wavelength or colour can also be used for sensing. Sensors based on colour tend to have a broad spectral response, which changes with the measured property. They may also operate at visible wavelengths. Sensors utilising wavelength have a very much narrower spectral response, which again changes with the measured property. As the intensity of the light returning from the sensor is not of primary importance, wavelength based sensors are largely unaffected by losses in the fibre connections or variations in the light source. Fibre Bragg gratings are based on wavelength change.

3.2.3 Fibre Bragg Grating Sensors

Fibre Bragg grating sensors are in fact spectral filters which utilise the principle of Bragg reflection. The gratings are a series of close parallel lines printed into the core of a fibre. This is usually achieved by the photosensitive effect; optical fibres are exposed to a periodic pattern of ultraviolet light (typical wavelength is less than 250 nm). The periodic pattern can be generated as a hologram or with a phase mask. Figure 3.1 shows an example

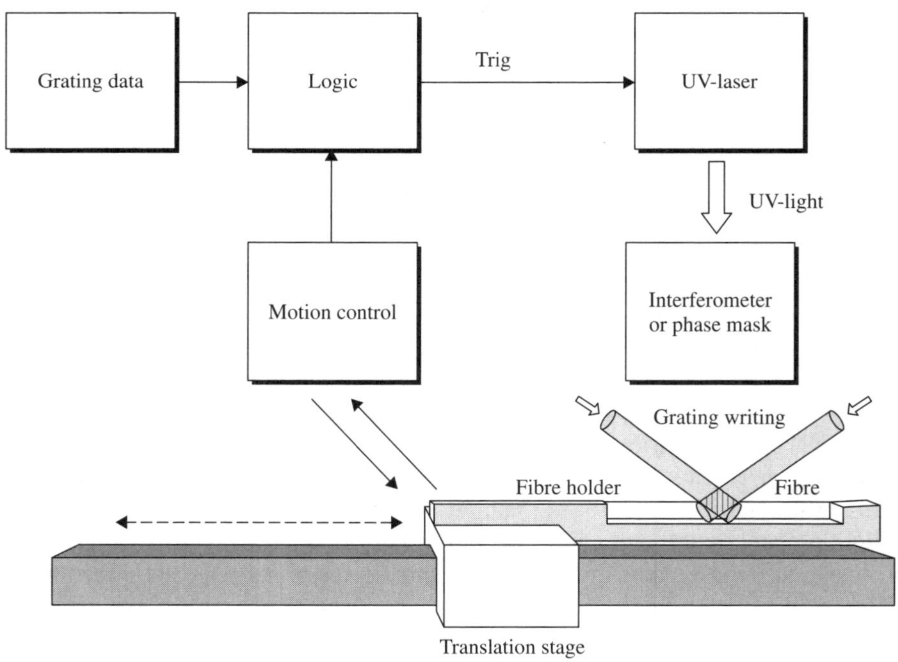

Figure 3.1 Schematic diagram of optical fibre grating process (Kersey, 1992)

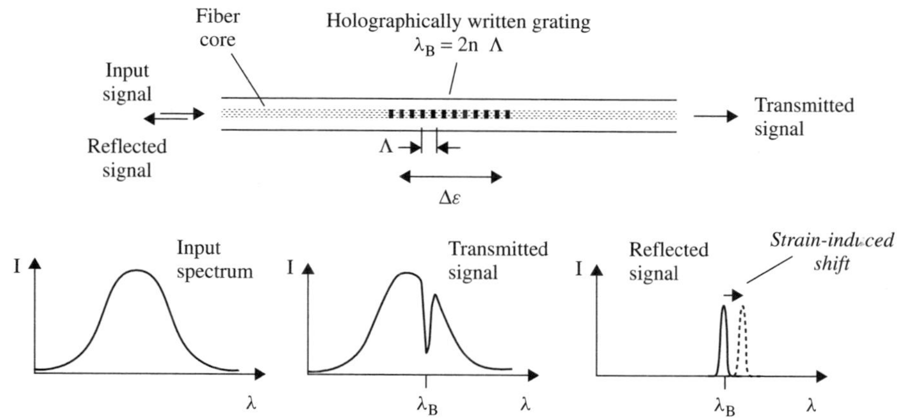

Figure 3.2 Sensing concept of a single FBG sensor (Kersey 1992)

of the holographic grating writing process. The grating, which is usually between 1 and 25 mm long, consists of periodic regions of higher and lower refractive indeces.

Broadband light passing through the grating is partially reflected from the grating. Interference between the individual grating planes results in a narrow wavelength range of light being reflected. The remaining wavelengths are transmitted through the grating. The reflection wavelength of the fibre Bragg grating is determined by the spacing between the individual grating planes. This spacing is set during manufacture of the Bragg grating. Thus the fibre Bragg grating's reflection wavelength can be set arbitrarily. Total reflection is possible in theory when the partial reflections from single grating planes add up in phase. The reflected component can be determined by the Bragg wavelength

$$\lambda_B = 2n\Lambda \tag{3.1}$$

where n is the average (effective) refractive index and Λ is the grating period. The reflection can be observed as peak in the spectrum. In contrast, the transmission exhibits a gap in the broadband spectrum. These effects can be used to measure the strain. When the load is applied to a structure, the grating is strained leading to a change of the Bragg spacing. This results in a change of the reflected wavelength and offers a robust measurement of strain. The relationship between the wavelength change and the strain is discussed in more details in Section 3.6.1. Figure 3.2 gives a sensing concept of a single element FBG sensor.

FBG sensors can be easily multiplexed and are widely used for strain measurement. The manufacturing of gratings and optoelectronic demodulation required for sensing significantly affect the costs of FBG sensors.

3.3 SENSOR TARGET SPECIFICATIONS

Operational load monitoring (OLM) systems can utilise optical fibre Bragg gratings for strain measurement. The sensors suitable for applications to metallic or composite structures will have to meet certain target performance specifications. These can be classified into system and environmental requirements. A set of system requirements for operational load monitoring, which underpin the design of the sensor system, includes: maximum sampling rate, maximum operating temperature range, maximum endurance temperature, maximum functional strain, design strain level, life time, operational acoustic noise level, number of sensors, location accuracy and load accuracy.

Sensors must be capable of achieving their full design performance within a certain region of temperatures. Similar requirements are needed for signal processors operating in unconditioned avionics bays. The load monitoring system should be additionally capable of achieving its full design performance when subjected to defined acceleration conditions and when within defined limits of ambient pressure.

The installed system should be of the minimum practical size and weight. Each signal processor, if implemented as a signal concentrator within a modular avionic environment shall be contained within an enclosure whose target dimensions are $100 \times 80 \times 20$ mm. The interface connectors will be on one of 100×20 mm sides. Power dissipation shall be significantly less than 10 W. If considered to be essential, then the Optical Signal Processor (OSP) may occupy more than one module or a multiple width module could be considered.

Other unspecified requirements are implicit in the system concept. The surface mounted sensors are aimed at retrofit installation on ageing, primarily metal skinned aircraft. The designs must therefore be compatible with installation on typical aircraft skin alloys of nonplanar topology in locations of restricted access. Application techniques must take this into account as well as required skill levels. It is also assumed that strain measurements must be fully, spatially resolved so that monitoring can occur at locations were principal strain directions are unknown or are subject to change under different operating conditions.

The costs of fibre sensors form an important element of target specification. These costs are significantly affected by manufacturing of Bragg gratings and optical signal processing. The higher costs of the OLM system based on optical fibre sensors must be compensated by the added value offered by the system.

Table 3.1 gives an example of the target performance specification of the optical fibre Bragg gratings OLM system developed within the *MONITOR* (Staszewski 2000) project.

Table 3.1 Optical sensor target specifications

Total number of Bragg grating sensors: 32
Number of sensors per patch: 4 (3 strain, 1 temperature only)
Number of strain patches per system: 8; arranged as 2 per channel from the OSP
Sensor type: Bragg gratings in single mode, 1300 nm fibre
Strain range per sensor: $\pm 3500\,\mu\varepsilon$
Spectral bandwidth per WDM sensor channel: 7.6 nm
Guard bandwidth: 1.3 nm
Bragg grating reflectivities: close to 100 %
Bragg grating reflection bandwidth: 0.4 nm
Peak reflection wavelength accuracy: ± 0.4 nm
Peak reflection wavelength interval: 8.9 ± 0.4 nm
Approximate gauge length: 3 mm
Operating wavelength regime: 1300 nm
Temperature sensor strain isolation: better than 0.3 % of backing material strain
Operating temperature range -54 to $+100\,^{\circ}$C.

Table 3.1 (*continued*)

Dimensions: to be determined
Inter-grating distance: to be determined
Fibre coating: to be determined
Materials: to be determined
Light source type: Superluminescent diode source pigtailed to single mode fibre
Source peak wavelength: 1300 nm (nom)
Source Spectral bandwidth: 70 nm (−3 dB selective)
Source power into fibre: 0.5 mW (integrated power)
Tunable filter type: Integrated optical acousto-optic tunable filter
AOTF tuning range: Across source spectrum
AOTF bandpass: 0.4 nm
AOTF scanning rate: up to 1.4×10^5 nm/s
AOTF tuning accuracy: ±0.011 nm
Resolution: 13-bit minimum. (40 dB dynamic range for a 70 nm tuning range)
Number of parallel output channels: 4
Signal detection: InGaAs PIN diode or equivalent
Number of detectors: 4 (one per channel)
Detection bandwidth: DC to 300 kHz (max)
Detector noise floor: 1 pW/$\sqrt{}$Hz (min NEP)
Signal processor bandwidth: 16.4 MHz (max) (13-bits at 2 kHz max)
Outputs (optical): 4 channel, single mode fibre connections
Output electrical: 4 parallel digital channels
Data type per channel: sequential, 13-bit (min) words at maximum rate of 16 000 words per sec
OSP target dimensions: *l*00 × 80 × 20 mm per module

3.4 RELIABILITY OF FIBRE BRAGG GRATING SENSORS

UV-induced fibre Bragg gratings have assumed significant importance in many fibre-optic applications. An important end-user issue is, of course, the reliability. A review some of the most important aspects that make fibre Bragg gratings and fibres prone to degrade or break and what measures that can be taken to prevent failure are discussed in this section. Both the fabrication process and the environmental conditions of the application determine the structural integrity and the lifetime of the device.

3.4.1 Fibre Strength Degradation

The mechanical strength of silica, determined by the chemical bonds of the glass matrix, gives in theory maximum stress values in the order of 20 GPa (Orowan 1949). However, defects and flaws, introduced during fabrication, on the surface of real fibres, lead to reduced fibre strength. The inert median breaking stress of pristine fibre is typically a few GPa, and Weibull plots have showed very narrow distributions (Limberger *et al.* 1996). The Weibull distribution is obtained by plotting the cumulative failure probability against the breaking stress. A narrow distribution means that all samples break close to the median breaking stress, a broad distribution means that some samples have failed at much lower stress. It should be pointed out, however, that the breaking stress of pristine fibres strongly depends on the conditions of drawing process and the chemical composition of the fibre.

The strength of fibres also depends on time and the environment. This can be explained by crack growth due to a stress enhanced chemical reaction that breaks the bonds or, as has been shown recently, that ageing produces a surface layer, presumably of hydrated silica, which is too weak to contribute to the strength of the fibre (Rondinella and Matthewson 1993). Whichever the process, water plays a major role in the degradation. A bare fibre is thus more susceptible to corrosion, but since most coatings are nonhermetic, coated fibre will degrade due to corrosion as well. In fact there is evidence that the patchy corrosion of a fibre with a nonperfect coating will cause a rougher surface and thus produce a faster degradation (Matthewson 1994).

The fibre coatings most commonly used today, such as UV-curable acrylates and thermal curing silicones, are opaque at the wavelengths where germanosilicate optical fibres are photosensitive. Thus, to be able to fabricate fibre Bragg gratings the coating has to be removed, mechanically or chemically, before the gratings can be written. Using a mechanical stripping tool usually damages the surface, which gives significant reduction in strength. When the fibre is stripped chemically, e.g. using hot sulphuric acid, it has been shown that the strength can be essentially kept the same as that of pristine fibre, (Limberger *et al.* 1996).

Recently the first Bragg gratings written through a UV-curable coating has been demonstrated (Espiniola *et al.* 1997). The results were encouraging; however, this jacket darkens and the transmittance degrades on exposure, which limits the effective grating writing time.

For efficient fibre Bragg grating fabrication, the fibre has to be significantly more photosensitive than standard telecommunication fibre. The photosensitivity can be enhanced in several ways. Standard fibre has a germania (GeO_2) concentration in the core of about 3 mol%. Increasing the germania concentration (Williams *et al.* 1992), codoping with suitable elements, like boron (Williams *et al.* 1993), or tin (Dong *et al.* 1995), or by 'loading' the fibre with hydrogen (Lemaire *et al.* 1993), will enhance the photosensitivity.

It appears that some of the photosensitive fibres, especially the boron-codoped fibre, appear to be more brittle than standard fibre when the coating has been removed prior to the grating writing. This could be due to increased internal stresses introduced by the dopant during preform fabrication and fibre drawing.

The grating writing process itself has been shown to degrade the fibre strength (Limberger *et al.* 1996; Lemaire *et al.* 1993). The decrease in mechanical resistance has been both attributed to surface damage (Limberger *et al.* 1996) and to increased internal tensile stress (Lemaire *et al.* 1993). The degradation depends on both the pulse energy density and accumulated dose, although the latter is more important (Limberger *et al.* 1996). A fibre showed degradation expressed as median breaking stress of almost a factor of two and the Weibull distributions were significantly broadened. The possibly reduced strength of photosensitive fibres may thus be compensated by the fact that they require a lower dose to accomplish the same index modulation.

The investigations in (Feced *et al.* 1997) have showed that the UV-induced degradation is wavelength dependent. Fibres with gratings written with a 193 nm source showed less reduction in strength than the same fibres with grating written with a 248 nm source.

3.4.2 Grating Decay

The formation of gratings generally involves the transition of the glass to a metastable state. This can either be the UV-induced displacement of an electron to a trapping site, or a local structural phase change in the exposed parts of the glass. At room temperature the gratings appear to be stable, i.e. the reflectivity and the Bragg wavelength remain unchanged over time, but when exposed to significantly higher temperatures the gratings start to decay and at sufficiently high temperatures they disappear completely. Thus, the UV-induced refractive index change is not a thermally stable state but reversible resulting in a decay of index. It can be assumed that the electrons or local structures are trapped in sites with varying 'depth', associated with different activation energies before they are reversed to the original state. The distribution between the different sites and the possible depths determines the thermal stability of the gratings. This distribution depends largely on the type of fibre, the pulse energy density, and the accumulated dose. An excellent review of the mechanisms involved is given in (Douay *et al.* 1997).

Comprehensive accelerated ageing tests have been carried out over the last few years. They are well summarized in (Douay *et al.* 1997; Baker *et al.* 1997; Kannan *et al.* 1997). In what follows, some typical results will be given. It is possible to distinguish between four types of gratings, type I, type II (Archambault *et al.* 1993), type IIa (Riant *et al.* 1996), and chemical composition gratings (Fokine *et al.* 1996).

Type I is by far the most common type. These grating are the ones most readily produced but they also prove to be the least stable. The results in the literature (see Figure 3.3) indicate that a higher dose will render more stable gratings and that boron doped fibres although very photosensitive give less stable gratings than those fabricated in highly Ge-doped or Sn-doped fibre (Dong *et al.* 1995). H_2 loading seems to affect the stability marginally.

In accelerated ageing tests (see Figures 3.4 and 3.5) at 85 °C, the gratings decay rapidly at first followed by a substantially decreasing rate of decay. After several thousand hours the gratings still show reflectivities of 80–95 % of the initial value. At 300 °C the initial decay is much faster than in the previous case and the gratings saturate at a value of about 20 % of the initial reflectivity after a few thousand hours. By pre-annealing the gratings they can be 'burnt in' or stabilized. The decay of the reflectivity after such an annealing process can then be predicted according to a power law (Erdogan *et al.* 1994). At modest temperatures grating lifetimes without significant degradation can be estimated well above 25 years.

In some applications the gratings may be exposed to substantially higher temperatures than room temperature for extended period of times. In some fibres when writing a type I grating, the grating reaches a maximum index modulation and then starts to disappear at further exposure. If the process is continued beyond the erasure point a new grating is formed. This grating turns out to be much more stable and is referred to as a type IIa grating. Gratings of this type written in germanium free oxynitride fibre have been shown to survive temperatures as high as 1200 °C for 30 min. (Dianov *et al.* 1997).

Another very stable type of grating is the type II grating (Archambault *et al.* 1993). This type of grating is formed when the pulse energy density exceeds a certain value, depending on type of fibre. Over this value the fibre gets physically damaged in the strongly absorbing core region. It thus takes very high temperatures to reverse this kind of gratings.

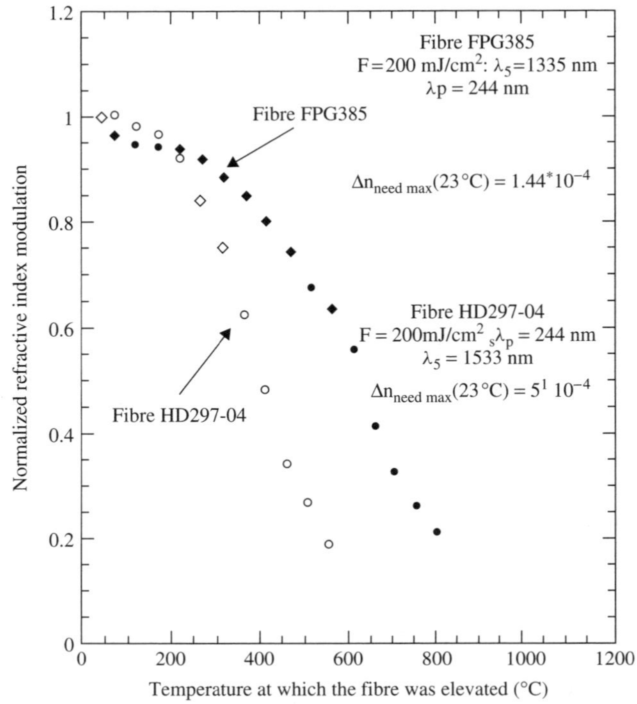

Figure 3.3 Normalised index modulation after a 30-minute annealing for gratings written in germanosilicate fibre (black symbol) and in Boron codoped germanosilicate fibre (open symbol) (Douay *et al.* 1997, ©1997 IEEE)

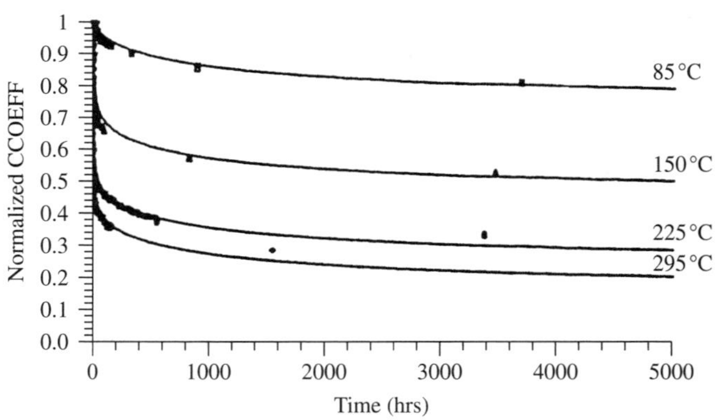

Figure 3.4 Accelerated ageing tests of hydrogen loaded fibre (Baker *et al.* 1997, ©1997 IEEE)

A new fibre, called *Tiger Fibre*, and grating fabrication method that makes it possible to produce gratings with so far unprecedented stability has been recently developed (Fokine *et al.* 1996). Gratings have, for instance been subjected to 800 °C for more that 500 hours without showing any sign of degradation, as shown in Figure 3.6. The grating fabrication procedure is more complicated than writing ordinary gratings, which of course will

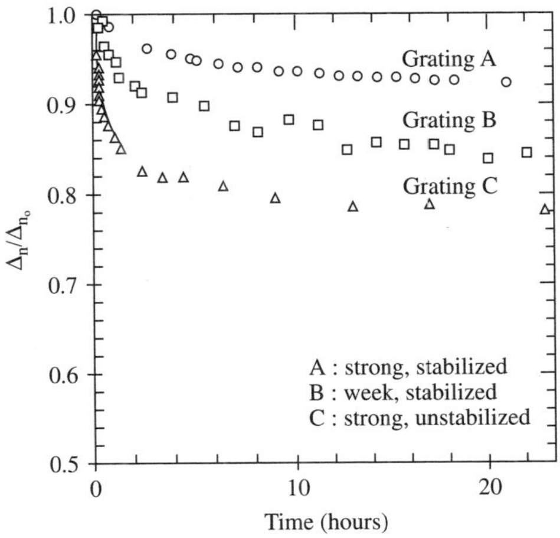

Figure 3.5 Decay of different types of gratings at 200 °C. Gratings A and B have been stabilised (Kannan *et al.* 1997, ©1997 IEEE)

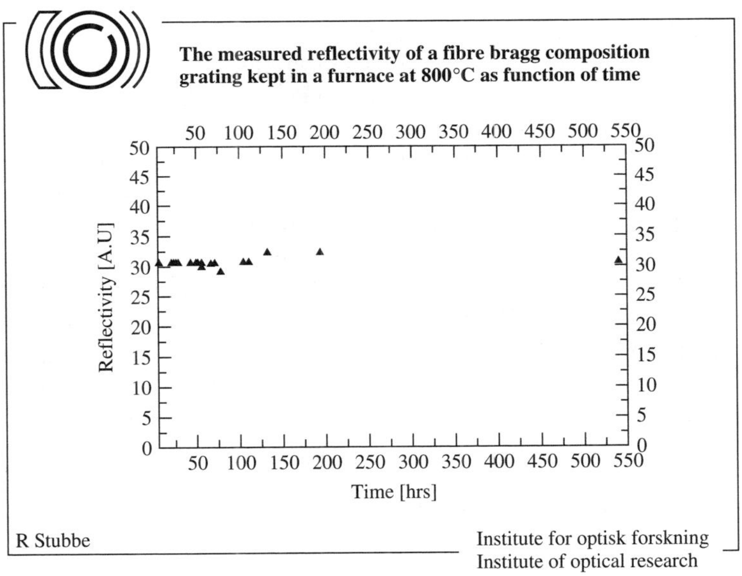

Figure 3.6 Ageing test at 800 °C of grating written into a *Tiger fibre* (Fokine *et al.* 1996)

make the gratings potentially more expensive. On the other hand their high temperature resistance makes it possible to involve coating and capsulation procedures that are not possible with other types of gratings.

3.4.3 Summary

The exposure of fibres to UV-light has a strength degrading impact. The median breaking stress may be reduced by a factor of two. A highly UV-sensitive fibre is thus a preferred choice in order to shorten the exposure time. Even more important, however, is the handling procedure during the grating manufacturing process. This and environmental factors may very well be what determines the structural integrity of the device. The fibre has to be chemically stripped and preferably kept under dry conditions while uncoated. To really ensure a high long-term reliability the fibre should, after the writing procedure, be recoated with a hermetic coating, such as a metal or carbon coating.

The fibre Bragg gratings themselves will, at room temperature conditions, not limit the lifetime of the device. A lifetime of at least 25 years is expected independently of fibre type. However, if higher temperatures are involved at any stage, e.g. during the manufacturing, special care has to be taken in the choice of grating type. B-doped fibres have shown the fastest decay and should in this case be avoided if type I gratings are to be used. If the fibre grating is going to be exposed to very high temperatures, gratings made in *Tiger Fibre* are a solution.

3.5 FIBRE COATING TECHNOLOGY

Fibre optic Bragg grating sensors for real time strain measurement in composite materials require an appropriate protective coating layer. The major design parameter for the coating material is high temperature resistance. The curing process of the composite, where the optical fibre sensors are embedded, requires a temperature of 190 °C for several hours. The fibre coating must be able to withstand this temperature without significant degradation. The operating temperature is specified at 100 °C, which the coating thus must withstand for the entire lifetime of 30 years (Stork 1992). Other important parameters for the coating material are: good coating- and film-forming properties, sufficient adhesion to the silica fibre, sufficient mechanical and chemical protection of the silica fibre and processing feasible in available fibre drawing equipment. A number of polymer coating materials for this application have been investigated in (FFA 1997). Of the material tested a polyimide (DuPont, Pyralin PI2525) was recommended due to ease of coating, processing, adhesion and high temperature properties. This section gives the background material related to the chemistry of coating silica with polyimide and the major parameters associated with the coating process.

3.5.1 Polyimide Chemistry and Processing

Polyimides constitute an important class of polymers due to their many desirable properties, some of which are (Gosh and Mittal 1996):

- good electrical insulating and HF properties;
- inertness to most solvents;
- low thermal expansion (for a polymer);
- wear resistance;
- radiation resistance;
- long-term stability;
- exceptional high temperature stability.

Due to these properties polyimides have found widespread use in many demanding high-technology applications and lot of effort has gone into tailoring polyimides for a

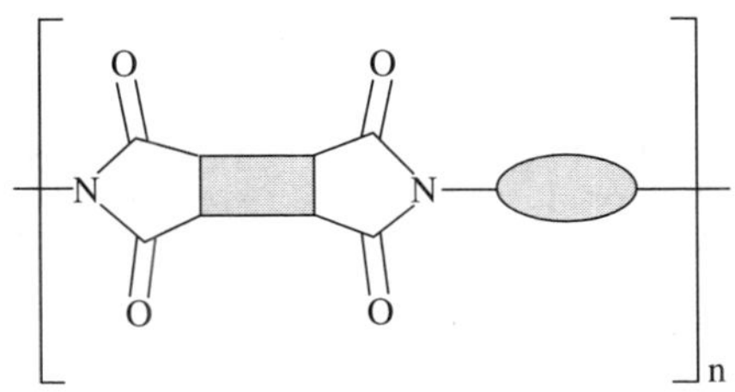

Figure 3.7 Generic structure of polyimide

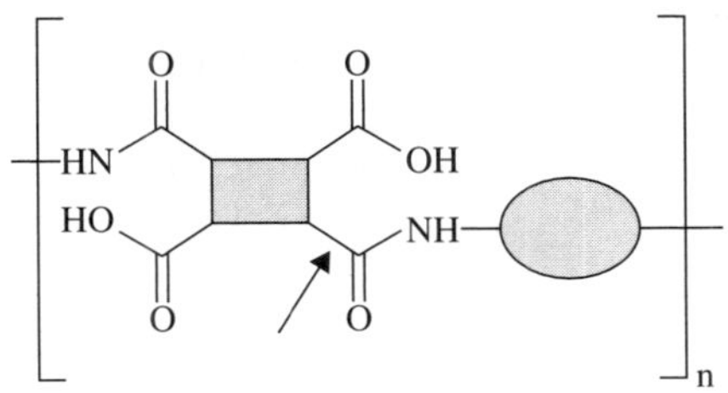

Figure 3.8 Generic structure of polyamic acid

widespread range of uses. There exists a multitude of different polyimides with vastly varying chemistry and different properties. It is clearly beyond the scope of this book to further expand on this.

The generic structure of a polyimide is shown in Figure 3.7. The backbone (shadowed objects in Figure 3.7) of the majority of the polyimides has a linear, planar aromatic or heteroaromatic structure. They are generally infusible and insoluble and therefore very difficult to process (e.g. apply as a coating on a fibre). One way past this obstacle is to process a soluble precursor and then convert it to the (insoluble) imide form once the desirable shape has been obtained. The most common precursor is polyamic acid (PAA) as illustrated in Figure 3.8.[1] The polyamic acid is usually dissolved in a suitable solvent such as N-methyl-2-pyrrolidone (NMP) or N,N-dimethylacetamide (DMAc) to a suitable viscosity.[2] After application of the solution the solvent is (partially) evaporated at elevated temperature and the polyamic acid is ready for ring closure (cyclisation, imidisation).

The preferred way for the subsequent cyclisation to an imide is by raising the temperature to above 130 °C[3] at which temperature the ring closure to imide (Figure 3.7) starts, obviously giving off a water molecule for each imide group formed. The ring closing

[1] Here and in the rest of this chapter the chemical formulae given are typical examples. For the polyimide actually used (DuPont Pyralin PI 2525), there might be differences in substituents and part of the backbone structure. This 'however' does not influence the major properties and behaviours discussed here. However, it can and will influence parameters such as absolute curing temperature, susceptibility to oxidative degradation, bonding strength to various surfaces, etc.

[2] These solutions are thermally unstable (the polyamic forming reaction between a diamine and a dianhydride is an exothermic equilibrium) and are also very sensitive to moisture. Moisture is present in the ambient air and is also formed by spontaneous ring closure. The water will react with the chain terminating anhydrides, eventually shortening the chain and lowering the average molecular weight. These solutions should be kept refrigerated in closed containers at all times.

[3] Various authors give figures ranging from 120 °C to 175 °C.

process, although it looks simple, is quite complicated and not fully understood in all details (Gosh and Mittal 1996). Initially it is fast as the chain is quite flexible and the viscosity is still low; the molecule can rotate round the indicated bond taking on conformations that bring the functional groups close together for reaction to take place. The water given off can easily diffuse from the reaction site. As the imidisation proceeds the chain becomes more rigid and cannot rotate freely, more solvent is driven off and T_g increases. When T_g reaches the reaction temperature there is usually a marked decrease in reaction rate. An increased temperature will close more rings but a ring closure degree in excess of 95 % is seldom attainable. During the imidisation there are also some competing side reactions (Gosh and Mittal 1996). It should also be noted that although polyimides show excellent thermal stability in inert atmosphere, withstanding temperatures >500 °C, oxidative breakdown in air can set in at much lower temperatures 250–350 °C. The final outcome of the imidisation process thus depends on the time and temperature history. The imidisation or curing process can thus, somewhat arbitrarily, be divided into three (overlapping) phases:

A. Solvent evaporation – Drying

This should take place at a temperature/time that effectively removes most of the solvent without causing formation of voids and bristles in the remaining material. This is normally done at temperatures below those at which significant ring closure takes place and for a period of up to 30 min. In fibre manufacturing only a few seconds are available for this step (below), which thus must be performed at higher temperatures.[4]

B. Closure of the majority of rings – Curing

During this step the majority of ring closures take place at the same time as the most of the remaining solvent is removed. This step usually (Gosh and Mittal 1996) takes place at the boiling point of the solvent for 10–30 minutes. For fibre coating, only about half a minute is available so this step should possibly be performed at a higher temperature.

C. Finalisation of ring closure – Post Bake

During this step the last possible ring closures are done. Normally this is done by heating up to 450 °C for up to 1 hour. In order to avoid oxidative degradation this post bake is often performed in inert atmosphere. For moderate fibre lengths this process can be done x-drawing tower if step B produces a dry, nonsticky coating.

For optimal results, the various parameters of the above mentioned phases must be experimentally determined.

3.5.2 Polyimide Adhesion to Silica

Although there are examples of the opposite, polyimides tend to have a poor adhesion to silica surfaces (Gosh and Mittal 1996; Plueddemann 1991). Published peel strength data for polyimides on SiO_2 are in the range 0–800 J/m^2. The higher values have been

[4] Pyralin PI2525 is dissolved in NMP with a boiling point of about 210 °C. Thus the ring closing reactions proceed with an appreciable speed at temperatures that effectively evaporates the solvent.

obtained for high temperature post bakes. There are several theories to explain adhesion between surfaces. Some of these are:

A. Chemical Bonding

A covalent bond has a high energy (up to several 100 kJ/mol) and is the preferred method for obtaining good adhesion. The prerequisite for this is the presence of functional groups in the two materials that can react with each other. These groups must also be in close proximity (<5) of each other. Ionic, hydrogen and van der Waals forces can also give rise to good adhesion but are generally more susceptible to attack from humidity.

B. Acid–Base Interaction

Strong acid-base interactions between the substrate and the overcoat will generally result in good adhesion. However, calculations on the acid base interaction between a polyimide and silica indicate very low adhesion (Gosh and Mittal 1996).

C. Diffusion

In this case the materials forming the interface diffuse into each other forming an inpenetrated network.

D. Mechanical Interlocking

The surface structure of one layer mechanically locks into the surface structure of the other.

The reasons behind the observation that some polyimides, after a high temperature post bake, can show a reasonable adhesion to silica is not yet fully accounted for (Gosh and Mittal 1996).

3.5.3 Silane Adhesion Promoters

In order to improve adhesion silane adhesion promoters (coupling agents) find a widespread use. One side of the coupling agent molecule binds covalently to the silica surface while the other can interact with the coating layer in different ways. The generic structure of such a promoter is shown in Figure 3.9.

The three RO groups in the silicon tetrahedron are alkoxy groups that eventually will couple to the glass surface, while the X group is the group that interacts with coating. X is given appropriate properties for a specific coating. Ideally X will be constructed so as to react covalently with appropriate parts of the coating molecules. Otherwise the interaction can be through ionic, hydrogen, van der Waals forces, etc. For promotion of adhesion

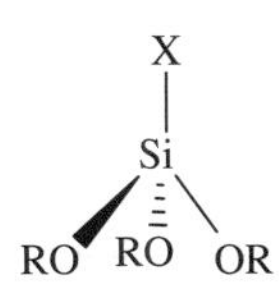

Figure 3.9 Generic structure of silane coupling agent

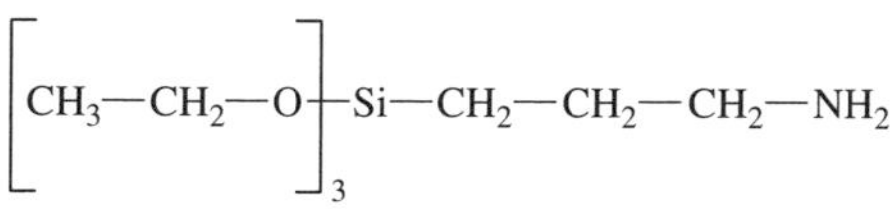

Figure 3.10 Structure of γ-APS

between polyimides and silica several silanes have been used (Plueddemann 1991). The most commonly used is γ-aminopropyl-triethoxysilane (APS) which also was used in this work, as shown in Figure 3.10.

The chemistry at both ends of the silane molecule is very complex and dependent on the conditions of its use. An in depth discussion of this is beyond the scope of this book. Below is given a short summary of the most important aspects in the present application.

The trialkoxysilanes are usually hydrolysed before application to the glass surface. This hydrolysis is done in water or a water/alcohol mixture. Two reactions take place: first the three step conversion to the corresponding silan(3)ol which then ultimately condenses to siloxane macromolecules, as shown in Figure 3.11.

The siloxane molecules are unsuitable for coupling to silica and will eventually precipitate when they get large enough. The solution becomes cloudy. The hydrolysis is fast (minutes) while the condensation is slow (several hours to days). At higher concentrations the condensation rate is faster. Also the silanol molecules tend to form aggregated monomers (Plueddemann 1991) through hydrogen bonding. The transition point for this behaviour is 0.15 % for APS. The silanol concentration should therefore be kept as low a possible.[5]

The silanol will react with the silica surface at appropriate sites. As the distance between the available reactive sites on silica surface is believed to be large, only one silanol group per molecule will bind to the glass. The rest of the silanol groups will take part in a condensation reaction ultimately leading to a continuous monolayer[6] of a siloxane with the organofunctional groups facing upwards, as shown in Figure 3.12.

This can be thought of as a two step process:

1. The dissolved silanol will react with the silica surface forming hydrogen and possibly also some covalent bonds to the glass at relatively low temperature.

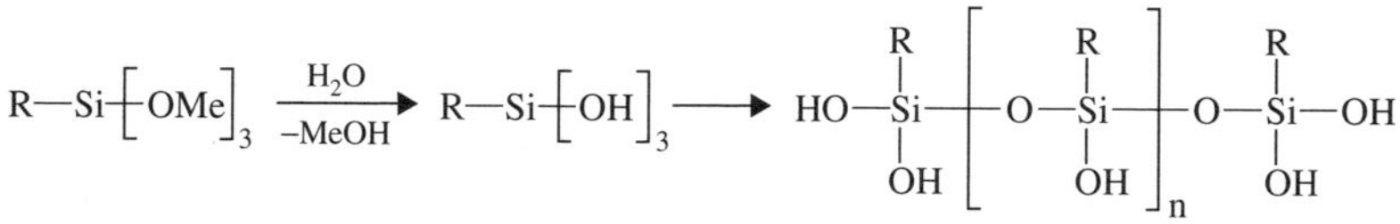

Figure 3.11 Hydrolysis of a trialkoxysilane

[5] It should, however, be noted that a low silanol concentration can come into conflict with the requirement of a high reaction rate as the available reaction time in fibre production is only a few seconds. The optimum concentration must be determined empirically.

[6] Actually a very thick layer will be formed with the bottom layer covalently bound to the silica. The top layers will be less well bounded. They could be used for adhesion promotion (through mechanical interlocking, etc...) if no covalent coupling between the coating and the promoter is possible. Otherwise the excess of silanol should be rinsed from the surface leaving preferably only the monolayer.

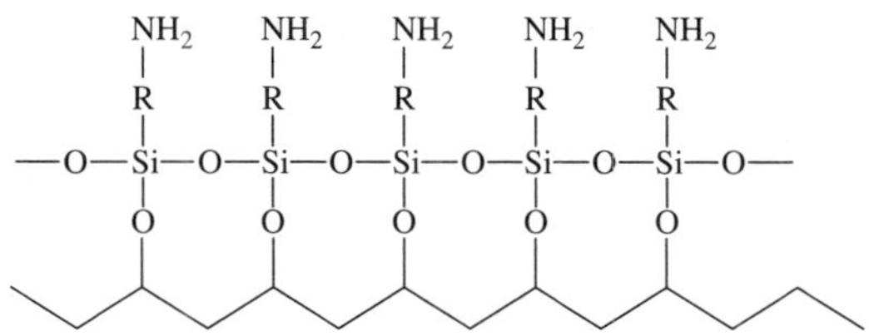

Figure 3.12 Amino organofunctional silane coupled to silica

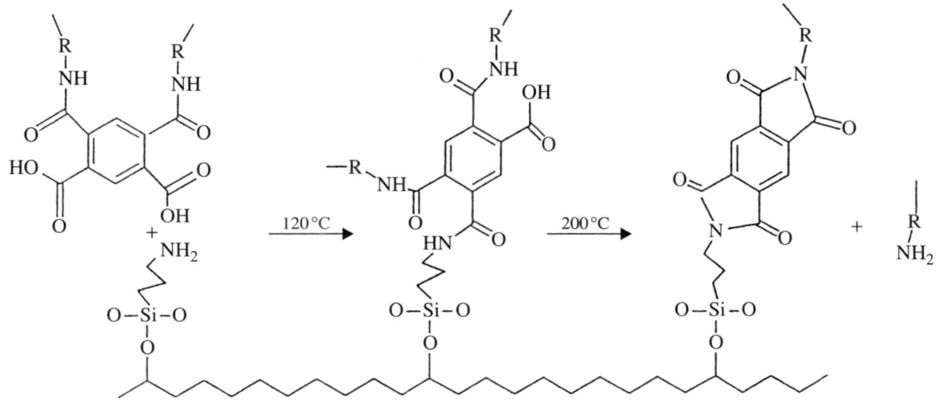

Figure 3.13 APS reaction with polyamic acid during thermal curing

2. At elevated temperature (100–120 °C) the solvent is driven off and the reactions with the silica surface and the condensation to a continuous siloxane layer are driven to completion.[7]

The amino groups are now available for reaction with the polyamic acid, as shown in Figure 3.13.

After coating the silanised silica with polyamic acid the solvent is driven at a comparatively low temperature (120 °C above). At this temperature the covalent bond between the APS and the PAA will also be formed. The temperature is then increased to increase the ring closure rate.

It should be noted that heating an organofunctional siloxane to too high temperatures (>160 °C) in air could cause oxidative degradation. Thus the curing and post bake (if used) should preferably be done in an inert atmosphere.

3.5.4 Experimental Example

This section gives an example of the experimental work performed in the laboratory and manufacturing conditions. The work done has been both laboratory tests on sheets of

[7] It should be noted that for the subsequent coating with polyamic acid the substrate must be effectively free from water. The PAA is rapidly degraded by any water present and will actually form a precipitate if the water concentration becomes too high.

glass and silica for evaluation of process parameters and actual attempts to coat during manufacturing of the fibre in draw towers.

All coatings were made with DuPont Pyralin PI2525 not more than a few months old, transported and stored refrigerated to $-25\,^{\circ}C$ or lower. The Pyralin 2525, containing $25 \pm 1\,\%$ of PAA in NMP was used undiluted. The first fibre coatings attempted utilised a silicone rubber (RTV) funnel with an opening of 0.2–0.3 mm filled with Pyralin at room temperature. The coatings thus obtained were very uneven. The temperature of the Pyralin was increased to $37\,^{\circ}C$ to adjust viscosity and the coating was performed in a very similar way to the method used for the more familiar acrylic fibre coating. The Pyralin is pressurised to just below 1 bar and the fibre passes through about 5 cm of the solution in a pressurised vessel, finally exiting the vessel through a diamond nozzle with a diameter ranging from 180 to 300 μm. This modification resulted in quite even and concentric coatings. The fibre, however, turned out to be quite brittle and this was attributed to pinholes in the imide. To overcome the effect of pinholes the fibre was double coated. A primary coating was applied in the draw tower during fibre manufacturing as described above. This coating was thermally cured to a nonsticky state and the fibre was wound onto a bobbin. The fibre then was given a second coat (using the same imide) by passing it through the appropriate parts of the draw tower. The adhesion of the coating to the fibre however turned out to be very poor and thus a DuPont adhesion promoter, VM652,[8] was utilised in the process. The promoter was used neat and applied through a silicon funnel as described above. The results however were quite disappointing: the coating now became non concentric and partly uneven. Parts of the coating were very thin indeed ($<2\,\mu$m). No significant improvement in adhesion was observed. In order to gain a better understanding of the processes and parameters involved, a series of laboratory experiments on sheets of glass and silica were performed aimed at finding approximate conditions for silanisation and imide formation. This will be described in more detail below. Finally the results from the laboratory test were carried over to the draw tower. This resulted in improved adhesion but coating concentricity still remains a problem.

During these tests various experiments were performed in order to gain a better understanding of some of the critical procedures outlined above. The first step was a series of initial investigations aimed at gaining a better understanding of the silanisation process. The coatings for unsilanised fibres made so far were examined in a stereo microscope. If the coating was severed with a scalpel it fell off the fibre virtually by its own weight. The adhesion obviously was very low indeed. The introduction of the adhesion promoter DuPont VM652 resulted in uneven coatings. Mixing of the Pyralin solution even with small amounts of the VM652 solution resulted in precipitation. This is a possible explanation for the uneven coatings obtained. Due to the high boiling point of the solvent ($112\,^{\circ}C$) some residual solvent can be carried into the Pyralin by the fast moving fibre thus causing precipitation. It is also common practice to evacuate polymer solutions before application in order to remove dissolved air that can cause bubble formation and voids in the finished product. Evacuation of Pyralin however did not reveal any dissolved gas and is thus unnecessary.

In the next step plane sheets of glass and silica were pretreated and silanised. Then they were manually coated with PAA and processed according to various schemes. At appropriate stages the samples were evaluated. The chemicals used were: Imide

[8] This is actually γ-APS dissolved in 1-methoxy-2-propanol.

1 – DuPont PI2525, Imide 2 – Epoxi Technology 0G105, Silane 1 – DuPont VM652 (γ-aminopropyl-triethoxysilane dissolved in 1-methoxy-2-propanol), Silane 2 – OSI Specialities γ-aminopropyl-triethoxysilane, neat. The evaluation of coating adhesion was made qualitatively under a stereo microscope utilising a scalpel. By careful scraping of the coating three different behaviours could be identified:

- The coating separates very easily from the substrate in the form of large continuous flakes–very poor adhesion;
- The coating can quite easily be separated from the substrate but will only form small flakes–some adhesion to the substrate;
- The coating can only be made to separate from the substrate with some difficulty. It will form only very small flakes–fairly good adhesion.

After the initial investigations and the study of the most important processing parameters, the experimental process was performed following the procedure shown in Figure 3.14. In order to obtain a contamination free surface, the sample ($\approx 3 \times 6\,\text{cm}^2$) was etched in 5 % hydrogen fluoride (HF) for 10 min, rinsed with deionised water and blown dry with N_2. In some cases the sample was dried at 900 °C for 30 min. The sample was immersed in the silane solution for 15 minutes, thus giving ample time for the reaction to go to completion (room temperature). Excess silane solution was removed by rinsing with

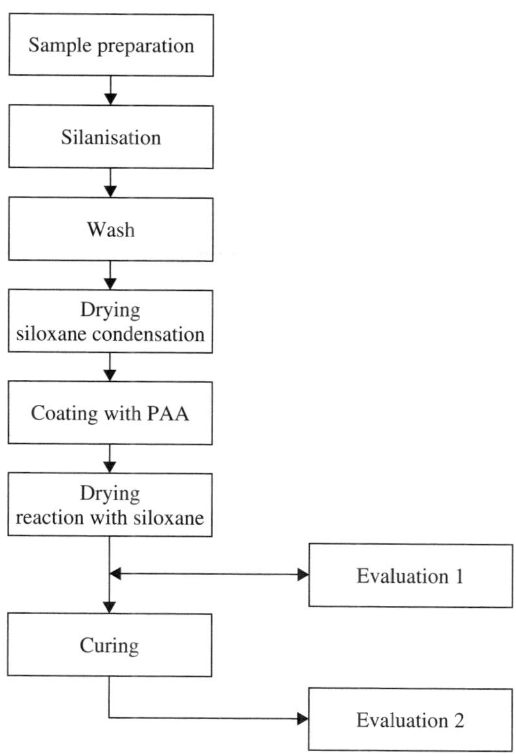

Figure 3.14 Schematic diagram of the experimental process

the appropriate solvent: water or dry methanol. The object was blown dry with N_2 and placed in an oven at 110 °C for 10 minutes in order to facilitate formation of the siloxane bridges. When the substrate had cooled down to room temperature, a thin (0.05–0.2 mm) layer of Pyralin was manually applied utilising a thin glass rod. The substrate was dried at 150 °C for 60 min. After this stage the coating was dry and non sticky. The coating was then evaluated for the first time following the procedure described above. The substrate was heated to 30 °C or 350 °C for 60 min and the coating was evaluated again. Tables 3.2–3.4 give examples of the experimental results. The results can be summarised as follows:

- Curing at elevated temperatures (300–400 °C) does not improve adhesion to the substrate. The results actually indicate a slight decrease in adhesion. This could be explained by two factors. Firstly, the drying step 150 °C for 60 min is enough to drive the ring closure close to almost completion. The extra closures obtained at the high temperature

Table 3.2 Fibre coating experiment – series 1

Sample	Imide	Silanisation	Evaluation 1	Cure	Evaluation 2
Al	1	1	2	350 °C, 60 min	2−
A2	1	none	1	350 °C, 60 min	2−
B1	2	1	2	350 °C, 60 min	1
B2	2	none	1+	350 °C, 60 min	1
C1	1	5 % of 2 in 1	2	350 °C, 60 min	1+
D1	2	5 % of 2 in 1	2	350 °C, 60 min	2−
E1	1	0.6 % of 2 in H_2O	1+	350 °C, 60 min	1+
F1	2	0.6 % of 2 in H_2O	2	350 °C, 60 min	1+

Table 3.3 Fibre coating experiment – series 2

Sample	Imide	Silanisation	Evaluation 1	Cure	Evaluation 2
G1	1	1	2+	300 °C, 60 min	2−
G2	1	none	1	300 °C, 60 min	2
H1	2	1	2	300 °C, 60 min	2+
H2	2	none	1−	300 °C, 60 min	1

Table 3.4 Fibre costing experiment – series 3

Sample	Imide	Silanisation	Evaluation 1	Cure	Evaluation 2
K1	1	1	2+	300 °C, 60 min	2−
K2	1	none	1−	300 °C, 60 min	1
L1	2	1	2	300 °C, 60 min	2
L2	2	none	1−	300 °C, 60 min	1
M2	1	0.6 % of 2 in H_2O	3	300 °C, 60 min	3−

cure will only effect adhesion properties marginally. Other important properties however might be more effected. Secondly, heating to high temperature in air will cause partial oxidation of the siloxane molecule's organic end, thus rupturing the covalent bond to the imide.

- There is no significant advantage of using the more expensive Epoxi Technology OG 105 imide instead of DuPont Pyralin 2525.
- Silanisation definitively improves adhesion in most cases.
- The silanisation procedure does not seem to be very sensitive to the silane solvent.

These results motivated the following changes to the fibre coating procedures:

- Silanisation by a 0.6 % w/w solution of APS in deionised water (less dangerous than an alcoholic solution) at room temperature.
- Removal of excess silane solution by thorough washing in deionised water.
- Drying and formation of siloxane bridges in inert atmosphere at about 120 °C.
- Drying and curing of the imide in inert atmosphere at as low a temperature as possible. There is a theoretical advantage if the primary coating is not completely cured before application of the second layer. In this case the molecules of layers can diffuse into each other before final curing thus possibly enhancing adhesion between the two imide coatings.

The previous fibre coatings experiments had shown that droplets from the silane applicator will glide along the fibre to the PAA applicator and eventually cause fibre jamming and/or very uneven layers. To prevent this, a special device was constructed to remove these droplets at the outlet of the applicators.

The modified process used for coating of fibre was performed in the next step. Figures 3.15–3.17 show a schematic diagram and photographs of the coating equipment, respectively. Altogether, a few hundred meters of imide-coated fibre were produced. Two samples of the fibre produced were post baked in N_2 for 18 hours at 220 °C and 3.5 hours at 285 °C, respectively.

Coating properties were evaluated under the microscope. Scraping the fibre with a scalpel indicated that the coating had a better adhesion than earlier coatings. There was no obvious difference in coating adhesion between the post baked and the not post baked fibre. Possibly the coating baked at 285 °C is somewhat harder. There were no signs of separation between the two coating layers. Parts of the fibre, close to the point were water contaminated the Pyralin, was brittle. That was probably due to poor coating quality as the water reacted with Pyralin. For the rest of the fibre the strength was very good. Parts of the fibre showed periodic variations in coating thickness with the amplitude of $\approx$12 μm and the period of 0.5–1 mm. The coating observed was not concentric with the fibre, as shown in Figure 3.18.

The devised methods for silanisation and polyimide curing have improved the adhesion to the silica fibre. The study shows that the adhesion is very important for coating. This requires appropriate utilisation of silane mixtures and process parameter optimisation. Parameters of interest in this respect are surface energies of fibre and coating, viscosity, nozzle diameter, fibre velocity and drying temperature.

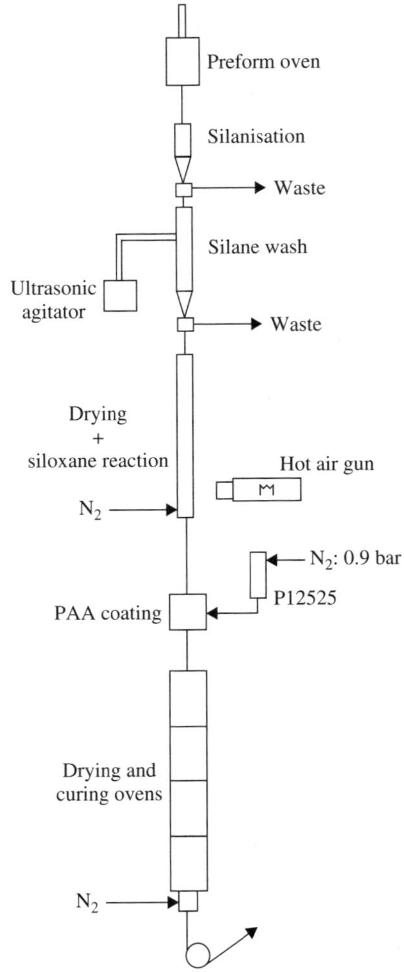

Figure 3.15 Schematic diagram of the coating equipment

3.5.5 Summary

Protective coating layers are important for a fibre optic Bragg grating sensor used to measure strain in composite materials. A polyimide seems to be the appropriate material for this application. The major process steps for coating a silica fibre with polyimide are summarised below. The major process parameters, which need to be optimised, are indicated in italics.

Silanisation

- Coat the fibre with the silanol from a water solution.
  - — *Type of silane or silanes (mixture) used*
  - — *Concentration*
  - — *Reaction time and temperature*

Below is a photograph of the draw tower:

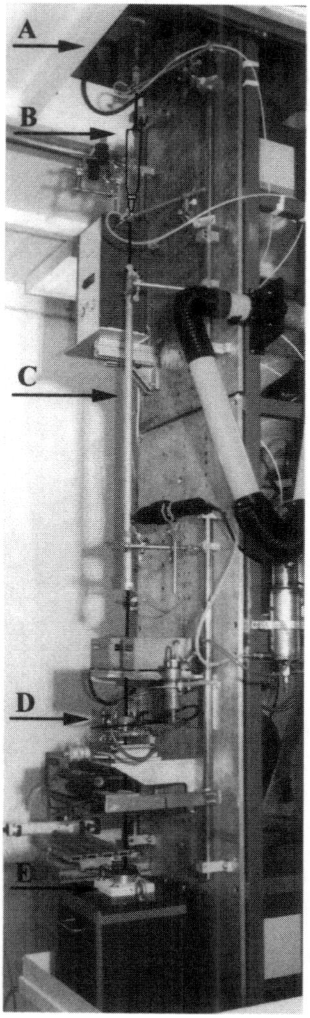

Figure 3.16 Silanisation and PAA coating equipment

- Remove excess silanol.
- Drive of the solvent and form the covalent bonds to silica and the continuous siloxane layer at elevated temperature, preferably in an inert atmosphere.
  — *Time and temperature*
  — *Gas flow*

Coating with polyamic acid

- Coat the silanised fibre with a solution of PAA in e.g. NMP.
  — *PAA concentration*
  — *Temperature*

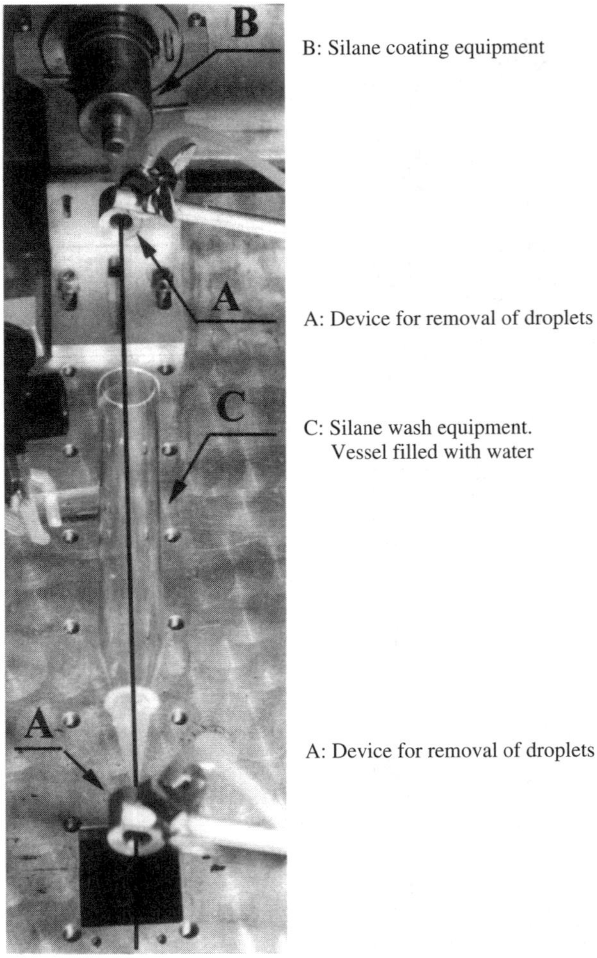

Figure 3.17 Close up of silanisation equipment

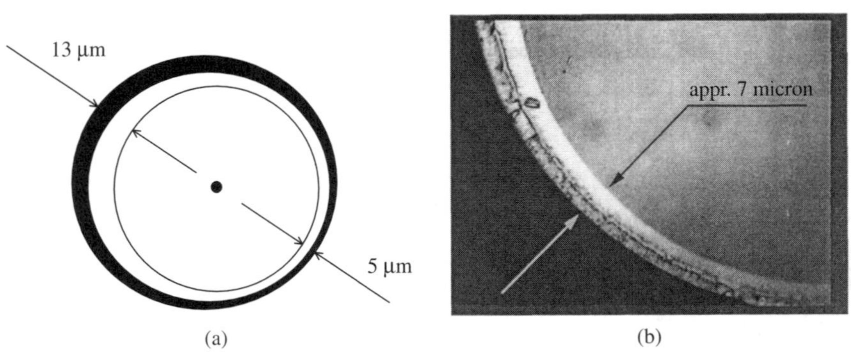

Figure 3.18 Coating unconcentricity: (a) schematic diagram; (b) microscopic photograph

- Evaporate the solvent at the same time forming covalent bonds to the organofunctional groups of the siloxane layer.
  — *Temperature and time*

Ring closure of polyamic acid

- Increase temperature to start ring closure of the PAA. At the same time most of the remaining solvent is driven off. To avoid oxidation of foremost the siloxane this step should be done in an inert atmosphere.
  — *Temperature and time*
  — *Gas flow*
- Possibly perform a post bake in inert atmosphere to enhance the degree of completion of the ring closures.
  — *Temperature and time*

There is thus quite a lot of empirical process optimisation to be done once selections of imide and promoter systems have been done.

3.6 EXAMPLE OF SURFACE MOUNTED OLM SENSOR SYSTEM

This section describes an example of a surface mounted optical strain gauge system for use in Operational Load Monitoring (OLM). The system comprises three major elements: sensors, Optical Signal Processor (OSP) and interconnections between sensors and the OSP. The architecture is illustrated in Figure 3.19. The OSP is a hybrid opto-electronic module, which provides a light source for a number of sensors, a set of detectors, a means of spectral analysis and signal processing to convert spectral measurements into meaningful strain information. Light is divided to supply, simultaneously, several optical fibres, each of which contains several sensors. Each fibre also provides a return path for reflected light from each sensor, which can then be detected for spectral analysis. The architecture shown in Figure 3.19 has four fibres, each containing eight sensors. The sensing elements themselves are in-fibre Bragg gratings which act as wavelength selective reflectors of incident light. The Bragg grating structure is a region of fibre in which the refractive index is spatially modulated in a sinusoidal pattern to form the grating. This distributed reflective structure, usually defined over a few mm of fibre length, has a peak reflection wavelength, λ_B, defined by

$$\lambda_{\mathrm{B}} = \Lambda/2\mathrm{n} \tag{3.2}$$

where Λ is the refractive index grating spacing or spatial period and n is the fibre's background refractive index. The reflection bandwidth depends on the uniformity and overall length of the grating (see later sections) but for a grating of approximately 3 mm length a typical reflection bandwidth would be 0.5 nm.

Since the reflection peak wavelength is dependent on a spacing parameter Λ, it is also sensitive to strain, i.e. a fractional change in Λ such that, for a given temperature, the fractional change in peak reflection wavelength, $\Delta\lambda$, is related to strain, ε, by

$$\frac{\Delta\lambda}{\lambda} = \frac{\gamma\Delta\Lambda}{\Lambda} = \gamma\varepsilon \tag{3.3}$$

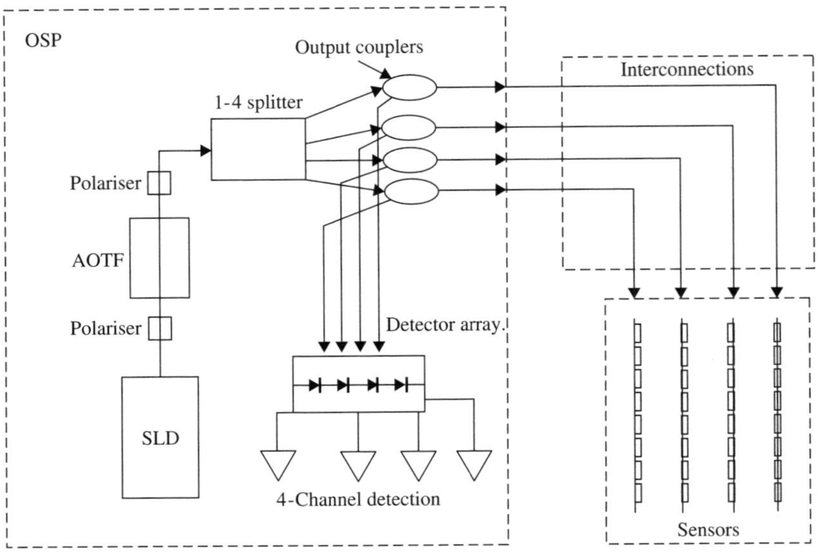

Figure 3.19 Schematic representation of the Bragg grating based strain sensing system

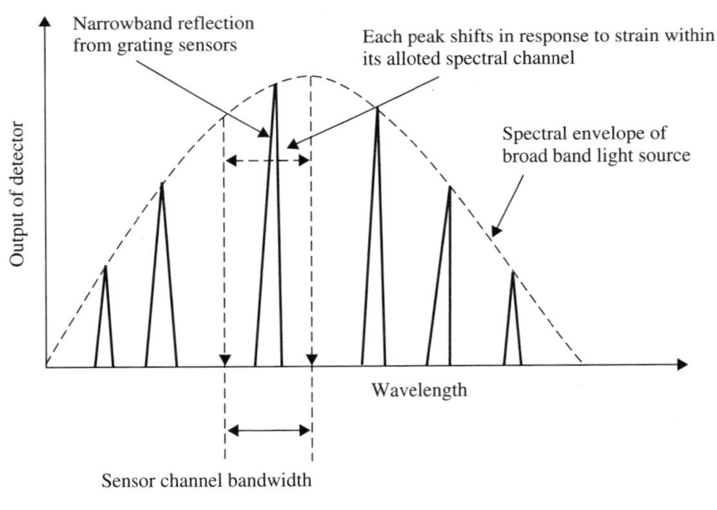

Figure 3.20 Wavelength encoding of Bragg grating strain sensor

where γ is a dimensionless gauge factor. This is the basis of the Bragg-grating-strain sensor. To determine the reflection wavelength of a grating or a number of gratings in a single fibre, light of broad spectral content must be shone down the fibre. For a given spectral envelope, a number of Bragg gratings each with a distinct, nominal, strain-free reflection wavelength can be illuminated as shown in Figure 3.20. By encoding individual gratings with different nominal reflection wavelengths in this way, a series of wavelength multiplexed strain sensors can be made in the same fibre. The reflected light can be spectrally analysed, the wavelengths associated with each reflection maximum determined and the relative shifts due to strain measured.

A number of design trade-offs for any sensor system are immediately apparent. From Figure 3.20 it can be seen that each Bragg grating must occupy a spectral channel sufficiently wide that at maximum strain, it will not overlap with a neighbouring (in the spectral domain) sensor. The application will therefore define the width of each spectral channel via the specified strain ranges.

The light source itself will possess a finite spectral bandwidth. One light source will therefore be able to accommodate a finite number of sensor channels, the total number being the ratio of light source spectral bandwidth and sensor channel bandwidth.

The total number of sensors can be increased by splitting the available source light between multiple parallel channels, as shown in Figure 3.19. Each channel will require separate parallel or sequential interrogation to recover the spectral information from Bragg gratings. The limit to the number of parallel channels will be determined primarily by the optical power budget for the system, i.e. to what extent source power can be split before the detected light levels from the sensors fall below the detectors sensitive thresholds.

The process of spectral analysis of the reflections from Bragg gratings can be performed in a number of ways. The design under discussion in this section uses a straightforward wavelength division multiplexing approach in which light in the system is spectrally scanned using an electronically tuned optical filter. The filter comprises a narrow bandpass wavelength filter, which can be tuned to any wavelength, with great precision, within the spectrum of the light source. When the tunable filter is set to a wavelength, which matches any one of the Bragg gratings' peak reflection wavelengths, a maximum response is detected in the light detection system. In this way a filter, which is repeatably scanned through the entire source spectrum, will produce at the detectors a series of peaks in a temporal sequence. These correspond to the reflection spectrum of the Bragg grating arrays. By timing the arrival of these peaks with respect to the start of a filter scan, the exact wavelength and hence state of strain of each Bragg grating can be deduced. The signal processing task then is one of timing peak signal events from a photodetector, with respect to a synchronised scan control waveform which modulates the tunable filter. The tunable filter can be placed in one of two positions within the sensor architecture: at the source end or at the detector end. At the source end the filter selects, in a tunable manner, a narrow portion of the source spectrum to be injected into the fibres. The combined filter and broadband source now constitutes a synthesised narrow band, tunable light source. As the filter is tuned the signals corresponding to the Bragg peaks will appear at the broadband detectors. A second position for the filter would be at the detectors where incoming light from the Bragg gratings is selectively transmitted to the detectors as the filter is tuned. This now constitutes a narrow band tunable detector. For the architecture adopted in Figure 3.19, the filter is best placed at the source end before the light is split into a number of paths. If placed at the detection end, the filter would be required to handle multiple input channels, which for the single fibre coupled device to be used in this chapter, would present some difficulties. The light source, tunable filter, optical path divider, detectors and signal processing electronics are all within the OSP, which is a self-contained module.

3.6.1 Sensors

The sensors for a surface mounted operational load monitoring system are based on optical fibre Bragg grating strain sensors. The sensor design is based on a patch concept and mirrors the design and function of conventional resistive strain gauge rosettes (Dally and Riley). This enables a general, planar strain measurement capability. Temperature compensation is achieved by a separate sensor, which is isolated from strain but in close proximity with strain sensors and sharing the same patch. The entire concept is illustrated in Figure 3.21. The patch consists of three strain sensors with principal sensitivities aligned along axes separated by an angle of 120°. A minimum of three independent strain measurements, at differing angles, are necessary to resolve the principal directions of planar strain and hence fully characterise the strain at a location on a structure's surface. Following the conventions of electrical strain gauges, θ takes the value of 45° or 60°. As with all strain sensors, electrical or optical, temperature effects can induce apparent strain readings from the sensor. For the case of Bragg gratings, sensor measurements are made by reading the peak reflection wavelength of the Bragg grating. A change in strain or temperature causes a proportional shift in this wavelength $\Delta\lambda$, where

$$\Delta\lambda = \lambda_0(\gamma\varepsilon_{xx} + (\xi\Delta\ T) \tag{3.4}$$

and (Dally and Riley)

$$\gamma\varepsilon_{xx} = (\varepsilon_{xx} - 1/2n^2(P_{12}\varepsilon_{xx} + 1/2(P_{11} + P_{12})(\varepsilon_{yy} + \varepsilon_{zz}))) \tag{3.5}$$

or (Sirkis)

$$\gamma\varepsilon_{xx} = (1 - 1/2n^2(P_{12} + v(P_{11} + P_{12})))\varepsilon_{xx} \tag{3.6}$$

and

$$\xi = (\Delta n/\Delta T) + \alpha \tag{3.7}$$

for a free fibre and λ_0 is an unshifted peak reflection wavelength, ε_{xx} is the axial strain, α is the thermal expansion coefficient of glass fibre (units/°C), $\Delta n/\Delta T$ is the thermo-optic coefficient (units/°C), ΔT the temperature change, γ is the strain optic gauge factor, P_{ij} is the strain optic tensor component and ν is the Poisson's ratio for the optical fibre.

Thermal effects for Bragg grating strain sensors typically amount to 6 ppm of fractional wavelength change per °C. For the required range of operation over 154 °C, this amounts to an additional apparent strain of up to 1000 $\mu\varepsilon$. For this reason, an additional temperature reference sensor is required for each strain sensing patch. This sensor must be isolated from strain so that temperature effects can be determined independently and the strain readings suitably compensated. This design, since it contains its own built in temperature referencing capability, would be suitable for attachment to any structure without the need to be thermally matched. This is analogous to the practice of including compensating gauges in bridge circuits for electrical strain gauges. Three possible sensor geometries are shown in Figure 3.21. The upper two geometries comprise two fibres, each with two Bragg gratings. One of the fibres in each case terminates in a strain relieving sleeve which envelopes a Bragg grating. This strain-relieved grating is sensitive only to temperature. The lower configuration uses a single fibre with all four gratings distributed at key locations to provide angularly differentiated strain response and again, terminates in a strain isolated temperature sensor.

In what follows various features, which need to be considered in the design of the sensor patches, are discussed in more details.

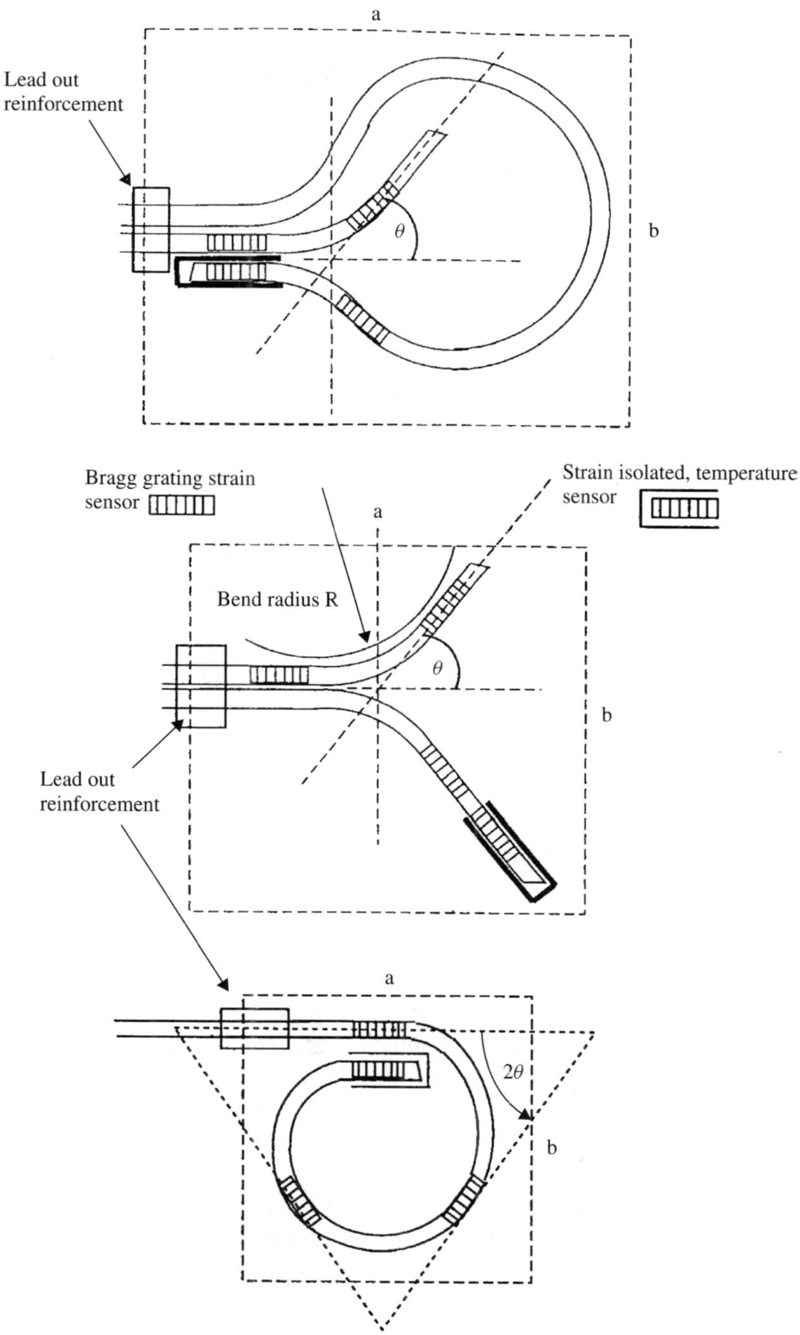

Figure 3.21 Sensor patch concepts using optical fibre Bragg gratings as strain and temperature reference sensors

3.6.1.1 Backing Patch

The role of this item is to act as a substrate holding the fibre sensor array together and transmitting strain from the underlying structure to the optical fibre itself. The important characteristics of the backing material are defined in terms of its shear modulus, creep, flexibility, and elongation capability.

The shear modulus must be sufficiently high that strain is transmitted faithfully from structure to sensors especially at elevated temperatures. Although rigid, the backing must also be sufficiently flexible to allow compliance with curved surfaces.

Much development work on electrical strain gauges has already been undertaken with polyimides successfully employed in a wide range of strain gauge designs. This material maintains its rigidity to high temperatures (in excess of 500 °C) and is also suitable at cryogenic temperatures.

Optical strain gauges do not require the electrical insulation properties of electrical strain gauge backing materials. Metal foil backing may also be suitable for optical strain gauge patch construction. This may widen the opportunities for bonding and encapsulation techniques. For example, aluminium and silica (from which optical fibres are made) have similar moduli. Aluminium coated fibres embedded in an aluminium matrix bonded to an aluminium structure may prove to be a highly desirable combination.

The ultimate dimensions of this component will be determined by end user requirements in terms of localisation of strain measurements. No immediately relevant requirements are yet known in this respect. A realistic target specification would be to match optical fibre strain gauge patches to electrical strain rosette packages currently available commercially. The constraints will also emerge from the configuration of optical fibres within the patch (see fibre bend losses).

3.6.1.2 Optical Fibre Bragg Gratings

The Bragg grating structure induced in the, optical fibre defines the sensor gauge region. The Bragg grating is a distributed structure within the core of the fibre. Typical fibre dimensions are 125 microns outer cladding diameter with a core of nine microns diameter (single mode at 1300 nm). A great deal of flexibility exists in the fabrication of Bragg gratings in terms of physical length, operating wavelength and reflection bandwidth. The last two factors are more important in the discussion of the OSP later in this document. In considering the design of the sensor patch itself, the physical property of gauge length is of prime importance.

The constraints on sensor gauge length arise both from the grating fabrication process and from the requirements imposed by the OSP. Systems described in this chapter utilise the acousto-optic tunable filter (AOTF) device, which operates as a scanning optical spectrum analyzer. The optical bandwidth of the device is 0.45 nm (FWHM) at wavelengths around 1318 nm. For optimal usage of light power budgets in the system, it was necessary to match optical sensor Bragg grating bandwidth to that of the filter device. Again, for power budget reasons, the peak reflectivity of each Bragg grating sensor was required to be a maximum of 100 %. For moderate values of the optical Bragg grating coupling constant, κ, the physical grating length L, and the grating FWHM bandwidth, $\Delta\lambda_g$ are related by

$$L = (\lambda_0^2/\Delta\lambda_g \pi n)\sqrt{\pi^2 + 1} \tag{3.8}$$

For a grating bandwidth of 0.45 nm and a peak reflection wavelength of 1318 nm an approximate grating length is 2.8 mm.

Strain sensitivity of Bragg grating is determined by a strain optic factor, γ (defined earlier), which will be fibre and temperature dependent and must be measured for each fibre type used in grating fabrication. Values for γ are typically 0.6 to 0.7 (Measures 1992).

In determining, unambiguously, the principal strains using a strain rosette, the transverse sensitivity to strain in each gauge must also be quantified. Ideally this sensitivity will be negligible for all but the most accurate strain measurement requirements.

3.6.1.3 Strain-Isolated Temperature Reference Sensor

In the earlier general design comments, the sensitivity of Bragg gratings to both strain and temperature effects was quantified. For accurate strain measurement over the range of temperatures set out in the requirements for the OLM system, some form of independent temperature measurement is required. With temperature information, all strain readings can be suitably compensated with the assumption, of course, that temperature conditions within a single sensor patch are uniform. Typically, a strain error of 12 $\mu\varepsilon$ (the required accuracy of the OLM system) would result from a temperature uncertainty of 1.4 °C, so provided spatial temperature gradients in the structure are less than 1.4 °C per maximum patch dimension, single point measurements will suffice (this assumes a value for $\xi = 6 \times 10^{-6}/°C$ (Measures 1992), $\gamma = 0.7$ and $\lambda_0 = 1300$ nm).

To isolate a Bragg grating from strain, the grating must be situated in a region of fibre with one end entirely undamped and free from axial strain. With the strain patch geometries shown in Figure 3.21, this will require the temperature sensing grating to be located at the tip of one of the fibres. The tip region containing the grating will be enclosed in a loose fitting sleeve such as a silica capillary. The fibre will require some buffer material within the sleeve to keep out moisture and contaminants. The sleeving will also be sealed around the fibre and fixed to the patch backing material. Any buffer material within the sleeving will be of extremely low modulus so as not to transmit stress from the patch to the enclosed Bragg grating. A non-setting, hydrophobic gel capable of remaining fluid at low temperature (up to −54 °C) will be suitable.

3.6.1.4 Optical Fibre Bending Radius

Bending in optical fibres induces loss of light from the guiding core region. Quantitative analysis of loss mechanisms is difficult for small bend radii (<5 mm for single mode fibre) but for larger radii the losses are governed by

$$\alpha_c = A_c R^{-1/2}\, e^{-UR} \tag{3.9}$$

where α_c is the optical electric field amplitude loss per unit length of fibre for bend radius R. The factors A_c and U are complex functions of the fibre parameters involving field mode radius, core-cladding index difference, propagation wavelength and fibre cut-off wavelength.

Possible sensor geometries, as shown in Figure 3.21, will entail differing intrinsic loss characteristics due to the length and radius of bends. Sensor placement along the fibre may also be optimised so as to offset effects of cumulative loss along a curved fibre path. The bend loss parameters will dictate allowable fibre geometries, which will in turn, constrain the minimum dimensions for a strain gauge patch.

Bend losses are also wavelength dependent and given that the Bragg grating sensor is a wavelength sensitive device, this factor could impact significantly on sensor performance. It is essential that the spectral bend loss characteristics of the chosen fibre are adequately characterised in the spectral region within which the sensor system operates.

3.6.1.5 Angle of Orientation of Bragg Grating Sensors

Bending losses may dictate the maximum angular changes in fibre path, which, for a given overall sensor patch dimension, will restrain the relative orientation of the individual Bragg sensors. The lower configuration in Figure 3.21 may contain sensor regions, which lie along curved fibre paths. Ideally, the three strain sensors in each patch would be oriented with the largest possible angle between their respective axes (i.e. 120°). This would allow the general application to structures with unknown strain field directions. Maximising the angle between sensor axes contributes to the accuracy of resolving strain in two dimensions (Foote 1995). To understand fully the consequences of curved sensor axes a general analysis of optical phase–strain relationships along arbitrarily configured fibre paths is needed.

The orientation angle may also be influenced by manufacturing processes. For example, during encapsulation and bonding to the patch backing material, the fibres will require support in a fixed configuration. The rigidity of the fibre when compared with that of the backing material may preclude excessive bends in the fibre path if the patch is to remain flat.

3.6.1.6 Encapsulation of Optical Fibre

Electrical strain gauges require encapsulation to protect from the effects of moisture, contaminants and mechanical damage. Similar requirements will apply to optical patches. Moisture penetration will adversely affect silica optical fibre by initiating fracture mechanisms in the fibre cladding. The mechanical robustness of the optical fibre sensors will be, to a first approximation, similar to electrical foil type strain gauges. As with electrical gauges, the encapsulant performs no function in the transmission of strain from the structure to sensor and so shear properties of this material are not as important as for the backing material.

A major distinction between the electrical and optical sensors is that in the optical case, contacting encapsulant can be metallic (or more generally electrically conducting). With this relaxed constraint, more choice in manufacturing methods and materials may ensue.

A desirable route to fabrication would be a one step process in which the backing material and encapsulant are applied to the optical fibre in one process. Casting processes would be most suited to this requirement. These could apply to both polymer and metallic patch materials. For Bragg grating sensors, temperature limitations must be respected which if exceeded, could lead to the thermal annealing of the induced Bragg structure. Ideally processes, which do not involve exposing the fibres to temperature in excess of 200 °C are to be preferred. If it is not possible to avoid excessive temperatures, the thermal stability of the chosen gratings and fibre should first be investigated.

3.6.1.7 Adhesive or Matrix Uniting Fibre and Backing Material

In the most complex realisation of the optical fibre strain sensor patch, a compound structure will result comprising a backing material, an adhesive layer bonding fibre to backing and an encapsulating layer covering the bonded fibre. This system presents five interfaces for the transmission of strain from the underlying structure into the optical fibre: the two interfaces either side of the adhesive layer bonding gauge backing to structure; the two interfaces either side of the adhesive bonding optical fibre to backing; and, finally, the interface between fibre and encapsulant. At each interface distortion of the strain field originating in the structure will occur. The ideal adhesive will form thin films with sufficient strength and ductility to allow strains up to the maximum design case. The preferred manufacturing route in which the optical fibre is cast into an encapsulant backing material would avoid this step in the process.

3.6.1.8 Lead-Out Reinforcement

Connections to optical fibres especially in a single mode, are far more difficult to achieve than electrical connections. Alignment tolerances for a fibre with core diameter of 9 μm must be at submicron levels for good transmission. To achieve this reliably, high precision and robust connector housings are required which at present are too bulky to consider as an integral part of the sensor patch. Their inclusion would dramatically compound the problems of encapsulation and gauge bonding.

The approach taken in this chapter is to equip sensor patches with flying leads or pigtails of optical fibre. The free ends of the fibre can be spliced into fixed optical fibre cable runs at convenient locations with little difficulty. They could even be terminated with rugged connectors at selected locations if demountable connections are required. This would be similar to existing surface mounted structural monitoring gauges which use line driving amplifiers with demountable electrical connectors at some point removed from the actual strain gauges.

The point where the optical fibres leave the sensor patch is a potential region of stress concentration and so must be suitably reinforced to inhibit fibre failure. Strain relief could be facilitated by extending a free region of fibre some way onto the patch before the first sensor. A low modulus encapsulant at this region would spread the concentration of stress. It would be convenient if the overall patch encapsulant material and lead-out reinforcement were the same material.

3.6.1.9 Strain Gauge Bonding

Convenience of installation and durability are prime factors for consideration when selecting adhesives for bonding gauges to structures. The approach in this chapter will be to make use of the many well-qualified adhesives currently available commercially for conventional strain gauge bonding. Many systems exist including epoxies, cyanoacrylates, solvent-set, polyimides and ceramics. The governing factors in selection will be the compatibility of adhesive with backing material and structural substrate material. Other application specific factors will also be of influence – for example, ultimate application requires a lifetime of up to 50 years. This would tend to rule out cyanoacrylate adhesives.

3.6.1.10 Termination of Fibre

The sensor patch designs of Figure 3.21 entail optical fibres, which terminate within the patch itself. To avoid back-reflections (Fresnel reflection from the fibre end) the fibres should be terminated in an angled cleave or index matching medium.

3.6.2 Optical Signal Processor

A top level description of the OSP was given in the introductory part of this chapter where the broad description of light source, tunable filter and detectors was given. The details of a viable system will now be discussed, based on these main components. The tunable filter to be used in the analysed systems is an acousto-optic tunable filter or AOTF.

To recap, with reference to Figure 3.22, the light source will comprise a semiconductor superluminescent diode operating in the 1300 nm wavelength range. The device has a spectrally broad output with a roughly Gaussian spectral power distribution with a FWHM bandwidth of 70 nm. The integrated power output is 0.5 mW into single mode fibre and is randomly polarised. The source must provide sufficient optical power to be distributed among a number of fibre sensor arrays. Sufficient power must also be available

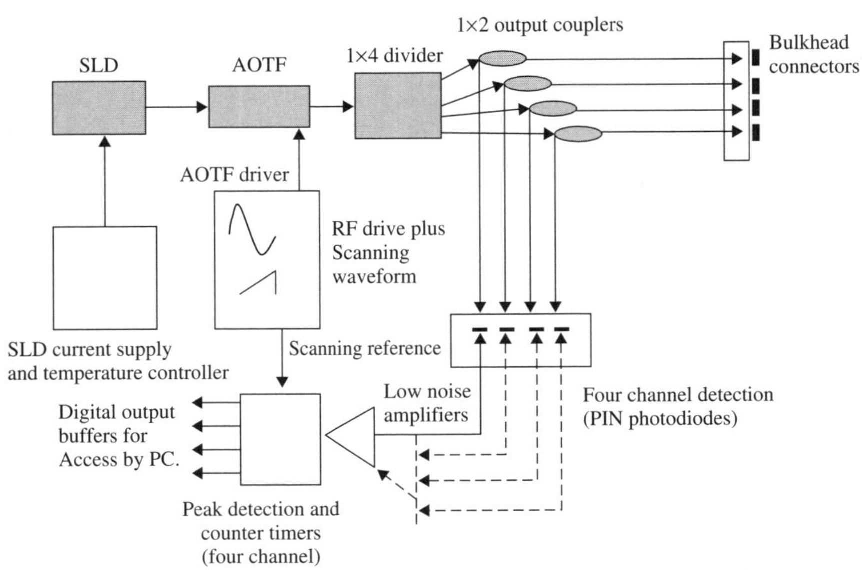

Figure 3.22 Optical signal processor architecture

Table 3.5 Insertion losses of the main components in the optical processing unit

Component	Insertion Loss
AOTF	13 dB
Divider network	6.5 dB + 3.3 dB
Connectors	0.2 dB each

to accommodate the insertion losses of other components such as the tunable filter and light dividing network. Some key component insertion losses are given in Table 3.5.

The total losses in any one complete optical circuit from light source through AOTF, divider network, sensor array and return path to detector can be calculated from:

$$\mathrm{Loss} = \mathrm{A} + \mathrm{D} + 2(\mathrm{C} + \mathrm{CON}) + \mathrm{BG} \tag{3.10}$$

in dB per sensing fibre, where A is the insertion loss of the AOTF equal to 13 dB, D is the insertion loss of the divider network equal to 6.5 dB, C is the insertion loss of the coupler equal to 3.5 dB, CON is the insertion loss of the connectors equal to 0.4 dB (2-off) and BG is the loss due to a single Bragg grating equal to 0.5 dB.

The losses through the output coupler and connectors between the OSP and the sensor array are doubled because of the round trip that light from the sensors is required to take. For this system, the total losses amount to 27.8 dB. A typical light source as described above will have an output with spectral power density of approximately 5 μW/nm. To match the characteristics of the Bragg gratings to the AOTF, they will be required to have a reflection bandwidth of 0.4 nm so the total effective loss of a single grating will be 0.5 dB × 0.4 per nm. The peak optical power associated with the detection of Bragg grating peak can now be calculated from the product of the SLD output power density, the round trip loss to a sensor grating and the spectral loss per grating which totals 3.3 nW.

The system in Figure 3.22 requires broadband detection (covering the spectral range of the SLD source) with sufficient sensitivity to provide usable signals for Bragg peak detection. The requirements call for a sensor system update rate of up to 2 kHz. This is interpreted as a maximum interval between sensor readings of 0.5 ms. To achieve this with the current architecture, the AOTF will be required to scan through the entire SLD spectrum at this rate which is equivalent to a scan rate of 140 000 nm/s. For a spectral resolution of 0.4 nm, the maximum frequency component of signals generated at this scan rate will be within a 300 kHz bandwidth. Signal to noise ratios of 10 are acceptable for peak detection. Combining the expected power levels at the detector with the temporal bandwidth requirements and a signal to noise ratio of 10 gives a required optical detector with a noise equivalent power (NEP) of approximately 1 pW/$\sqrt{\text{Hz}}$. This should be possible with InGaAs PIN diode and transimpedance amplifier combinations. Avalanche Photodiode detectors are also capable of this type of performance.

Peak detection and location takes place in the temporal domain which, when referenced directly to the AOTF scanning waveform can be mapped to wavelength. To achieve an equivalent strain resolution of $\pm 12\mu\varepsilon$ as specified in the requirements, a Bragg peak wavelength the measurement accuracy of ±0.011 nm is required (assuming a value for

γ of 0.7 at 1300 nm). If the full SLD spectral width is to be used, this would require a Bragg peak location system with at least 13-bit dynamic range. To achieve this it is necessary to make relative timing measurements between the occurrence of peaks in the detector signal (corresponding to Bragg grating maxima) and the AOTF scanning control signal. This can be realised by employing a 13-bit (min) up/down counter to generate, via a D/A converter, the AOTF scan waveform. The same counter can then be used in a synchronous fashion, to perform the timing measurement of Bragg grating peak events in the detector signals. The overall *digital* bandwidth of the system will be governed by the required accuracy and the scanning frequency of the AOTF. With 13-bit accuracy, bandwidth will be a minimum of 16.4 MHz per channel.

For 1300 nm single mode fibre, standard products exist for performing 1×4 splitting. Similarly, the output couplers can be realised using standard 1×2 or 2×2, 3 dB couplers. In both cases, the components must be free from polarisation sensitivity. The style of connector at the bulkhead of the OSP unit is not important provided, again, the components display minimal polarisation sensitivity.

The operation of the AOTF requires randomly polarised light from the SLD to be polarised at the input. The AOTF will also output linearly polarised light. The operation of the sensor system beyond this point does not depend on the polarisation state of the light in the fibre. However, some components may display a degree of polarisation sensitivity, which will be detrimental to the operation of the system as a whole. In sourcing components for this system, careful screening of polarisation properties will be necessary. Any polarisation dependent effects in the Bragg gratings must also be quantified.

The system illustrated in Figure 3.22 shows four independent channels. Each channel will connect to an array of Bragg grating sensors, which will be arranged in terms of peak reflection wavelengths, to utilise fully the available source spectral bandwidth. Returning to the original requirements, the maximum strain and temperature ranges will imply a maximum expected wavelength shift $\Delta\lambda$, for any Bragg grating sensor. Using the expression

$$\Delta\lambda = \lambda_0(\gamma\varepsilon_{xx} + \xi\Delta\,T) \tag{3.11}$$

produces a value of $\Delta\lambda = 7.6$ nm where ε (absolute) is equal to 7000 $\mu\varepsilon$, and ΔT (absolute) is equal to 154 °C. The other quantities are as previously defined. This value for $\Delta\lambda$ defines the spectral bandwidth for each sensor channel. With an available source spectral bandwidth of typically 70 nm, this would allow eight sensor channels to fit within the source spectrum with guard bands of 1.3 nm for each sensor.

The discussion of sensor patch design focused on a four-sensor design (three for strain and one for temperature). With this degree of multiplexing, a single output fibre from the OSP could supply two such sensor patches giving a total system capability of 32 sensors or eight complete sensor patches.

3.6.3 Optical Interconnections

It was mentioned in Section 3.6.1 that the design approach taken in this chapter with regard to optical connections will avoid integral patch connectors. Although laboratory realisations could be readily achieved using existing low profile multifibre devices such as the *Europtics MT* connector, an engineered integral connection system capable of reliable operation during the anticipated flight trial is beyond the scope of this book. However, during the course of sensor patch development, the specification for integrated on-patch connectors could be more clearly defined for future development in cooperation with commercial suppliers.

The demonstration system described in this book will use existing in-line connectors already qualified to a level of environmental tolerance. These will be based on either multiple, single fibre connectors or multifibre connectors such as the *Europtics* device mentioned above. This choice of interconnection method still leaves a great deal of flexibility in sensor placement on structures. Patches constructed with flying lead connections can be custom terminated with qualified connectors either using termination equipment *insitu* or by splicing to factory terminated cables. This will guarantee rugged connections for the evaluator systems tested in Chapters 4 and 6. For suitable sleeving material and reinforcing for fibre cables, in-house optical harness studies can provide background for materials selection and supply. Harness and cable routing requirements will only become clear at the evaluator design stage in Chapters 4 and 6.

3.7 OPTICAL FIBRE STRAIN ROSETTE

The concept of the optical fibre rosette discussed in Section 3.6 was implemented to develop a surface mounted patch for strain measurement. The patch is equivalent of a *delta* electrical strain gauge rosette. The concept was to duplicate the function of a three axis, $3 \times 120^\circ$ strain measurement gauge using a single optical fibre. The fibre was constrained into triangular loop geometry by a process of lamination between thin polymer films. Use of these materials ensured that the same bonding procedures and materials developed with many years experience for electrical strain gauge could be used unchanged, for these developmental devices. A significant extra feature of these surface mounted optical devices was the inclusion of a fourth sensing element, isolated from the strain, but sensitive to temperature. An example of the optical strain gauge rosette with equivalent electrical device is shown in Figure 3.23. The advantage of the optical rosette is very straightforward – only one fibre is required instead of 12 wire connections for the classical strain gauge. Figure 3.24 shows an output example from 3-axis optical strain rosette with strain isolated temperature sensor when bonded to an aluminium test coupon in compressive load. The measurements were made at constant temperature.

3.8 EXAMPLE OF EMBEDDED OPTICAL IMPACT DETECTION SYSTEM

The effects of impacts on composite structure in aircraft are deemed to be a significant threat to structural integrity. These events lead to BVID (Barely Visible Impact Damage), which challenges existing inspection methods. Typical causes of BVID are low velocity impacts such as bird strike, runway stones and tool-drop (hangar rash) during maintenance and standby. BVID is a prime source of delamination in composite structures and an automated inspection system (one of the components of a Health and Usage Monitoring System) would need to record and quantify such damage. Damage could be quantified either directly by using an active inspection probe technique or, as in the case of impact detection, indirectly by passively 'listening' for damaging, events.

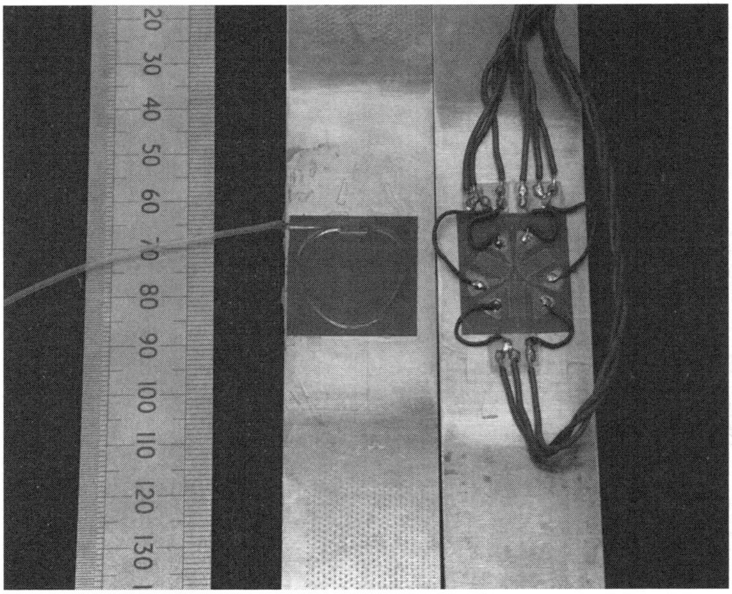

Figure 3.23 Optical fibre strain rosette

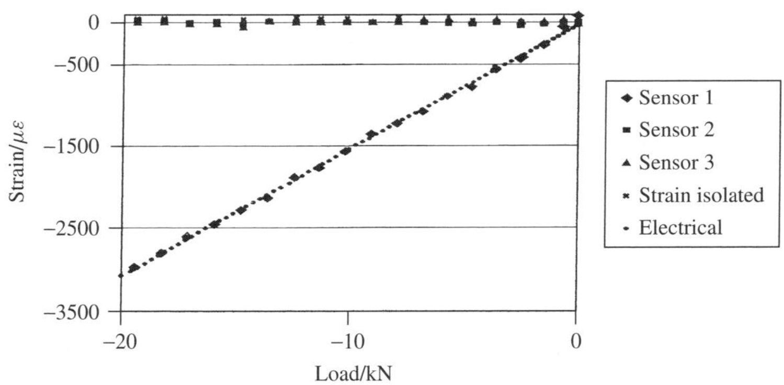

Figure 3.24 Output from 3-axis optical fibre strain rosette

The optical sensing technology described in this section is designed to be complementary with strain sensing system for use in aerospace OLM systems. The technology is based on optical fibre Bragg grating sensors, both surface mounted and embedded approaches have been pursued. Although the details of the opto-electronic interface equipment will be different for the load monitoring and damage detection systems, synergy exists on the issues of structural implications of optical fibre embedment in carbon fibre composites and fibre sensor endurance to manufacturing and handling conditions. The sensors must be capable of responding to impact events of a potentially damaging nature, on the CFC structure. From the sensor data the required information is the location of the impact plus the energy and duration of the impact event. Knowledge of these parameters will enable an assessment of the likelihood of the occurrence of damage and its location.

The sensors must be capable of operating under the different load cases likely to be encountered during the operating lifetime of the structure or component. The sensors must also be of sufficient sensitivity to detect damaging events without an unrealistically dense sensor deployment. Further, the signal bandwidth of the sensors should be commensurate with the task of locating the event by a temporal triangulation technique.

The technology for impact detection described in this section is based on optical fibre Bragg grating sensors. These devices are fabricated within the cores of otherwise standard telecommunications optical fibre by an ultraviolet laser exposure process. Details of the fabrication process, its variants and optical properties of these devices are described in the previous sections. The nature and operation of Bragg grating sensors can be summarised in the following:

- Bragg gratings are localised regions in a length of optical fibre. Many gratings can be imprinted along a single length. Each grating is typically 2 or 3 mm in length. These gratings constitute the sensors.
- The gratings have the property of reflecting light that is shone down the fibre, in a predetermined band of wavelengths. The grating themselves are periodic ripples in the refractive index in the fibre's core.
- As the fibre, and hence the grating is strained, the band of wavelengths at which the Bragg grating reflects is shifted. This strain can be induced quasi-statically, in which case the grating acts as a strain sensor, or can be dynamic such as the stress wave event caused by an impact.
- The sensor system operates by shining a broad range of wavelengths simultaneously down the fibre. The wavelength of the reflected light is detected in an opto-electronic module where wavelength shift is converted to an electrical signal that constitutes the raw sensor signal. This can then be captured and analysed using standard data capture equipment.

The sensor system, based on a concept of a patented innovation (Zhang and Bennion), is illustrated schematically in Figure 3.25. Light from a broadband superluminescent light emitting diode (SLD) source with centre wavelength around 1300 nm is divided four ways by a passive fibre splitter. Each fibre from each arm of the splitter illuminates a chirped Bragg gratings sensor via a 3 dB fused fibre coupler (CI). The light reflected from the sensing grating then enters another 3 dB fused fibre coupler (C2). On one output arm of each of C2 there is another chirped Bragg grating, whose centre wavelength at zero strain is different from that of the sensing grating; this second grating, is termed the filter grating. The light from the output arms of each of C2 is then detected on a pair of PIN photodiodes (D1 and D2). The ratio of these two detected signals will be linearly related to the strain applied to the sensing grating. The filter can be used in either a transmission or a reflection mode. In this case the reflection mode is used in which light from the sensor is simultaneously filtered and reflected by this filter grating. In the transmission mode, where sensor light is filtered but passed through the filter grating, the detected light would include an additional background component from the residual light not rejected by the filter. If this transmitted component were used for recovering the sensor information, the overall signal-to-noise ratio of the system would suffer. Transmitted light also suffers from short wavelength loss where intrinsic fibre properties cause light to be coupled out of

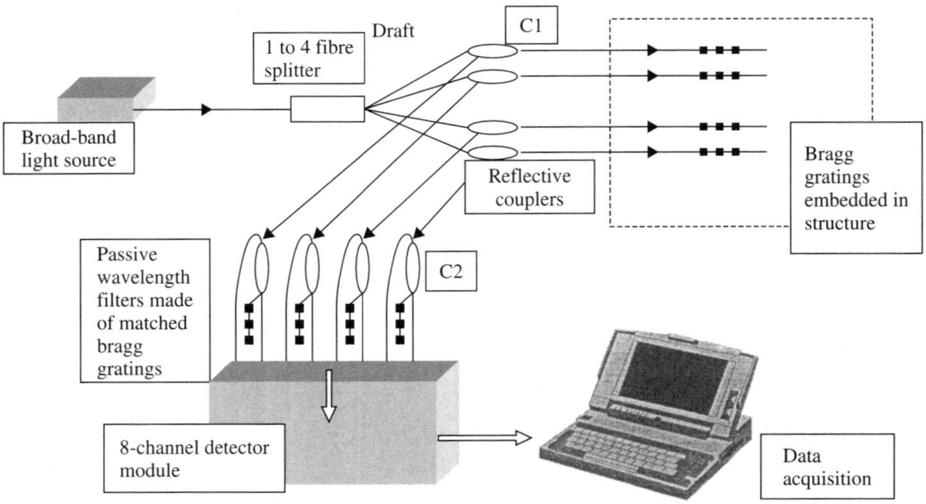

Figure 3.25 Schematic diagram of the optical fibre impact detection system using Bragg grating sensors

the fibre at selected wavelengths. These effects would combine to distort the wavelength response of the filter grating in transmission and hence the sensor's response to static and dynamic strain. The use of a reflective geometry for the system eliminates these disadvantages.

The characteristics of these Bragg gratings are distinct from the type used in Chapter 6 for the Operational Load Monitoring in-flight application. The term *chirped* refers to the fabrication method for the particular gratings used in this case. Instead of containing regular, periodic refractive index ripples, the cores in chirped gratings have a spatially varying period of index modulation. This results in a broader reflection spectrum than for unchirped gratings. Figure 3.26 shows a typical reflection from one of the sensor gratings. The spectral bandwidth is approximately 20 nm. For each sensor grating, the system contains a filter grating at the detector module, which acts as a wavelength-to-amplitude converter for the optical signal returning from the sensor. The principle is illustrated in Figure 3.27. The spectral intensity of the broadband SLD source will be

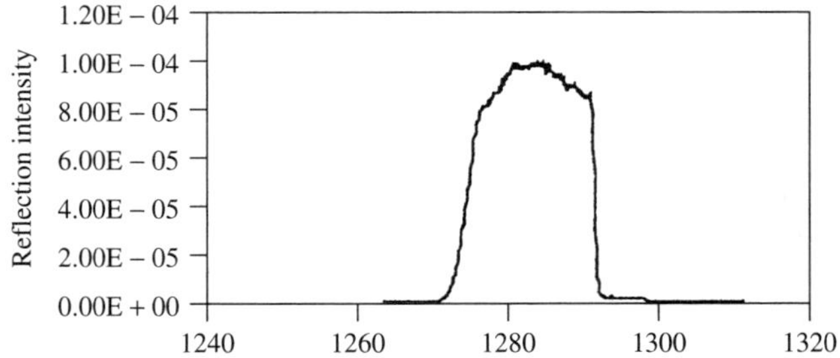

Figure 3.26 Reflection spectrum of sensor S1′

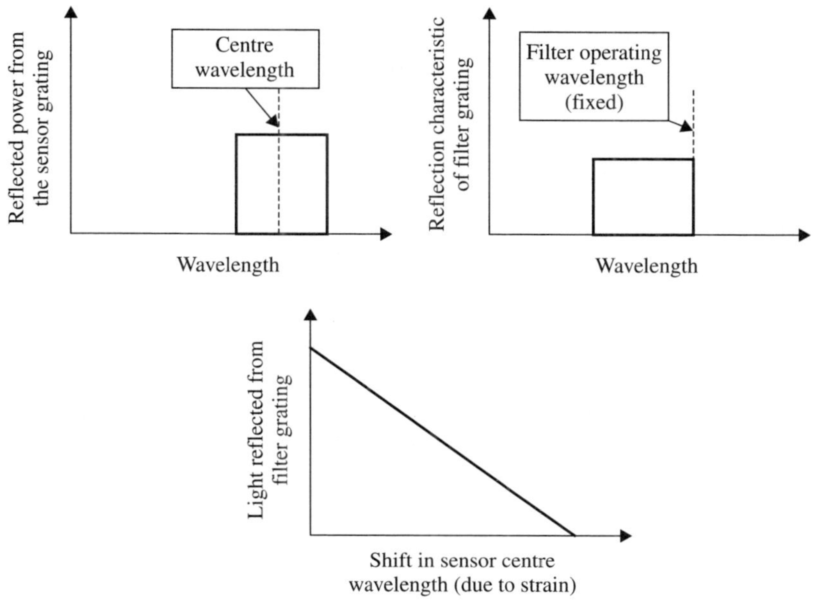

Figure 3.27 Principle of sensor system operation

approximately Gaussian and is given by

$$I_{SLD}(\lambda) = I_0 \exp \; - (2(\lambda - \lambda_0)/\Delta\lambda)^2 \quad (3.12)$$

where I_0 is a constant, λ is the wavelength, λ_0 is the centre wavelength of the source and is the full width at half of the source. The intensity reflected from the sensor will be given by

$$I_s = \int_0 I_{SLD}(\lambda)\ R\ (\lambda)\ d\lambda \quad (3.13)$$

where $R(\lambda)$ is the reflectivity of the sensing grating and is a function of both wavelength and strain, ε. The intensity after the filter grating will be given by:

$$I_F = \int I_{SLD}(\lambda)\ R\ (\lambda)\ F\ (\lambda)\ d\ \lambda \quad (3.14)$$

where $F(\lambda)$ is the reflection coefficient of the filter grating.

In this sensor architecture the strain experienced by the grating is found from the ratio I_F/I_S. For the ideal case of a perfect 'top-hat' sensor reflection profile and filter transmission function, the ratio above will be a normalised convolution of two square functions yielding a linear dependence with shift in sensor centre wavelength. In practice, there is always structure to the reflection and transmission spectra as can be seen from Figure 3.26. The spectral profile of the light source also influences the linearity of output.

The sensor system comprised eight independent optical fibres, each containing a single chirped Bragg grating. These gratings were nominally identical and were produced in standard, single mode (1300 nm) telecommunications grade optical fibre.

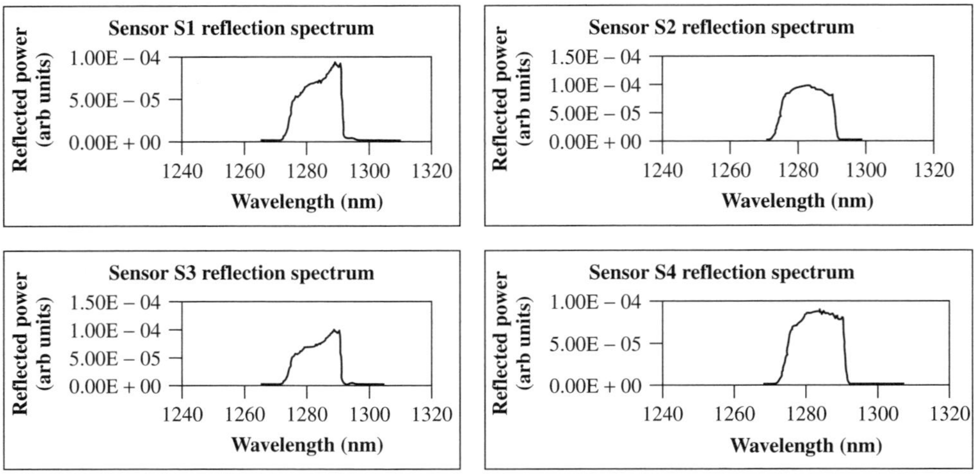

Figure 3.28 Reflection spectra of sensors S1 to S4

Figure 3.28 shows reflection spectra of four of the eight sensor gratings. Some of the sensors (e.g. S1 and S3) display asymmetry in their spectral profiles. This is due to short wavelength loss mechanisms and is dependent on which end of the fibre is used to address the grating. The fibre ends available for addressing the sensors were those left undamaged after the composite panel manufacture and test item installation.

The nonideal spectral profiles of sensor, gratings and filter gratings in the sensor interface equipment impose a calibration requirement on the system. There are a number of permutations of interconnection between sensors and detector channels all of which must be calibrated if flexible use of the system is to be achieved. Each calibration entailed straining the sensor gratings (prior to embedment in the composite test item) in a calibrated straining clamp. The optoelectronic interface module comprised a set of four, fused 2×2 fibre couplers that split the output from each sensor into two arms. For each coupler, one output arm was fused to a fibre containing a filter grating, nominally identical in every channel. The fibre without the filter grating was then connected to a custom made photodetector module (built by *Opto-Sci Ltd.*). The filtered light passed back through the coupler and was routed to an identical detector module. For each sensor channel therefore, there are a pair photodetector modules. To eliminate the effects of unquantified changes in the reflected power from the sensors due to, say connector loss or fluctuations caused by fibre bending, the outputs from each pair of detectors were fed to the numerator and denominator inputs of four analogue divider modules. These have the effect of dividing out common-mode optical intensity variations in the sensor channels. Each photodetector module has a -3 dB bandwidth of 0 to 1 kHz. Calibration curves for sensors S1 to S4 are shown in Figure 3.29, for the particular detector permutations used in the experimental trials. The calibration could only be performed using tensile strain on the unembedded sensors. The reasonable assumption is made that performance is linear over compressive and tensile strains. Calibration was performed over a strain range of approximately 0 to 4000 $\mu\varepsilon$. The sensors are capable of measurement over an absolute range of approximately 10 000 $\mu\varepsilon$. The response of all sensors over this strain ranges is very linear and the calibration factors are given in Table 3.6.

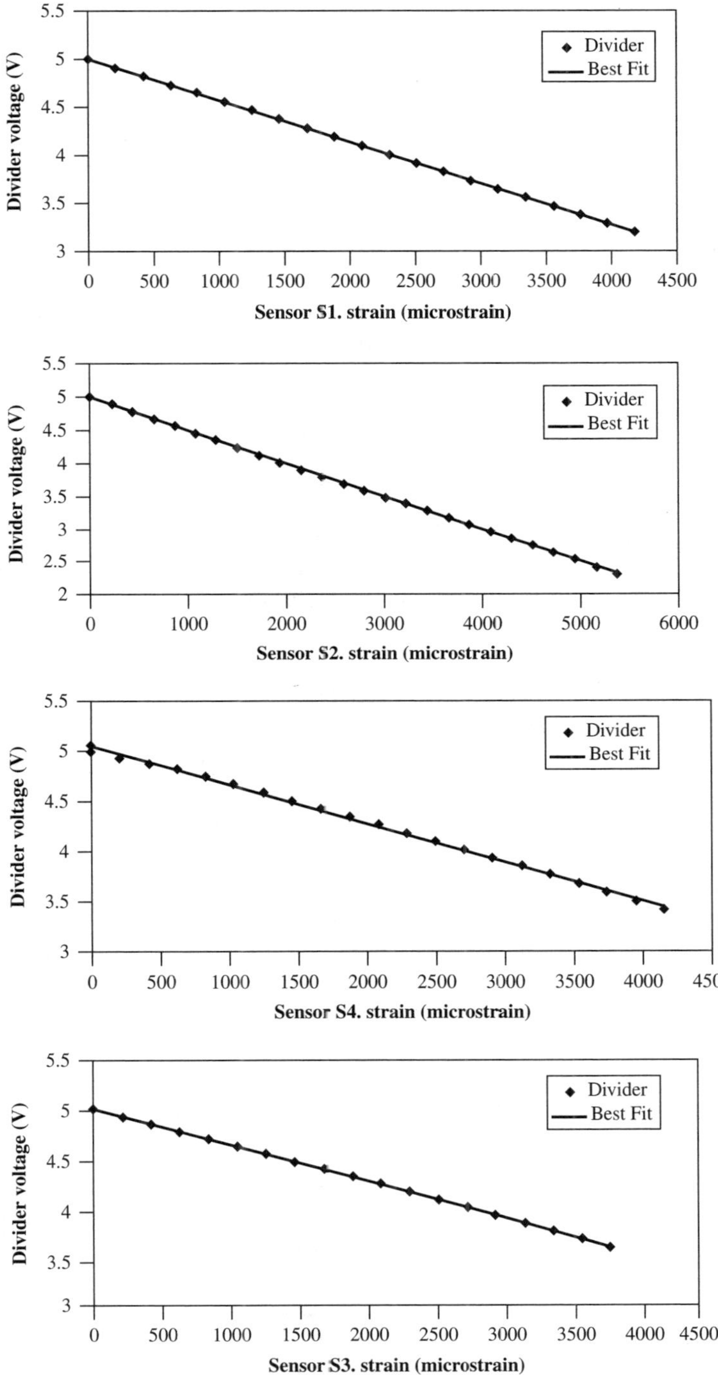

Figure 3.29 Calibration curves for sensors S1 to S4

Table 3.6 Optical fibre sensor calibration factors

Sensor I.D.	Calibration factor (mV/$\mu\varepsilon$)
S1	0.433
S2	0.503
S3	0.364
S4	0.383

A composite panel was fabricated from T300/914 unidirectional prepreg. The lay-up was a 48 ply, quasi-isotropic design of nominal thickness 4 mm. A ply build-up region to simulate reinforcing was included during fabrication. A stiffener was bonded to the panel after cure. Eight optical fibres were incorporated each containing a single Bragg grating sensors labelled SI to S4 and SI′ to S4′. S3, S4, S3′ and S2′ were incorporated between layers 4 and 5, the remainder between layers 44 and 45. These layers are all 0-degree plies with the optical fibre predominantly in the 0-degree direction. The regions of fibre containing sensors S4, S3, S2′ and S3′ were, however, routed in a 90-degree direction (Figure 3.31). This allows the effect of sensor orientation (if any) with respect to impact location to be determined during data analysis. The remaining sensors were all oriented in the 0-degree direction.

Optical fibres were jacketed with acrylate and were protected at the ingress/egress regions at the edges of the panel using short lengths of Teflon tubing. The fibres were laid up to allow access to the sensors from both ends of the fibre. This builds in a level of redundancy should one end of the fibre become damaged during manufacture or one side of the panel damaged during installation into the test rigs.

The panel was also built up with a stiffening frame to take fasteners for attachment to the metal sub-frame, as shown in Figure 3.31. After fabrication and cure the panel was C-scanned (Figure 3.30a) and measured. The C-scan indicated good uniform consolidation and the main panel thickness was to design (4 mm). Sensors were checked for functionality and all eight were deemed functional although not all fibres were accessible from both ends. Further stiffening elements were bonded to the panel and then the panel was drilled and fastened to the metal sub-frame as shown in Figure 3.31.

The test item was installed in a loading and vibration frame. An instrumented impactor was used to impact at sites P1 to P5 (see Figure 3.32). Impacts were made at differing energy levels, with applied vibration and with external heating. For each impact, signals from a group of four sensors were digitised and logged. The experimental set-up for taking sensor measurements during impact trials is shown in Figure 3.32. For all impacts at positions P1 and P2, sensor S1′, S2′, S3′ and S4′ were monitored. For impact positions P3, P4 and P5, sensors S1, S2, S3 and S4 were monitored.

Figure 30b shows a C-scan example of the panel after impact trials. Damage is clearly visible in the region around impact site P5 that experienced the highest impact energies. Figures 3.33 and 3.34 show the sensor signals after conversion to strain amplitude using the sensor calibration data. Two impacts events represented here were of 2 J energy at location P5 and of 12 J energy at location 4, respectively. The impacts were taken at room temperature with no load. From these data an indication of the relative time of arrival of the stress wave at the sensors can be given. It is possible to compare the relative timing of the signal impulses to the positions of the sensors relative to the impact site.

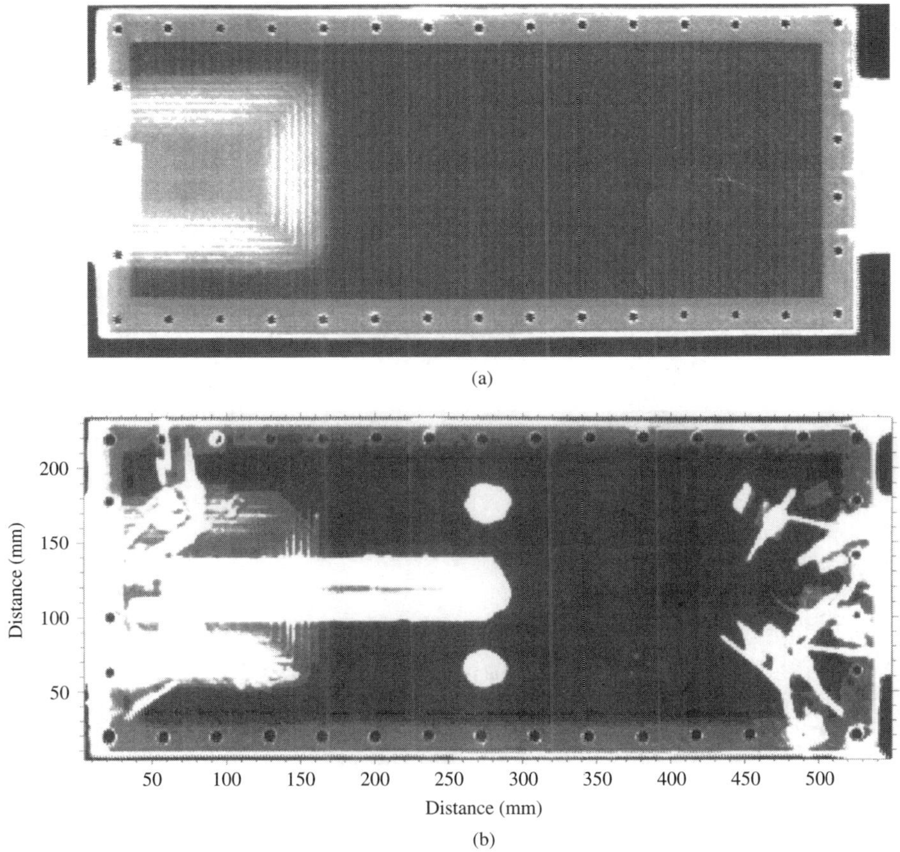

Figure 3.30 C-scan images of the manufactured test panel: (a) before impact tests; (b) after impact tests

Figure 3.31 Composite structure used for impact tests

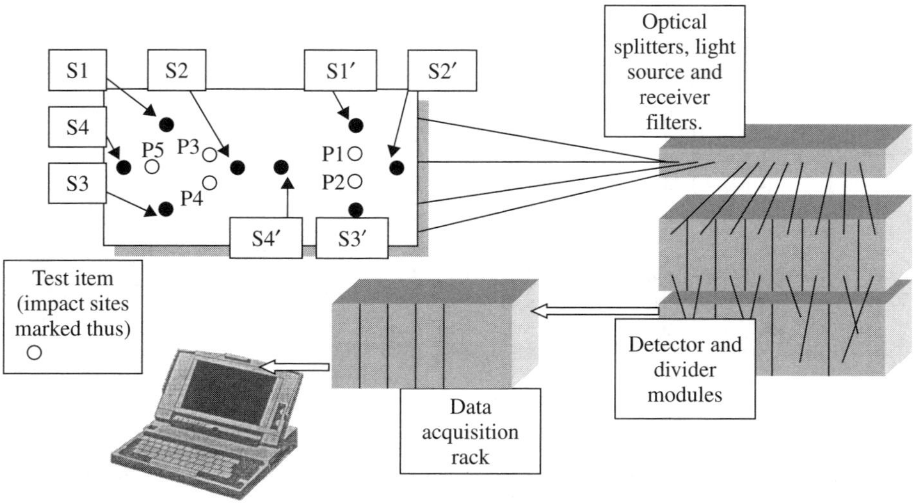

Figure 3.32 Experimental test layout

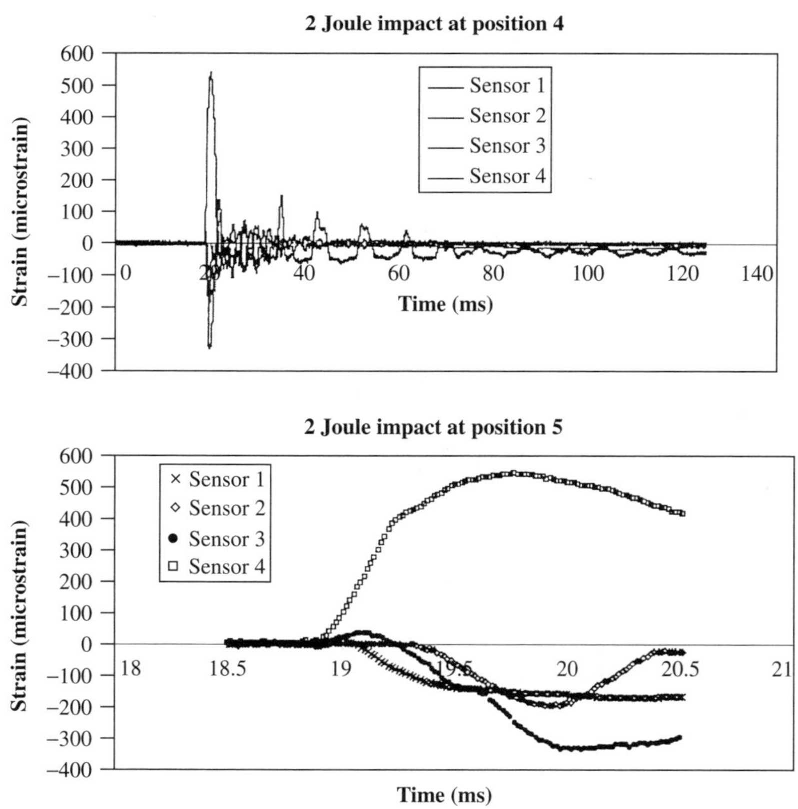

Figure 3.33 Sensor data for 2 J impact

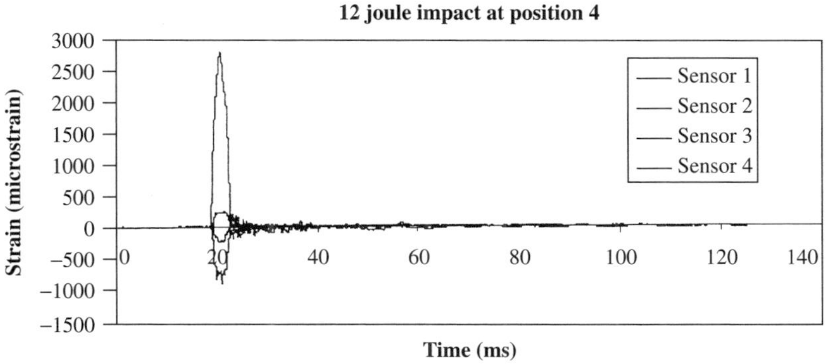

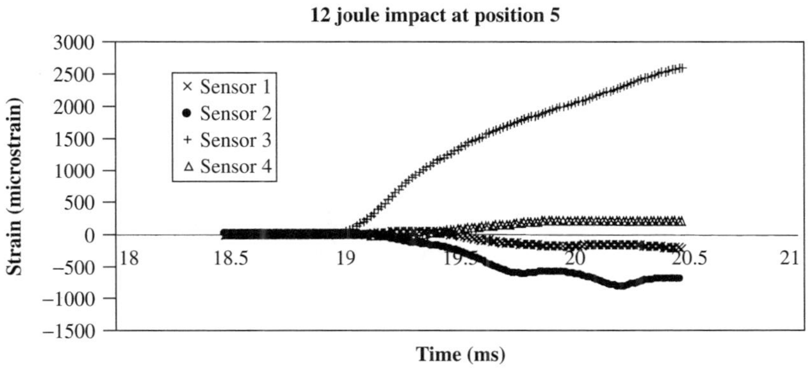

Figure 3.34 Impact data for 12 J impact

Figure 3.35 displays time versus distance information taken from the two impact events. The timing information was taken from the signals for each individual sensor. From the initial rise of each signal, the event time was taken as the intercept between the gradient of the leading edge and the time axis. The distance information was taken directly from the impact locations on the composite impact test item.

The example presented in this section shows that the system based on chirped fibre Bragg gratings, is capable of producing linear signals in response to static strain and dynamic stress wave events such as those resulting from impact. The sensors were successfully embedded into a test panel manufactured in aerospace type composite material and underwent trial impacts up to and beyond damaging energies. The sensor signals are of good signal to noise ratio and display no evidence of saturation over impact energies ranging from 2 to 24 J. More examples related to strain monitoring and on-line flight tests are presented in Chapter 6.

3.9 SUMMARY

Optical fibre sensors have a great potential for load monitoring in aerospace structures. Recent years have shown many research and laboratory developments in this area. It is very likely that optical fibre sensors will replace strain gauges in OLM systems in the near future. However, various financial and technical aspects need to be addressed in

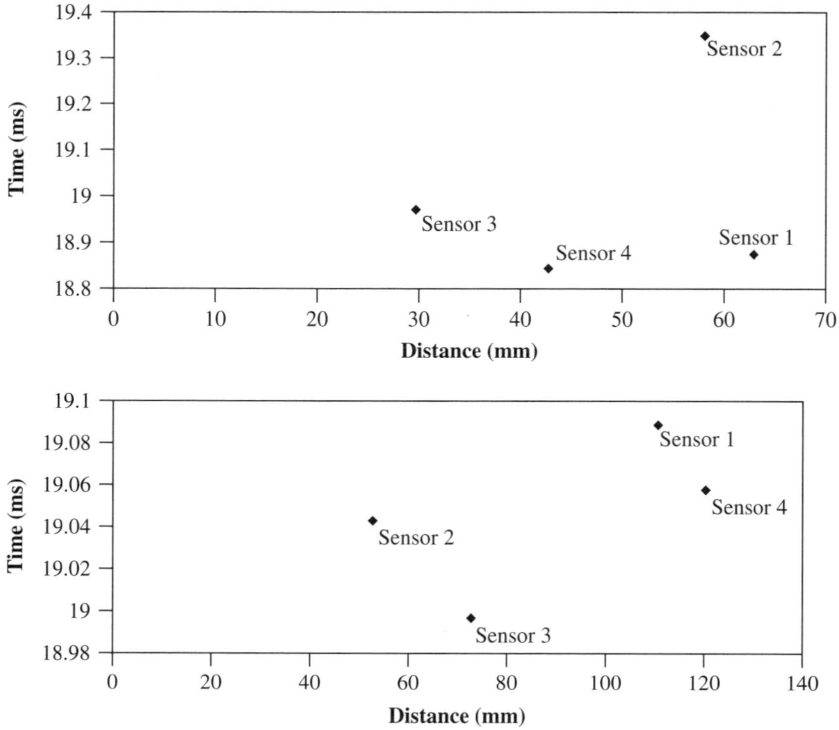

Figure 3.35 Time vs distance for impact events detected by sensors S1 to S4

the development of such systems. This includes: costs, manufacturing specifications, performance specifications, sensor installation, system integration, optical signal processing, calibration, redundancy, survivability, repairability and many other important elements. All these aspects have been discussed in this chapter. An example of the system performance will be shown in Chapter 6.

REFERENCES

Archambault, J.L., Reekie, L. and Russel, P.S.J. 1993. 100 % reflective Bragg reflectors produced in optical fibres by single excimer laser pulses, *Electron. Lett.*, Vol. 29, p. 453.

Baker, S.R., Rourke, H.N., Baker, V. and Goodchild, D. 1997. Thermal decay of fiber Bragg gratings written in boron and germanium codoped silica fiber, *Journal of Lightwave Technology*, Vol. 15, p. 470.

Betz, D., Thursby, G., Culshaw, B. and Staszewski, W.J. 2003. Acousto-ultrasonic sensing using fibre Bragg gratings, *Smart Materials and Structures*, Vol. 12(1), pp. 122–128.

Claus, R. (ed.). 1992. *Proceedings of the Optical Fibre Sensor-Based Smart Materials and Structures*, Virginia Polytechnic.

Culshaw, B. 1988. *Optical Fibre Sensing and Signal Processing*, Peter Peregrinus, London.

Dally, J. and Riley, W. (eds). *Experimental Stress Analysis*, McGraw Hill.

Dianov, E.M., Golant, K.M., Khrapko, R.R., Kurkov, A.S., Leconte, B., Douay, M., Bernage, P. and Niay, P. 1997. Grating formation in germanium free silicon oxynitride fibre, *Electron. Lett.*, Vol. 33, p. 236.

Dong, L., Cruz, J.L., Reekie, L., Xu, M.G. and Paynem, D.N. 1995. Enhanced photosensitivity in tin copdoped germanosilicate optical fibres, *Photon. Technol. Lett.*, Vol. 7, p. 1048.

Douay, M., Xie, W.X., Taunay, T., Bernage, P , Niay, P., Cordier, P., Poumellec, B., Dong, L., Bayon, J.F., Poignat, H. and Delavaque, E. 1997. Defensification involved in the UV-based photosensitivity of silica glasses and optical fibres, *Journal of Lightwave Technology*, Vol. 15, p. 1329.

Erdogan, T., Mizrahi, V., Lemaire, P.J. and Monroe, D. 1994. Decay of ultraviolet-induced fiber Bragg gratings, *J. Appl. Phys.*, Vol. 76, p. 73.

Espiniola, R.P., Atkins, R.M., Simoff, D.A., Nelson, K.T. and Pacxkowski, M.A. 1997. Fiber Bragg gratings written through a fiber coating, *Proc. ORF '97*, post-deadline paper PD4.

Feced, R., Roe-Edwards, M.P., Kanellopoulos, S.E., Taylor, N.H. and Handerek, V.A. 1997. Mechanical strength degradation of UV exposed optical fibres, *Electron. Lett.*, Vol. 33, p. 157.

FFA. 1997. Private communication, June.

Fokine, M.A., Sahlgren, B.E. and Stubbe, R. 1996. High temperature resistant Bragg gratings fabricated in silica optical fibres, *Proc. Australian Conf. on Optical Fibre Technology (ACOFT)*, post-deadline paper PD-1.

Foote, P. 1995. Optical fibre Bragg grating sensors for aerospace smart structures, *IEE Coll. Proc. on Optical Fibre Gratings and Their Applications, London, January*.

Gosh, M.K. and Mittal, K.L. (eds). 1996. *Polyimides: Fundamentals and Applications*, Marcel Dekker, New York.

Kannan, S., Guo, J.Z.Y. and Lemaire, P.J. 1997. Thermal stability analysis of UV-induced fiber Bragg gratings, *Journal of Lightwave Technology, Journal of Lightwave Technology*, Vol. 15, p. 1478.

Kersey, A.D. 1992. Multiplexed fibre optic sensors. In: J.E. Pearson (ed.), 1992, *Optical Technologies for Aerospace Sensing*, Critical Reviews of Optical Science and Technology, Vol. CR47, SPIE, Bellingham, USA, pp. 200–225.

Krohn, D.A. 1988. *Fiber Optic Sensor Fundamentals and Applications*, Instrument Society of America, Research Triangle Park, North Carolina.

Lemaire, P.J., Atkins, R.M., Mizrahi, V. and Reed, W.A. 1993. High-pressure H_2 loading as a technique for achieving ultrahigh UV photosensitivity and thermal sensitivity in GeO_2 doped optical fibres, *ELectron. Lett.*, Vol. 29, p. 1191.

Limberger, H.G., Varelas, D., Salathé, R.P. and Kotrotsios, G. 1996. Mechanical degradation of optical fibres induced by UV light, *Proc. of SPIE*, Vol. 2841, p. 84.

Matthewson, M.J. 1994. Optical fiber reliability models, *Proc. of SPIE*, Vol. CR50, p. 3.

Measures, R. 1992. Advances towards fiber optic based smart structures, *Optical Engineering*, Vol. 31, No. 1, p. 35.

Orowan, E. 1949. *Fracture and Strength of Solids, Rep. Prog. Phys.*, Vol. 12, p. 49.

Pearson, J.E. (ed.). 1992. *Optical Technologies for Aerospace Sensing*, Critical Reviews of Optical Science and Technology, Vol. CR47, SPIE, Bellingham, USA,

Plueddemann, E.P. 1991. *Silane Coupling Agents*, 2nd edition, Plenum Press, New York.

Riant, I., Borne, S. and Sansonetti, P. 1996. Dependance of fibre Bragg grating thermal stability on fabrication process, *Proc. of ORC '96*, Vol. 2, OSA Tech. Dig. Series, paper TuO5.

Rondinella, V.V. and Matthewson, M.J. 1993. Effect of chemical stripping on the strength and the surface morphology of fused silica optical fiber, *Proc. of SPIE*, Vol. 2074, p. 52.

Staszewski, W.J. 2000. Monitoring on-line integrated technologies for operational reliability – MONITOR, *Air & Space Europe*, Vol. 2, No. 4, pp. 67–72.

Stork, Y. 1992. Beläggning av Temperaturtålig Polymer på Optisk Fiber, Examensarbete vid Ericsson Calbeles-Onerc, Autumn, 92 12 23.

Udd, E. 1991. *Fiber Optic Sensors – An Introduction for Engineers and Scientists*, John Wiley & Sons, Inc., New York.

Udd, E. (ed.). 1992. *Fiber Optic Sensors*, Critical Reviews of Optical Science and Technology, Vol. CR44, SPIE, Bellingham, USA.

Williams, D.L., Ainslie, B.J., Armitage, J.R. and Kashyap, R. 1992. Enhanced photosensitivity in germanium doped silica fibres for future optical networks, *Proceeding of the 18th European Conference on Optical Communication*, Berlin, paper We B9-5.

Williams, D.L., Ainslie, B.J., Armitage, J.R., Kashyap, R. and Campbell, R.J. 1993. Enhanced UV photosensitivity in boron codoped germanosilicate fibres, *Electron. Lett.*, Vol. 29, p. 45.

Zhang, L. and Bennion, I. Apparatus for sensing temperature and/or strain in an object, UK Patent Application No. 9705976.0.

4

Damage Detection Using Stress and Ultrasonic Waves

W.J. Staszewski[1], C. Boller[1], S. Grondel[2], C. Biemans[3], E. O'Brien[4], C. Delebarre[2] and G.R. Tomlinson[1]

[1] *Department of Mechanical Engineering, Sheffield University, Sheffield, UK*
[2] *IEMN, Valenciennes, France*
[3] *DaimlerChrysler, Berlin, Germany*
[4] *Airbus UK, Filton, UK*

4.1 INTRODUCTION

Elastic waves and their propagation have been used for many years to analyse impact response problems, mechanical properties of various materials and structural damage. Various types of methods based on sound and ultrasound are applied for nondestructive testing (NDT). Acoustic Emission and Ultrasonic inspection are the most widely used techniques in industrial applications. The first technique is passive and does not require any external signal excitation; stress waves are structure-born and produced internally by defects. The second approach requires high-frequency external excitation. The maturity and proven damage detection applications are the major advantages of these techniques. The Acousto-Ultrasonic approach combines elements of Acoustic Emission and Ultrasonic inspection. Although the method has been around for more than thirty years, it has not been used widely in practice. Recent advancements in this area include applications of guided ultrasonic waves. Lamb waves are particularly attractive for damage detection in aerospace structures; a vast amount of literature has been published over the last ten years.

This chapter briefly discusses damage detection methods based on sound and ultrasound. Various elements related to wave propagation mechanism, monitoring strategy and

Health Monitoring of Aerospace Structures – Smart Sensor Technologies and Signal Processing.
Edited by W.J. Staszewski, C. Boller and G.R. Tomlinson ISBN: 0-470-84340-3

transducer schemes are discussed. The focus is on definitions, basic principles, similarities and dissimilarities between different techniques. For more details the reader is referred to the appropriate literature. Engineering applications are illustrated using damage detection case studies. Further application examples of Acoustic Emission and Lamb wave based damage detection are given in Chapter 6.

4.2 ACOUSTIC EMISSION

4.2.1 Background

Acoustic Emission (AE) is one of the first and most widely used NDT techniques for structural damage detection. This is supported by a number of literature references related to the subject (e.g. (Scott 1991; Muravin 2000; Holroyd 2001)). The technique relies on transient sound waves propagating in the analysed material. Most of these waves are short-time transient events (burst signals) of significant energy between 100–1000 kHz. The waves can propagate long distances in circles, i.e. in all possible directions (Figure 4.1). Therefore AE testing can cover large, often inaccessible, monitored areas. The distance of propagation depends on material properties, geometry, frequency and environment.

Acoustic events at their origin are high-frequency (in MHz), wideband impacts emitted internally by microcracks and/or inclusion de-cohesion (e.g. metallic inclusions, bubbles) under external loading applied to monitored specimens. These material defects release elastic energy due to rapid local stress redistribution as a result of loading. The energy results from growing cracks, rubbed surfaces of cracks or dislocations. Other sources of AE are also possible such as phase transformation or melting. The AE process is similar to seismic waves propagating as a result of dislocations of various rock formations leading to earthquakes. In contrast to other sound/ultrasound-based techniques it is a passive as well as static method of damage detection, when using the terms mentioned in Chapter 1 before; AE transducers listen to the monitored structure. A relatively small number of sensors is required to cover large monitored areas.

4.2.2 Transducers

AE waves propagating in the monitored specimen can be detected by appropriate transducers which convert mechanical energy into electrical signals. There exist various types

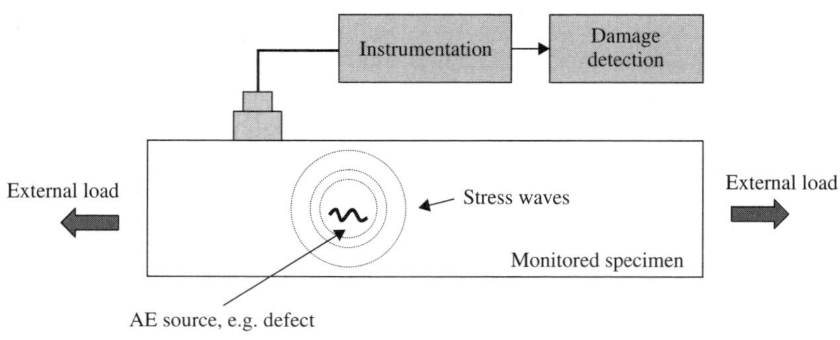

Figure 4.1 Acoustic Emission principle

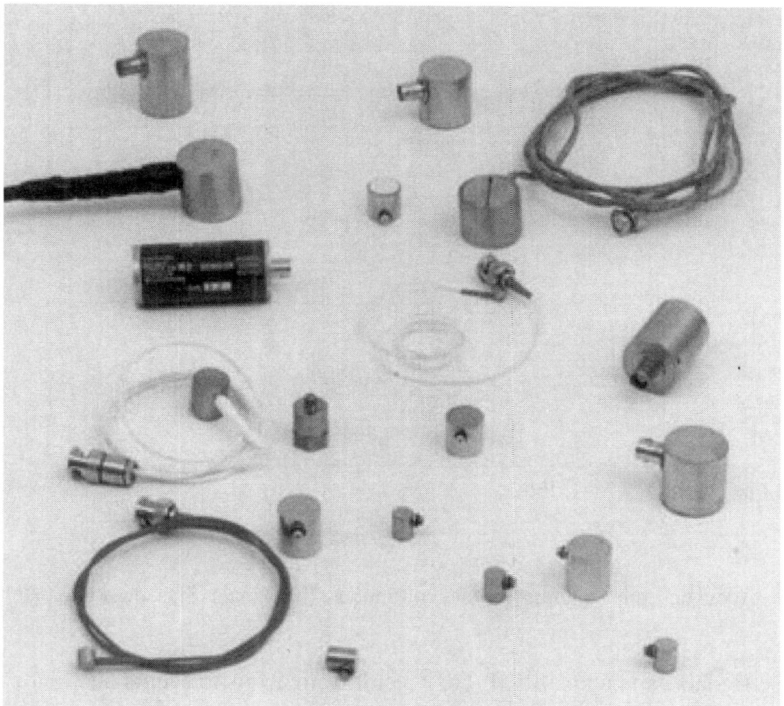

Figure 4.2 Acoustic Emission transducers (reproduced by the permission of MISTRAS Holdings Corp.)

of AE transducers such as piezoelectric, electrodynamic, laser and capacitance sensors. Piezoelectric transducers are well-proven and by far the most widely used for AE testing. Figure 4.2 shows examples of classical piezoelectric transducers. The sensitivity of AE transducers is determined by the bandwidth and the resonance frequency. The sensitivity in general is defined using a logarithmic scale given in dB as

$$A = 20 \times \log \frac{U}{U_{ref}} \tag{4.1}$$

where U is the output voltage from the sensor and U_{ref} is the reference voltage equal to 1 μV. Sensors data need to be amplified before further signal processing. Often pre-amplifiers are integrated with some types of sensors. Sensors are usually mounted using magnetic holding devices, clamps or glue. Good acoustic contact between the sensor and the monitored surface is essential for AE testing. When the sensor is not permanently bonded on the surface, often special types of coupling agents (e.g. silicon grease, oil) are required.

4.2.3 Signal Processing

There are two types of AE signals encountered in practice: these are transient (or burst); and continuous signals. Transient signals can be separated in time, i.e. the beginning and

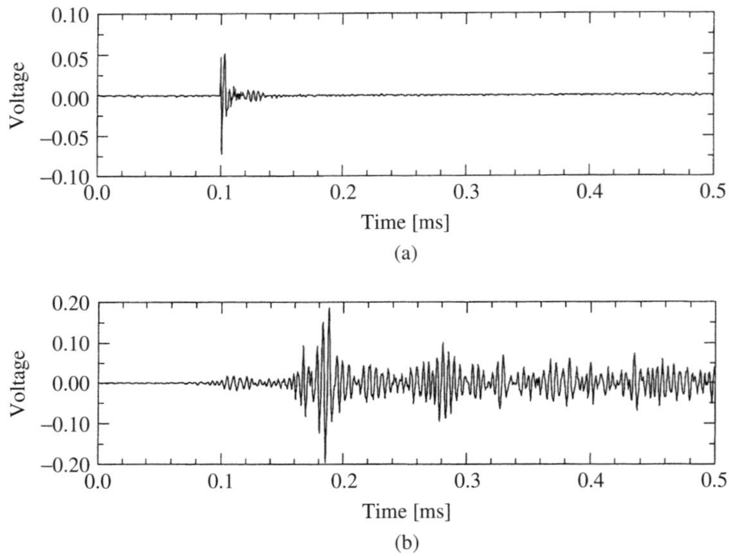

Figure 4.3 Acoustic Emission burst (a) and continuous; (b) signals (Staszewski and Holford 2001)

end of these signals can be identified. They result from acoustic events due to local defects. Continuous signals cannot be separated in time. They are produced by various unwanted phenomena, e.g. plastic deformation or friction. These signals often include background noise which can include both mechanical and electrical disturbances. Figure 4.3 gives examples of burst and continuous AE signals.

Various signal processing techniques are used to analyse AE signals and extract features related to defects. Usually, a threshold level is defined. The AE threshold needs to be exceeded before further analysis. A number of well-defined standard parameters are then used in order to identify defects. This includes:

- Peak amplitude – maximum amplitude of the signal;
- Arrival time – absolute time when the signal first crosses the threshold level;
- Duration time – time interval between the first and last crossings of the threshold;
- Rise time – time interval between the first threshold crossing and the peak amplitude;
- Ring down count – number of threshold level crossings.

Figure 4.4 summarises AE parameter used for damage detection. Often the amplitude values of the AE signal are integrated over the duration time in order to estimate the energy of the burst signal. The level of the background noise can be estimated calculating the Root Mean Square (RMS) value of the signal before the arrival time. Various clustering techniques of AE signals are used in order to mark and locate the highest AE activities. The information obtained from AE tests is displayed using numerical (cumulative or differential characteristics giving total number of events, total energy or pressure) and/or graphical (cluster graphs giving source location) diagrams. Signal waveforms can also be analysed using the time and frequency domains. Source location is possible using arrays

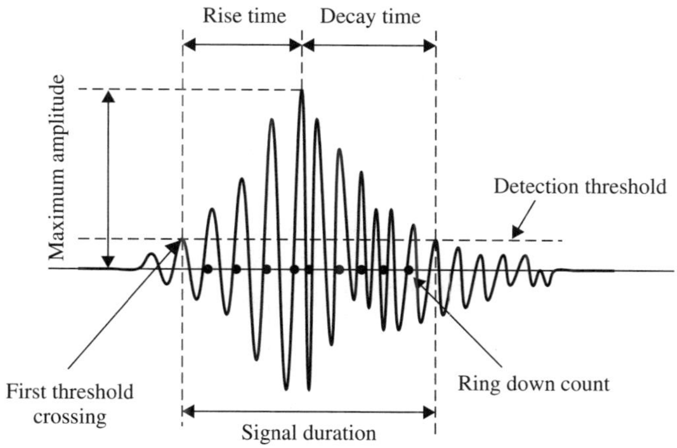

Figure 4.4 Summary of Acoustic Emission parameters used for damage detection

of sensors. Once the sound velocity and signal arrival time differences to these sensors are known, damage location can be estimated using classical triangulation procedures.

4.2.4 Testing and Calibration

AE testing comprises three major steps: detection of AE activities (damage detection), estimation of the AE burst rate (damage severity) and AE source location (damage location). All steps require the AE equipment to be calibrated. This involves an electronic waveform generator (often called a pulser) which produces pre-defined transient signal or a special device which simulates an AE event using the break of the pencil graphite lead. The later device is commonly known as the Hsu–Nielsen source, after the developers of the technique. The break of a 2H 0.5 mm diameter pencil lead approximately 3 mm from its tip generates an acoustic signal which is similar to a natural AE burst.

AE detects damage in the moment of its occurrence. Defect characteristics are unique events and once the event is missed signals cannot be reproduced. Therefore successful damage detection often requires continuous monitoring in service. Also, defects (e.g. cracks) that do not grow are not detected. Burst signals depend on loading and environmental conditions. It is important to monitor these effects for the reference.

The AE testing offers relatively rapid inspection for damage and what is important it can be used for a wide variety of materials. The method offers global monitoring and does not require access to whole monitored areas.

4.3 ULTRASONICS

4.3.1 Background

There exist various damage detection techniques based on ultrasound (Buirks *et al.* 1991; Rose 1999; Schmerr 1998). Classical ultrasonic techniques utilise various phenomena

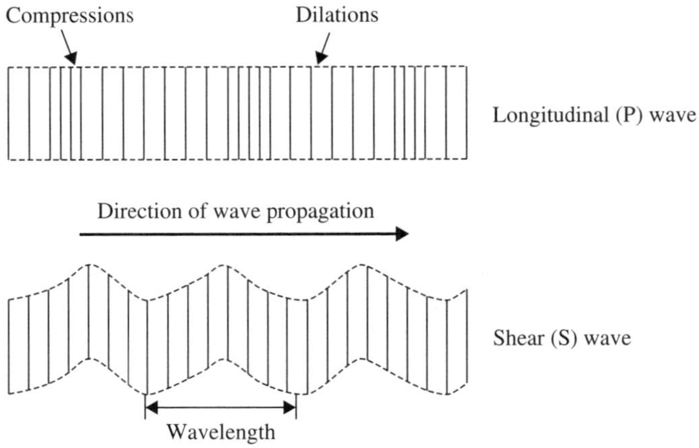

Figure 4.5 Longitudinal (a) and shear (b) bulk waves

and/or properties of ultrasonic waves propagating in the material in order to detect defects. It is well known that various types of waves can propagate in solids. Longitudinal (often called compressional, dilatational pressure or P-wave) and shear (often called transverse or S-wave) waves are the most commonly used wave modes for ultrasonic testing. Both types of waves are represented graphically in Figure 4.5. The direction of particle motion is either in the direction (longitudinal waves) or perpendicular (shear waves) to the direction of wave propagation.

The velocity of propagating waves is one of the most important parameters in ultrasonics. The relationship between the velocity of the longitudinal ultrasonic wave and elastic properties of the material is often used to evaluate the condition of the structure. This relationship can be given as

$$\nu^2 \approx \frac{E_L}{\rho} \tag{4.2}$$

where ν is the velocity of the longitudinal ultrasonic wave, E_L is the longitudinal modulus of elasticity and ρ is the density of the monitored specimen. Similarly, the square velocity of the shear wave is proportional to the shear modulus E_s over density.

Ultrasonic waves can travel long distances in solid materials. However, the energy of sound decreases with the distance of propagation. Propagating waves are additionally scattered (or reflected) and absorbed by different material/structural boundaries. As a result, the amplitude of propagating waves is attenuated. When a wave passes between different media, the velocity of propagation changes. Additionally, various mode conversions can occur. Ultrasonic testing utilises wave attenuation, reflection and refraction for damage detection. Successful NDT requires an understanding of the ultrasonic field. This involves not only wave propagation principles but also problems related to pressure variations, directivity analysis and beam angle of divergence, as discussed in (Rose 1999).

It is important for successful damage detection that the wavelength of ultrasound used for testing is of the order of the defect's size. The wavelength λ can be calculated as

$$\lambda = \frac{c}{f} \tag{4.3}$$

where c is the wave velocity and f is the input frequency. This formula is valid only for continuous waves (or pulse waves with a large number of oscillations). The wavelength value for short-time pulse waves (pulse wave with a small number of oscillations) varies due to the broadband nature of the pulse signal.

4.3.2 Inspection Modes

Ultrasonic testing procedures are based on two major inspection modes. These are normal beam inspection and angle beam inspection. The ultrasonic pulse travels through a thickness of material in a normal beam inspection. The so-called *pulse–echo* method requires two transducers. The pulse is generated by the transmitter and captured by the receiver. In contrast, the *pitch–catch* mode involves only one transducer. The pulse generated by the transducer passes trough the material, gets reflected by a boundary and is captured by the same transducer. The angle beam inspection mode introduces refracted shear waves to the monitored specimen. The inspection modes are shown graphically in Figure 4.6.

4.3.3 Transducers

There exist various types of ultrasonic transducers. Most acoustic transducers use piezoelectric ceramics which convert electrical energy into the mechanical energy and vice versa. Figure 4.7 shows a diagram of a typical ultrasonic transducer. The type, size and frequency bandwidth of transducers are important for damage detection resolution. There are various types of transducers used in practice. This includes contact and noncontact transducers. Contact transducers require appropriate coupling (gel, water) for the ultrasound energy transfer. Special types of transducers are used when specimens are immersed in water. Noncontact transducers do not require any coupling but show significant attenuation at high frequencies.

The geometry, frequency and size of the transducers determine the ultrasonic beam near field (Figure 4.8). The following formulas can be used in practice (Rose 1999)

$$\sin\frac{\alpha}{2} = \frac{0.6\lambda}{r} \quad \text{and} \quad N = \frac{r^2 f}{c} \tag{4.4}$$

where α is the angle of divergence, r is the transducer radius and N is the near field distance defined as the point on the axis of the transducer which separates intense oscillations from a smooth decay. A small angle of divergence, required in practice, is a

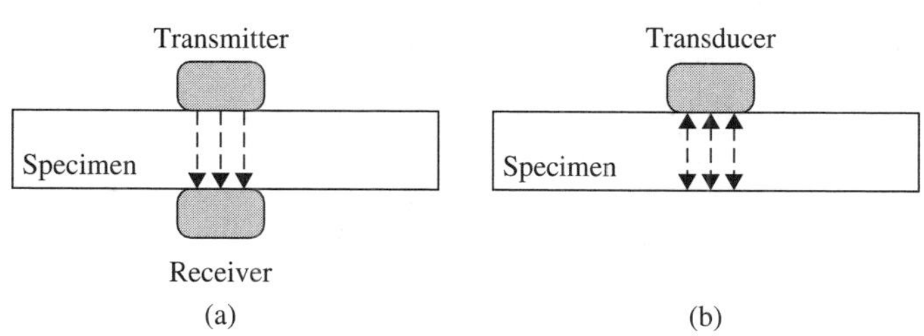

Figure 4.6 Ultrasonic inspection modes: (a) pulse-echo; (b) catch-pitch

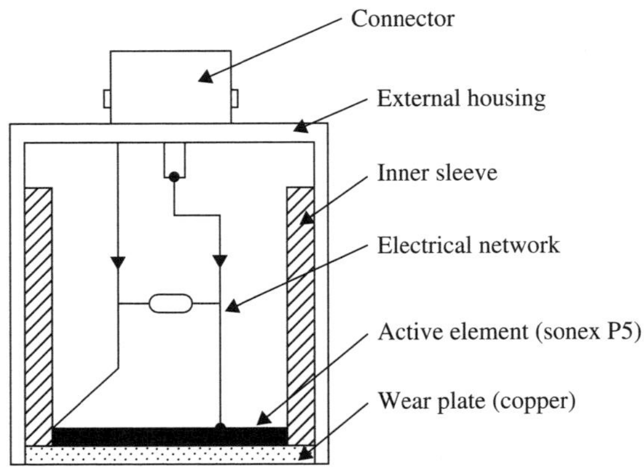

Figure 4.7 Schematic diagram of a typical transducer used for ultrasonic testing

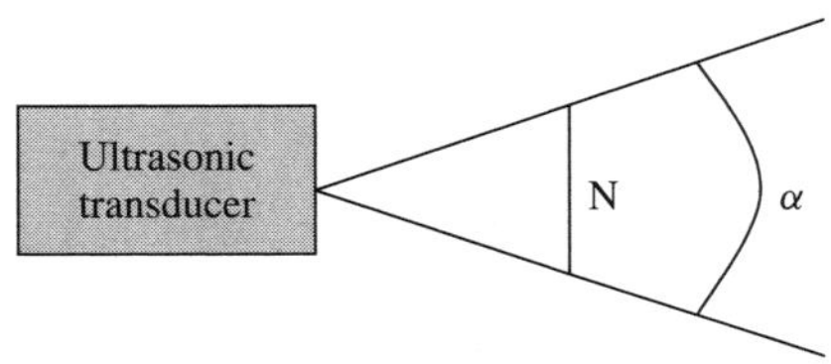

Figure 4.8 Ultrasonic beam near field

compromise between the transducer radius and the input frequency. Often both parameters are decreased in order to reduce destructive interferences in the near field.

Frequency bandwidth is also important since it affects the penetration of the material and damage detection sensitivity. Low-frequency transducers (below 2 MHz) offer better penetration whereas high-frequency transducers (above 15 MHz) provide better sensitivity to small defects.

4.3.4 Display Modes

Ultrasonic data is presented in a form of ultrasonic scans representing damage detection results. Four different types of scans are used in practice. These are:

- *A-scan*: pulse amplitude presented as a function of travel time;
- *B-scan*: two-dimensional representation giving travel time of an ultrasonic pulse versus transducer position;
- *C-scans*: signal echoes are displayed in a top view of the test surface giving defect location maps;
- *D-scans*: a modified C-scan in which time-of-flight values are displayed in a top view on the specimen surface.

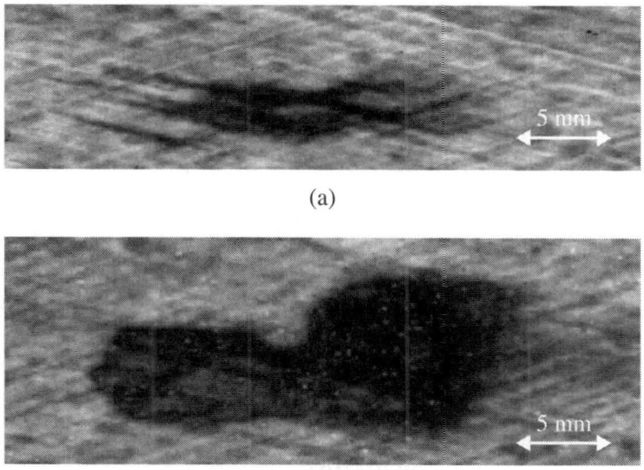

Figure 4.9 Ultrasonic C-scan examples from a composite plate: (a) delamination after the 2 J impact; (b) delamination after 7 J impact (Pedemonte *et al.* 2001)

Figure 4.9 shows an example of the ultrasonic C-scan for the composite plate with a delamination. The damage is exhibited as the dark area on the plot.

Ultrasonic inspection can be used for a wide range of materials. It is highly sensitive to small defects. However, it requires direct access to inspected surfaces and good inspection skills. Full defects maps (C-scan, D-scan) require scanning and are relatively time consuming. Nevertheless, ultrasonic testing has been used for many years as one of the most successful damage detection techniques.

4.4 ACOUSTO-ULTRASONICS

The acousto-ultrasonic technique is based on stress waves introduced to a structure by a probe at one point and sensed by another probe at a different position. The acousto-ultrasonic method, introduced in the late 1970s (Vary 1988), combines elements of Ultrasonics, guided wave Ultrasonics and Acoustic Emission. All these four techniques utilise stress waves propagating in structures for damage detection. However, Acoustic Emission is the only technique which does not require any external excitation; stress waves are produced by material defects. The other three techniques require an external source of high-frequency excitation to produce stress waves. However, paths of classical ultrasonic waves are well defined and traceable. Similarly, guided wave Ultrasonics is based on well-defined propagation of waves, as described in Section 4.5. In contrast, propagation of acousto-ultrasonic waves is difficult to analyse. This is due to the fact that the method uses high-frequency (usually above 0.5 MHz) impulse excitation which results in a large number of mixed modes. Alternatively, the broadband excitation (Gaussian white noise or sweep sine) could also be used, as reported in (Biemans *et al.* 2001) and shown in Section 4.8.1. Actuating and receiving transducer probes are often not in a line-of-sight position regarding the damage. As a result, the actual transducer responses include not only directly propagating wave modes but also reflected and scattered modes. The rich content of the stress wave energy of complex modes carries a lot of information regarding possible material and/or structural damage. Figure 4.10 shows an example of an acousto-ultrasonic stress wave from damage detection studies in a composite-metal joint (Kwon *et al.* 2000).

There exist a number of different parameters to describe the stress wave energy. Often AE parameters described in Section 4.2.3 are used in practice. Alternatively, the Stress

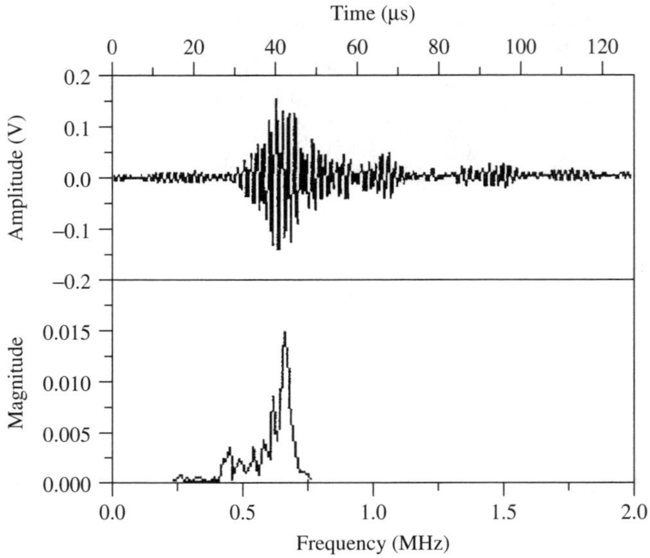

(a)

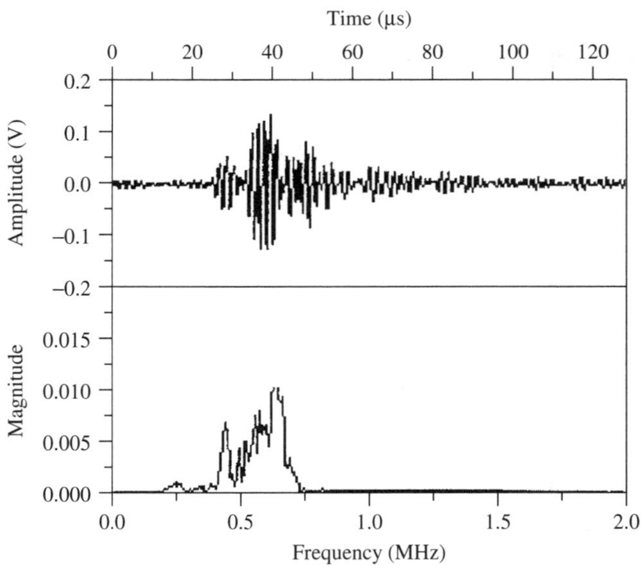

(b)

Figure 4.10 Acousto-ultrasonic stress waves from damage detection in a composite-metal joint: (a) unfatigued specimen; (b) fatigued specimen (Kwon *et al.* 2000)

Wave Factor (SWF), based on power spectral density, is applied. The SWF can be defined using spectral moments given as (Kiernanand and Duke 1988)

$$M_n = \int_0^{f_{\max}} S(f) f^n \mathrm{d}f \tag{4.5}$$

where $S(f)$ is the power spectral density, $f_{\max}$ is the maximum frequency of the analysed spectrum and $n = 1, 2, \ldots$.N. The M_0 moment is the most widely used parameter in practice. Figure 4.11 gives an example of this parameter for the acousto-ultrasonic data presented in Figure 4.10.

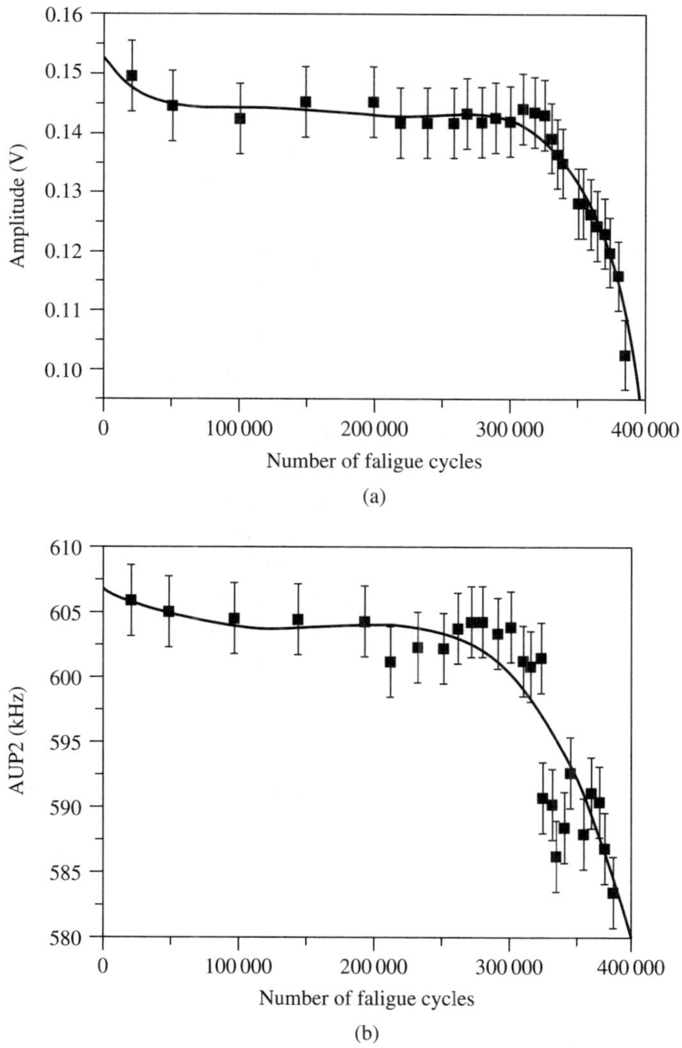

Figure 4.11 Stress wave factors for the data presented in Figure 4.10: (a) peak amplitude; (b) M_1/M_0 (Kwon *et al.* 2000)

4.5 GUIDED WAVE ULTRASONICS

4.5.1 Background

Guided waves are governed by the same wave equations as bulk waves. However, in contrast to bulk waves, they have an infinite number of modes associated with propagation. Guided waves are able to interact with defects in structures due to their propagation properties that are highly sensitive to any discontinuities in materials. This section defines various types of guided waves and briefly introduces Lamb waves which are the most widely used guided waves for structural damage detection.

4.5.2 Guided Waves

Propagating wave packets which are superpositions of various modes are often called guided waves. There are various types of guided waves available in practice. Wave packets, resulting from appropriate stress and strain boundary conditions, which travel on the surface of a solid body, are known as surface waves. Surface waves usually exhibit large amplitudes and travel slower than other types of guided waves.

Rayleigh waves are the best known surface waves (Viktorov 1967). They are nondispersive for uniform material properties. However, their mechanism of propagation is very complex; waves are polarised and surface particles are moved around an ellipse. The components of Rayleigh waves can couple with a medium surrounding the surface of the body. This coupling affects the amplitude and velocity of the wave. The amplitude of the wave decreases rapidly with depth. The rate of decrease depends on the wavelength. Therefore inspection methods based on Rayleigh waves are used mostly to detect surface defects. Other examples of surface waves include Stonely and Love waves. These waves are not commonly used for damage detection and are better known to seismologists than to damage detection experts. Stonely waves occur at an interface between two media. They are closely related to Rayleigh waves. The existence of Stonely waves depends on the density and shear modulus ratios of the neighbouring media. Love waves are horizontally polarized shear waves which also exist on the surface. However, in contrast to the other two types of surface waves they are highly dispersive. Love waves are the fastest propagating surface waves.

Lamb waves (Viktorov 1967) are dispersive plate waves that occur for traction-free forces on both surfaces of the plate. The velocity of these waves depends on the product of frequency of excitation and thickness of the plate. They can propagate long distances and are used for damage detection of plate-like structures. Lamb waves are the most widely used guided waves for damage detection. They are discussed in more details in the next section.

4.5.3 Lamb Waves

Lamb waves refer to elastic perturbations propagating in a solid plate with free boundaries for which the displacements correspond to different basic propagation modes, with symmetric and antisymmetric vibrations. For a given plate thickness d and acoustic frequency

f, there exists a finite number of such propagation modes specified by their phase velocities. A complete description of such propagation characteristics for plates is normally given in the form of a set of dispersion curves, illustrating the plate-mode phase velocity as a function of the frequency–thickness product. Each curve represents a specific mode, which is conventionally called A_0, S_0, A_1, S_1, A_2, S_2, etc. where A_n denotes antisymmetric modes and S_n symmetric modes.

When an excitation is applied at some point on the plate, the excitation energy encounters the upper and lower surfaces of the plate. Longitudinal waves (P) are polarised parallel to the plate whereas shear horizontal waves (SH) form a series of modes. After some time when longitudinal waves are polarised in the direction perpendicular to the surface, shear vertical waves (SV) form modes in connection with P waves; these combined P+SV guided waves are known as Lamb waves.

The most common approach for solving the Lamb wave problem is the method of potentials. For a solid medium bounded by two parallel planes a distance $2d$ apart (see Figure 4.12), the equation of motion, which contains only the particle displacement, can be given as (Achenbach 1984)

$$\mu w_{i,jj} + (\lambda + \mu) w_{j,ji} + \rho f_i = \rho \ddot{w}_i \tag{4.6}$$

where w_i are displacements, f_i are body forces, λ, μ are Lamé constants and ρ is the density. The boundary conditions for the surface tractions can be defined as

$$t_i = S_{ij} n_j \tag{4.7}$$

where t_i are traction forces, S_{ij} are stresses and n_j are cosine directions. The longitudinal and shear waves of plain strain are governed by the well-known wave equations that can be obtained from Equation (4.6) using the Helmholtz decomposition as (Achenbach 1984)

$$\frac{\partial^2 \phi}{\partial x_1^2} + \frac{\partial^2 \phi}{\partial x_3^2} = \frac{1}{c_L^2} \frac{\partial^2 \phi}{\partial t^2} \tag{4.8}$$

$$\frac{\partial^2 \psi}{\partial x_1^2} + \frac{\partial^2 \psi}{\partial x_3^2} = \frac{1}{c_T^2} \frac{\partial^2 \psi}{\partial t^2} \tag{4.9}$$

respectively. Here, ϕ and ψ represent decomposed displacement variables, c_L indicates the velocity of longitudinal wave whereas c_T is the velocity of shear (transverse) waves.

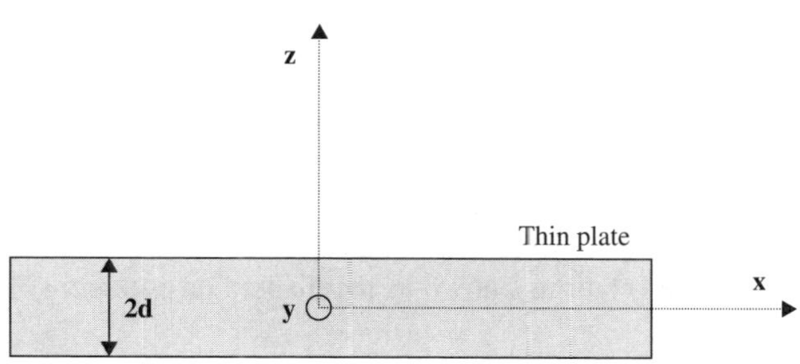

Figure 4.12 Geometry of the plate for Lamb wave propagation

The actual longitudinal and transverse displacements of the plate can be obtained as

$$w_1 = \frac{\partial \phi}{\partial x_1} + \frac{\partial \psi}{\partial x_3} \tag{4.10}$$

$$w_2 = \frac{\partial \phi}{\partial x_3} - \frac{\partial \psi}{\partial x_1} \tag{4.11}$$

respectively. Similarly, stresses can also be expressed in terms of the field variables but are not analysed here. The solutions of Equations (4.8–9), representing travelling waves in the x_1 direction and standing waves in the x_3 direction, can be assumed in the form

$$\phi = \Phi(x_3)\exp[i(kx_1 - \omega t)] \tag{4.12}$$

$$\psi = \Psi(x_3)\exp[i(kx_1 - \omega t)] \tag{4.13}$$

where ω is the frequency and k is the wavenumber. Substituting Equations (4.12–13) into Equations (4.8–9) gives the unknown amplitude functions Φ and Ψ as

$$\Phi(x_3) = C_1 \sin(px_3) + C_2 \cos(px_3) \tag{4.14}$$

$$\Psi(x_3) = D_1 \sin(qx_3) + D_2 \cos(qx_3) \tag{4.15}$$

where

$$p^2 = \frac{\omega^2}{c_L^2} - k^2 \quad \text{and} \quad q^2 = \frac{\omega^2}{c_T^2} - k^2 \tag{4.16}$$

and C_1, C_2, D_1, D_2 are arbitrary constants. Since both field variables involve sine and cosine functions, which are odd and even respectively, the solutions are often split into two symmetric and antisymmetric modes. The displacements for the symmetric modes are

$$w_1 = ikC_2 \cos(px_3) + qD_1 \cos(qx_3) \tag{4.17}$$

$$w_2 = -pC_2 \sin(px_3) - ikD_1 \sin(qx_3) \tag{4.18}$$

whereas the solutions for the antisymmetric modes can be given as

$$w_1 = ikC_1 \sin(px_3) - qD_2 \sin(qx_3) \tag{4.19}$$

$$w_2 = pC_1 \cos(px_3) - ikD_2 \cos(qx_3) \tag{4.20}$$

These equations, known as Rayleigh–Lamb equations for guided waves in plates, show that Lamb wave propagation is generally complex due to the coexistence of at least two modes at any given frequency and the strongly dispersive nature of these modes at high frequency. The traction-free boundary conditions for the plain strain need to be additionally applied in order to obtain the constants C_1, C_2, D_1 and D_2. This leads to Rayleigh–Lamb frequency relations known as the dispersion equations (Achenbach 1984)

$$\frac{\tan(qh)}{\tan(ph)} = -\frac{4k^2pq}{(q^2 - k^2)^2} \tag{4.21}$$

for symmetric modes and

$$\frac{\tan(qh)}{\tan(ph)} = -\frac{(q^2 - k^2)^2}{4k^2pq} \tag{4.22}$$

for antisymmetric modes, where $h = d/2$. The above equations can be solved numerically in order to predict velocities of a propagating Lamb wave of frequency f in a plate of thickness d. The results are presented as a function of the fd frequency–thickness product. Figure 4.13 gives an example of the dispersion characteristics for an aluminium plate. This clearly shows how many complex modes can propagate in the plate. Single S_0 and A_0 Lamb wave modes are possible only for small values of the frequency–thickness product ($fd < 2$).

Furthermore, a single and pure Lamb wave mode may generate a variety of other modes either by interacting with a surface or subsurface flaw or by crossing the interface between two materials of different boundaries. As a result, the problem becomes very difficult to solve analytically. Often numerical analysis is employed to study wave propagation. Figure 4.14 shows a simulated example of the Lamb wave propagating in the aluminium plate with a damage slot positioned in the middle of the plate. Here, the scatter around the damage can be clearly observed in the contour plots of the propagating wave. The study was performed using the Local Interaction Simulation Approach (LISA) (Delsanto *et al.* 1992, 1994, 1997). The method has been proven to perform very well for wave propagation models with complex boundaries, imperfect material interfaces and heterogeneous materials (Agostini *et al.* 2000; Lee and Staszewski 2002, 2003a, 2003b, 2003c).

4.5.4 Monitoring Strategy

Real engineering structures under inspection are usually quite complex when compared with simple plates studied in laboratory conditions and reported in the literature. The

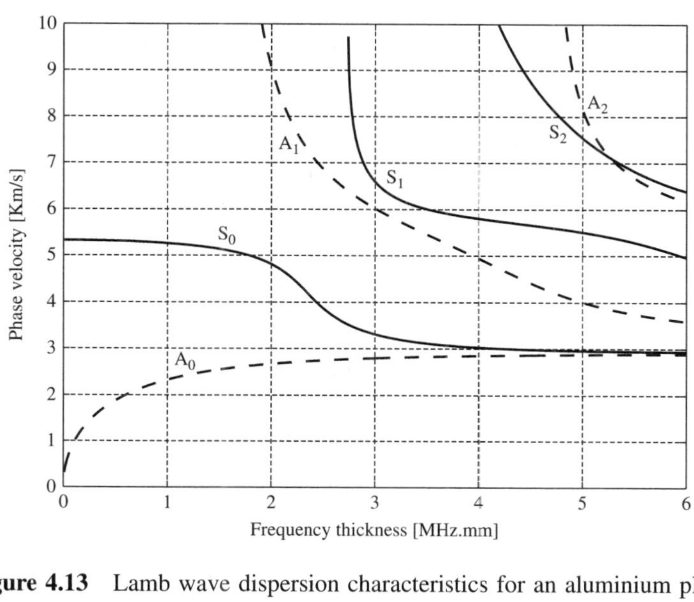

Figure 4.13 Lamb wave dispersion characteristics for an aluminium plate

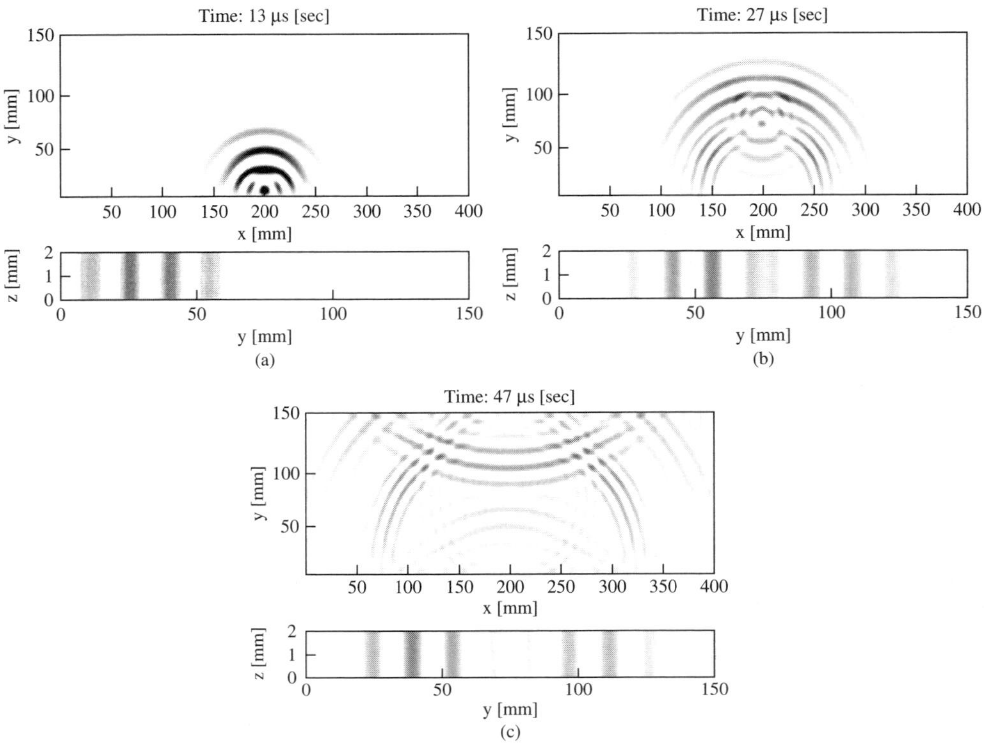

Figure 4.14 Snapshots of the S_0 Lamb wave mode propagation in an aluminium plate with a damage slot after: (a) 13 μs; (b) 27 μs; (c) 47 μs. The figure shows out-of-plane (top parts) and in-plane (bottom parts) vibration (Lee and Staszewski 2003a)

complexity is determined by various types of joints, stiffeners, rivets, complicated shapes or varying thickness. This causes the entire analysis to be much more complicated and requires an appropriate monitoring strategy.

The methodology or strategy of monitoring is extremely important for successful damage detection. The basic factors, which determine the Lamb wave based damage detection analysis are related to properties of the structure under inspection, transducer schemes, choice of excitation input signal, and appropriate signal processing, as reported in (Wilcox *et al.* 1999; Staszewski and Boller 2002). Other elements include various aspects related to transducer coupling methods, optimal sensor locations and sensor validation procedures. The last but not least is the hardware used for monitoring, graphical interface and data storage organisation.

The dispersive nature of Lamb waves and also the finite number of modes at a given frequency makes long-range inspection very difficult. To overcome these problems, low frequency–thickness products are often utilised for damage detection. In this case only two fundamental modes A_0 and S_0, are used. It is important to limit the bandwidth of the excitation to a range over which there is little dispersion (i.e. the phase velocity does not change significantly with frequency).

The first applications of Lamb waves for damage detection used bulky wedge (angle-beam) transducers. It appears that piezoceramic elements are now the most widely used transducers due to the fact that they can be used as sensors and actuators at the same time. Often piezoceramics can become an integral part of the monitored structure, as shown in Section 4.6.1. Recent advancements in this area include optical fibre sensors (Betz *et al.* 2002) and MEMS (Micro-Electro-Mechanical Systems) sensors (Khuri–Yakub *et al.* 2000). Both types of sensors can be used not only for Lamb wave detection but also for strain measurements. MEMS devices can additionally generate Lamb waves. A number of different practical aspects need to be considered once choosing transducers for Lamb wave detection. This includes coupling, connectors and environmental protection. The cold bonding is usually preferred to hot bonding. The process of bonding must be as easy (if not easier) as the procedure for bonding strain gauges. Also, bonded sensors are better than embedded due to possible sensor failures and replacements. Reliable connectors and environmental protections are required to prevent sensor failures. Wireless applications are possible with piezoceramic and MEMS sensors. Coupling, connectors and environmental protection are particularly important in the case of optical fibre and piezoceramic sensors.

Different types of signals are used as input excitation. It is considered that the simpler the input signal the simpler the output signal for damage detection. The choice of input excitation is often a compromise between the amplitude and the mode generation. Low-voltage signals are possible when the input frequencies are within transducer resonance frequencies (see Section 4.6.3). This is often associated with intelligent signal processing to remove undesired modes and extract features related to damage. In practice transducers resonance frequencies do not coincide with single Lamb wave mode frequencies.

Previous studies (Biemans *et al.* 2001; Staszewski *et al.* 1999a) show that even simple input signals can lead to complex output signals due to various attenuation and dispersion effects which are not related to damage. This clearly shows that intelligent signal processing is one of the most important elements of the Lamb wave based damage monitoring strategy. Examples of various signal processing techniques for damage detection are discussed in Chapter 5.

Once the transducers excitation signals are chosen the question is where to put sensors for optimal damage detection. Recent years have shown considerable progress on the problem of determining the number and location of sensors in engineering structures. Some of these techniques are discussed in Chapter 5.

4.6 PIEZOELECTRIC TRANSDUCERS

4.6.1 Piezoelectricity and Piezoelectric Materials

Piezoelectric transducers are the most widely used sensors for damage detection based on sound and ultrasound. Piezoelectricity is an electric polarisation effect due to mechanical forces. In other words an electric charge is collected on the surface of the piezoelectric material when it is squeezed.[1] Often the converse effect is possible, i.e. the material generates a mechanical strain in response to the applied electric field. Both effects are illustrated graphically in Figure 4.15. Piezoelectricity was discovered by Jacque and Pierre

[1] *Piezo* is in fact a Greek term for *to squeeze*.

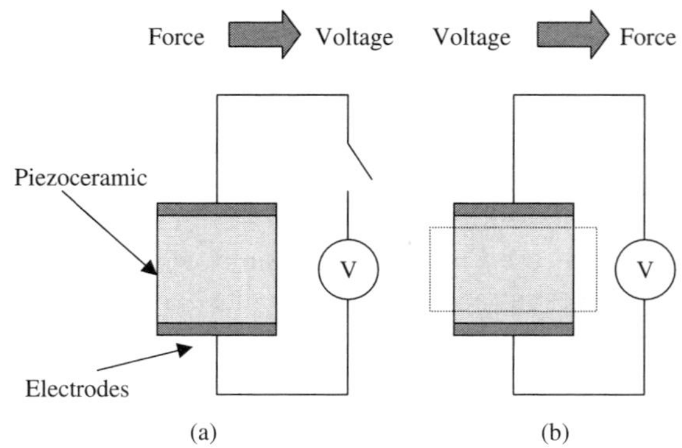

Figure 4.15 Direct (a) and inverse; (b) piezoelectric effects

Curie in 1880s. It is an anisotropic property of crystalline materials and results from nonuniform charge distributions within a crystal's cells. There are number of different materials which exhibit these effects. Natural materials include quartz (SiO_2), Rochelle salt and tourmaline. The quartz crystal exhibits stiffness comparable to steel and shows high voltage sensitivity. Manufactured piezoelectric ceramics were introduced in the early 1950s. The first material developed with piezoelectric properties was barium titanate ($BaTiO_3$). The two most widely used manufactured materials are lead zirconate titanate (PZT) and polyvinylidene fluoride (PVDF). The former is a ceramic and the later is a polymer film. Piezoelectric ceramics (often called piezoceramics) are quite brittle and need to be handled with care. Piezoelectric polymer films are, in contrast, very flexible and easy to handle for sensor applications. PVDF exhibits the strongest known piezoelectric behaviour of all polymers. PVDF offers better direct piezoelectricity and worse inverse piezoelectricity than the PZT, and therefore it is more often used for sensing applications.

Piezoceramic materials have been used for various types of transducers for many years. Recent years have shown extensive progress in developing sensing devices. The adaptation and integration of piezoceramic sensors onto, or into, structures has become more feasible. Low-profile piezoceramic sensors with wrap-around electrodes are available as thin plates and discs which can be bonded, and/or embedded, in monitored structures. Other important developments in this area include piezoceramic paints (Egusa and Iwasawa 1993), Smart Layers™ (Chang 1998) and piezoceramic fibres (Yoshikawa *et al.*). Smart Layers™ are Kapton dielectric films with an embedded network of distributed piezoceramic PZT sensors. These layers can be fabricated in various sizes and shapes, as shown in Figure 4.16.

Piezoceramic sensors are particularly attractive for structure integrated damage detection since they exhibit simultaneous actuator and sensor behaviour. This allows for both passive and active damage detection. In fact the majority of damage detection applications based on guided wave ultrasonic utilise PZT type sensors. There exist a vast amount of literature related the properties and behaviour of piezoceramic elements.

4.6.2 Constitutive Equations

Piezoelectric materials exhibit both mechanical and electrical properties. It is well known that mechanical properties of any linear elastic material can be described by the Hooke's

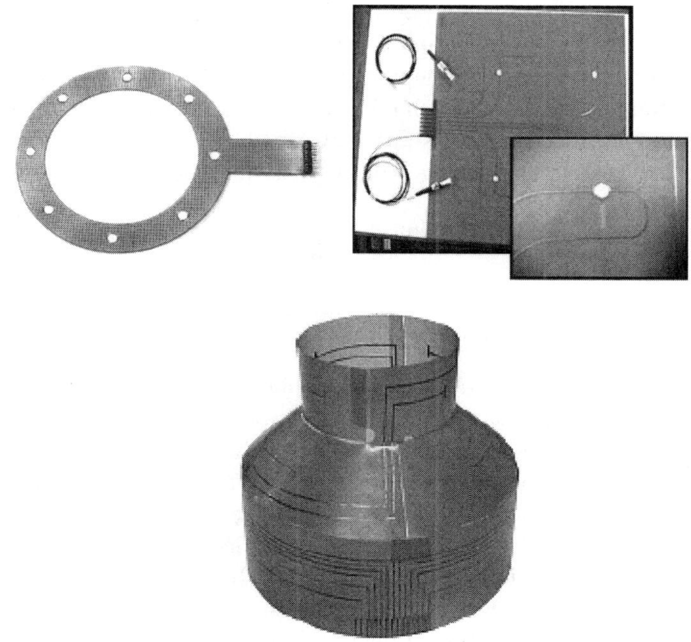

Figure 4.16 Smart Layer™ sensors. (Courtesy of Acellent Technologies Ltd, California)

law which gives the relationship between strain S and stress T

$$S = cT \tag{4.23}$$

where c is the compliance characterising the material. By analogy, a similar relationship exists to describe electrical properties of dielectrics. It shows how the electric displacement D changes as a result of the electric field E applied

$$D = \varepsilon E \tag{4.24}$$

where ε is the permittivity. The electric field E is equivalent to a force field in mechanics and represents the work done against the electric field in order to move a charge. In contrast, the electric displacement D gives the redistribution of charge when the material is subjected to an electric field. Equations (4.23) and (4.24) are often called constitutive mechanical and electrical equations, respectively. Piezoelectric materials exhibit coupled mechanical, electrical and piezoelectric properties. The constitutive equations describing these materials can be given as

$$\begin{aligned} S &= S_E T + d^t E \\ D &= \varepsilon_T E + dT \end{aligned} \tag{4.25}$$

where S is the mechanical strain, E is the electric field, T is the mechanical stress, D is the electrical displacement (all these state variables are second-order tensors), d is the piezoelectric coefficient and ε is the permittivity. Here, subscripts E and T indicate that the quantities are under constant (usually zero) electrical and stress fields, respectively. Additionally, superscript t indicates the transpose. Equation (4.25) gives the strain–charge form of the constitutive equations which relate the electric and mechanical fields. The other three forms give rearranged relationships between the state variables. The constitutive equations describe how voltage changes in the material when the charge moves, and the other way around.

Piezoelectric coefficients d_{ij} defined as

$$d_{ij} = \frac{\textit{strain}}{\textit{applie delectric field}} = \frac{\textit{charge density}}{\textit{applied stress}} \tag{4.26}$$

characterises either the strain produced by an applied electric field or the charge density developed from the applied stress field. The subscript i indicates the direction of electric field or displacement, whereas the subscript j gives the direction of strain or stress. The axis notation can be chosen as shown in Figure 4.17; $i, j = 1$, 2, and 3 indicate the shear strains associated with directions 1, 2 and 3, respectively. Figure 4.18 illustrates the physical interpretation of the piezoelectric coefficients. Here, the electric field E_3 is applied to the transducer of thickness t and width w. The transducer expands vertically (in the direction indicated by the subscript 3) by the factor equal to $1 + \varepsilon_3$; the new thickness of the transducer becomes $t(1 + \varepsilon_3)$. At the same time the transducer shrinks horizontally (in the direction indicated by the subscript 1) by the factor equal to $1 - \varepsilon_1$; the new width of the transducer becomes $w(1 - \varepsilon_1)$. The piezoelectric coefficients are then equal to

$$d_{33} = \frac{\varepsilon_3}{E_3} \quad \text{and} \quad d_{31} = \frac{\varepsilon_1}{E_3} \tag{4.27}$$

Note that d_{31} is in this case a negative number.

Often the conversion between the electrical and mechanical (or mechanical and electrical) energies is described using the electromechanical coupling coefficient k_{ij} which is defined as

$$k_{ij}^2 = \frac{\textit{stored electrical energy}}{\textit{applied mechanical energy}} = \frac{\textit{stored mechanical energy}}{\textit{applied electrical energy}} \tag{4.28}$$

where the subscript notations are identical to the convention used for piezoelectric coefficients. This parameter describes the efficiency of the energy conversion.

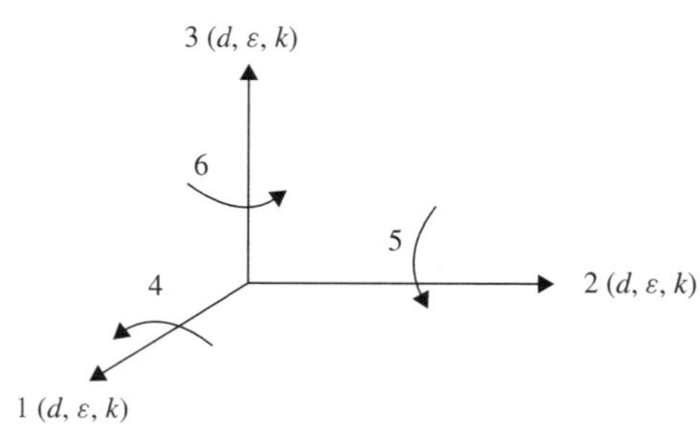

Figure 4.17 Axis notation for the strain/stress and electric fields

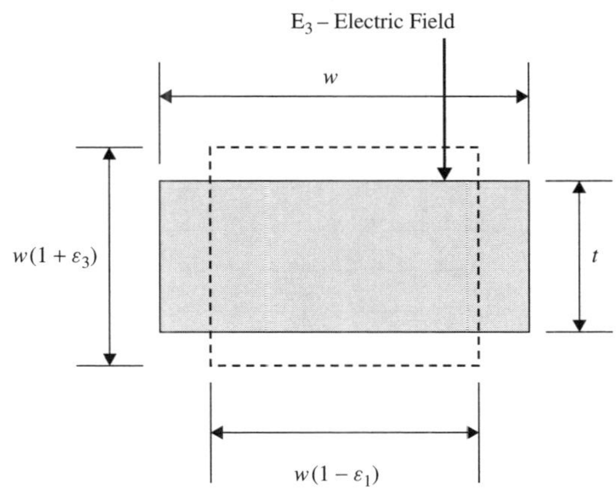

Figure 4.18 Physical interpretation of piezoelectric coefficients

4.6.3 Properties

The manufactured piezoceramics are initially isotropic. They are composed of randomly oriented cells which are polarised when the strong DC current is applied. This process is called poling due to the fact that electric dipoles are permanently aligned by the electrical field. In contrast to natural piezoelectric materials, piezoceramics show high charge sensitivity.

The voltage–strain behaviour of piezoceramics is in general nonlinear. Only the linear part of the voltage–strain curve is used in practice for sensing. Most piezoceramics also exhibit a typical hysteretic behaviour which exhibits the energy loss.

All piezoceramic materials have operating limits associated with temperature and voltage. The piezoelectric properties are exhibited below the so-called Curie temperature. When this temperature is reached or exceeded the material is not piezoelectric anymore, i.e. electric dipoles change their orientations. Some piezoceramics can operate up to 550 °C. The properties of piezoceramic materials also change as a result of temperature. Figure 4.19 shows an example of the experimental peak-to-peak voltage amplitude as a function of the temperature. The excessive voltage (usually above 1000 V) has a similar depolarising effect. Piezoceramic materials are also very brittle. High level of tensile stress can lead to mechanical damage and depolarisation.

The impedance can be used in order to describe dynamic properties of the piezoceramic transducers. The maximum performance of the transducer is achieved when the excitation frequency is chosen as one of the natural resonance frequencies of the transducer. This enables a very efficient conversion from the electrical to mechanical energy. Resonance frequencies are associated with different vibration modes of the transducers. These modes can be used for the generation of different types of waves. Often Frequency Response Functions (FRF) and transfer characteristics are obtained for the transducers using theoretical and experimental studies. Various piezoceramic elements available in the standard Finite Element (FE) codes can be used for the theoretical analysis. Figure 4.20 gives an example of such analysis for a simple disc-type transducer. Here, the strongest radial and thickness modes are identified. It is clear that resonance frequencies of the

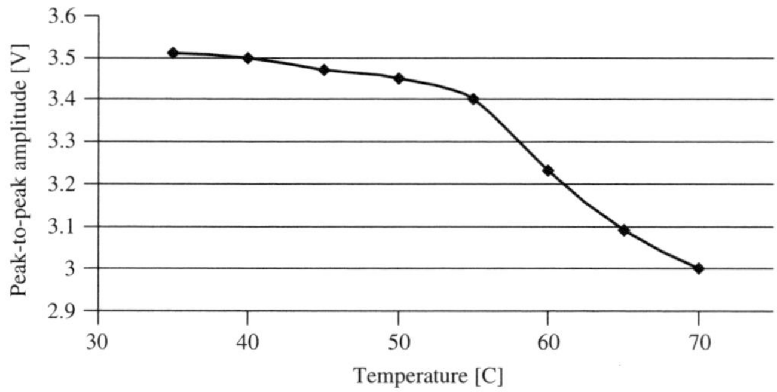

Figure 4.19 Temperature effect on Lamb wave responses when monitored with piezoceramic sensors

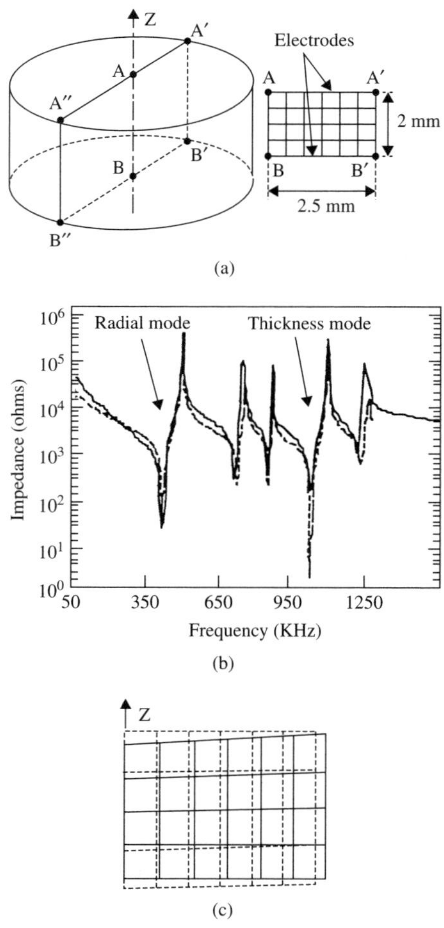

Figure 4.20 Piezoceramic transducer characterisation using modal analysis: (a) model geometry; (b) transfer functions: solid line – experimental curve, dashed line – numerical simulation; (c) vibration modes (Moulin *et al.* 1997, reproduced by the permission of The American Institute of Physics)

embedded or bonded transducers will be influenced by the acoustical loading conditions associated with the boundaries.

4.7 PASSIVE DAMAGE DETECTION EXAMPLES

4.7.1 Crack Monitoring Using Acoustic Emission

Acoustic Emission analysis was performed on a full scale *Airbus A320* aircraft inner wing as a part of the certification procedure. Figure 4.21 shows a schematic diagram representing the specimen on a fatigue test rig. The location of Acoustic Emission sensors is given in Figure 4.22 together with the three dimensional geometry of the analysed part of the wing. The data from sensors located in different planes allow for estimating the true position of the acoustic source due to fatigue crack damage. The signal amplitudes from the source varied between 55 and 100 dB with the majority of results around 90 dB after distance amplitude correction. Figure 4.23 gives an example of the variation of Acoustic Emission amplitude with distance in metres. All noncrack noises that come

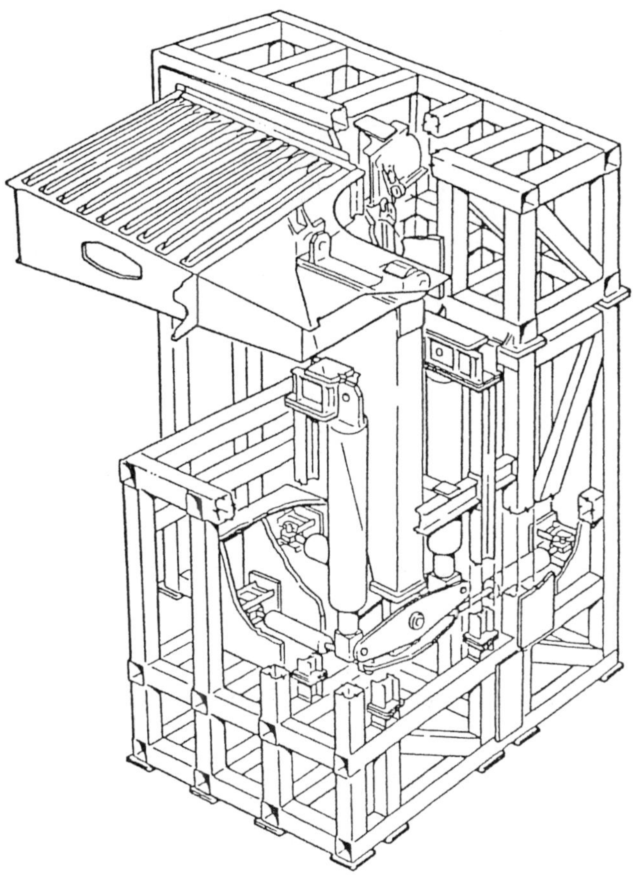

Figure 4.21 Schematic diagram representing the wing specimen on a fatigue test rig

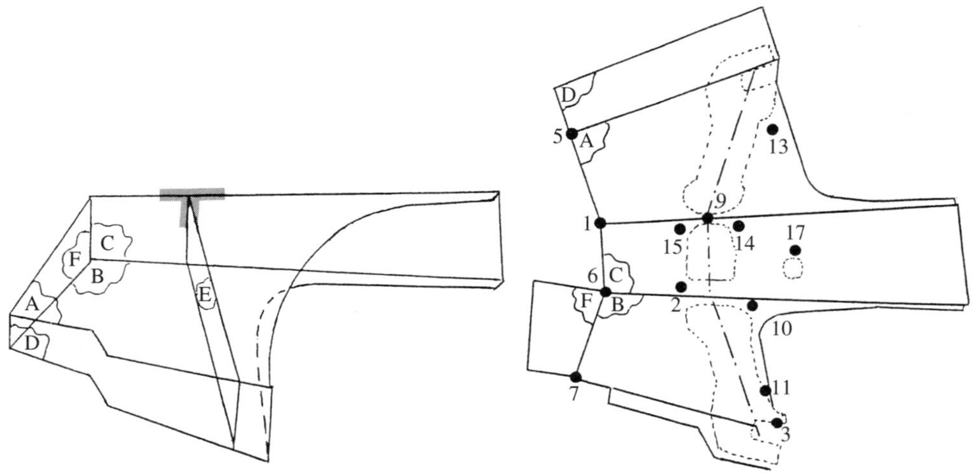

Figure 4.22 Three-dimensional geometry of the analysed part of the wing together with the Location of Acoustic Emission sensors

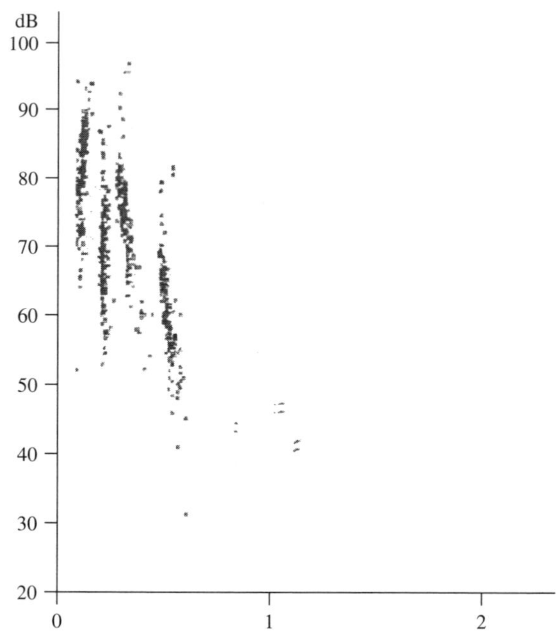

Figure 4.23 Example variation of Acoustic Emission amplitude with distance in metres

from background, fretting, etc., were eliminated before any source location studies. The analysis of the data established within 50 cycles showed the conclusive evidence of the presence of cracks and was used to direct conventional NDT to the locations. The growth of the length of the cracks was also monitored and showed that the five cracks that were detected could be monitored contemporaneously.

The results showed that Acoustic Emission could detect stress waves that emanate from growing cracks that occurred during full-scale fatigue tests.

4.7.2 Impact Damage Detection in Composite Materials

This section presents impact damage detection results from the passive monitoring approach utilising piezoceramic sensors (Staszewski *et al.* 1999b). The experimental tests involved a series of impacts on a simple composite structure shown in Figure 4.24.

The two specimens used in the experimental investigations were rectangular composite panels (530 × 225 mm) fabricated from the T300/914 carbon/epoxy unidirectional prepreg using nominal thickness of 0.15 mm/ply. They were made from 32 plies, except for the area at the root which was 48 plies, and the lay-up sequences were $[+45, -45, 0_4, +45, -45, 0_4, +45, -45, 0_2, 90_2, 0_2, -45, +45, 0_2]_s$ and $[-45, +45, 0_3, -45, +45, 0_2, +45, -45, 90_2, -45, +45, O]_s$, respectively. A ply build-up region was fabricated to simulate a reinforcement. A composite stiffener was bonded to the root side of the panel. The padded area allowed for the effects of ply drop-offs on the detection procedure to be investigated. The stiffener allowed for a disbond to be analysed. The composite panels were representative of aircraft wing skin parts. The panels were instrumented with one piezoceramic *PZT Sonox P5* plate (15 × 15 × 1 mm) bonded to the root side, as shown on a schematic diagram in Figure 4.25. The composite panel was mechanically fastened to a stiffening aluminium subframe using screws. This allowed for the quick replacement of the analysed plates. The composite panel was attached to the metal loading frame, as shown in Figure 4.26.

The composite/aluminium structure was statically loaded (0.2 kN, displacement equal to 18 mm) as a cantilever beam using a *Schenck* 250 kN servo-hydraulic test machine. The functionality of the sensor was checked using low energy 2 J impacts at random positions

Figure 4.24 Composite structure representative of an aircraft wing skin

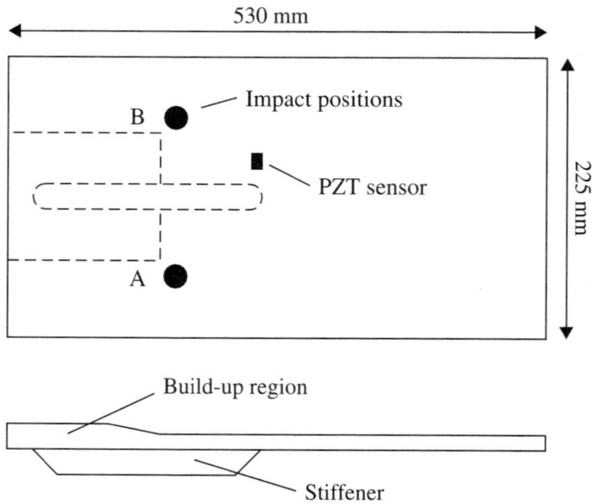

Figure 4.25 Schematic diagram of the composite panel (Staszewski *et al.* 1999)

Figure 4.26 Loading arrangements for the wing-box structure (Staszewski *et al.* 1999)

on the plate. A mobile impactor was used in the tests. This initial study was followed by a series of test impacts at positions A and B (Figure 4.25) with the energy levels equal to: 2, 4, 8, 12, 20 J for impact location A; and 2, 4, 8, 12, 16, 24 J for impact position B. The panel was A-scanned after each impact to check for possible damage. The presence of clear delamination was observed after the 20 and 24 J impact at position B.

The impact strain data were acquired from the piezoceramic sensor using a digital 2-channel *HP* oscilloscope with a sampling frequency of 400 kHz. An example of the strain data for the 4 J impact at position B is given in Figure 4.27. The absolute maximum values of the strain data are given in Figure 4.28. The results show that in general the

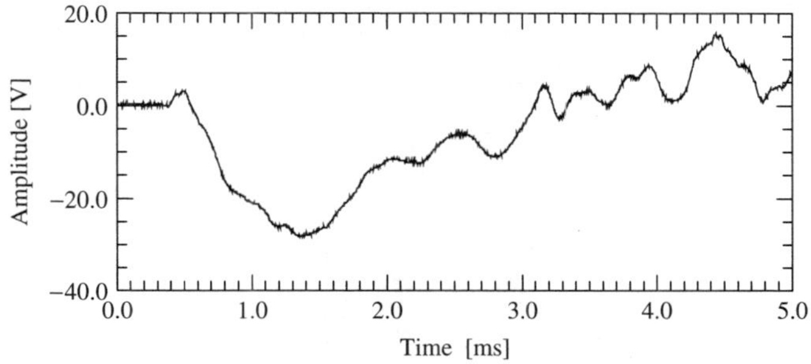

Figure 4.27 Impact strain data for the 4 J impact at position B (Staszewski *et al.* 1999)

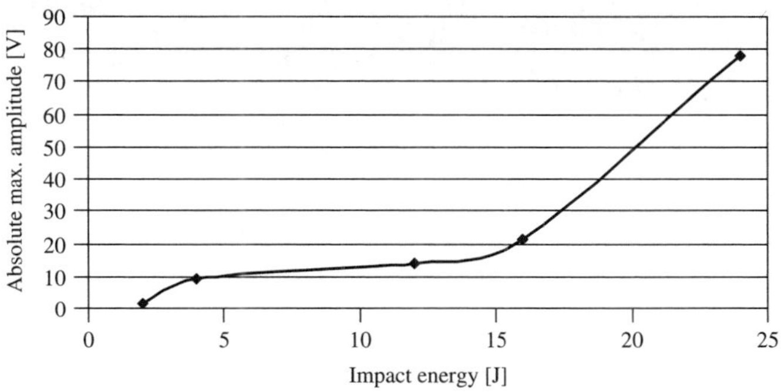

Figure 4.28 Absolute maximum values for the impact strain data

characteristics increase with the impact energy. Thus impact energy levels can be estimated analysing the strain data obtained from the piezoceramic sensor. These energy levels can be related to possible damage severities in the composite panels. Further analysis examples using advanced signal processing techniques are given in Chapter 5.

4.8 ACTIVE DAMAGE DETECTION EXAMPLES

4.8.1 Crack Monitoring in Metallic Structures Using Broadband Acousto-Ultrasonics

A simple fatigue experiment was performed in order to obtain the strain data from metallic specimens with growing cracks (Biemans *et al.* 2001). The specimens used were two rectangular (400 × 150 × 2 mm) aluminium plates with cracks initiated by spark erosion in the middle of the plates. Figure 4.29 shows the experimental set-up. The fatigue test was performed on a *Schenck* 250 kN servo-hydraulic test machine running in load control. The aluminium specimens were subjected to three different load conditions:

- *no load* – close to 0 kN;
- *static load* – at 12.5 kN mean load;
- *dynamic load* – static load plus sinusoidal tension–tension cycling loading with a frequency of 6 Hz and maximum amplitude of 11.5 kN.

Figure 4.29 Experimental set-up for crack monitoring using acousto-ultrasonics (Biemans *et al.* 2001)

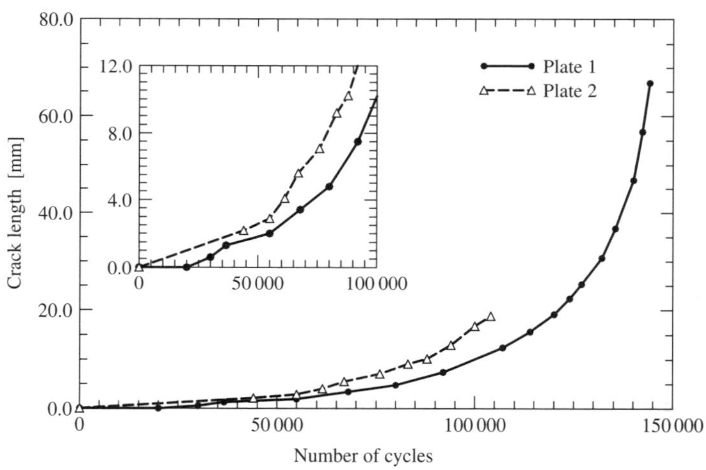

Figure 4.30 Crack propagation curve (Biemans *et al.* 2001)

Crack lengths were determined using a microscope to an accuracy of 0.01 mm. Figure 4.30 shows an example of a crack propagation curve for the analysed specimens.

Each plate was instrumented with six piezoceramics (PZT material *Sonox P5*; dimensions: $15 \times 15 \times 1$ mm) bonded in a symmetrical configuration 20 mm below and above the initiated crack, as shown in Figure 4.31. The piezoceramic no. 2 on each plate was used as an actuator in order to perform the acousto-ultrasonics damage detection.

The study involved two different types of excitation:

1. Gaussian white noise with the maximum frequency of 16 kHz;
2. sine sweep from 10 to 50 kHz with the ramp time of 0.01 s.

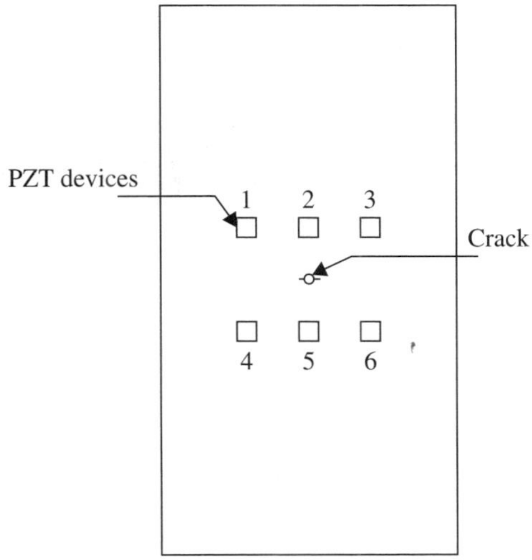

Figure 4.31 Sensor locations on the aluminium specimen used in crack monitoring tests (Biemans *et al.* 2001)

The maximum amplitude of excitation was equal to 20 V. The signals were generated using the *DIFA SCADAS II* and *Philips PM5138* generators.

The piezoceramics 5 and 6 were used to monitor the growing cracks. The response data, from the piezoceramic sensors no. 5 and 6, were recorded by a *DIFA SCADAS II* 24 channel measuring system running the *LMS 3.4.04* data acquisition software and by a digital 2-channel *Tektronix TDS 210* oscilloscope. Figure 4.32 shows an example of the piezoceramic response to the sine sweep excitation for different types of load used in the experiment. Here, a clear drift can be observed for the dynamic load due to the 6 Hz tension–tension cycling loading.

Various signal parameters were estimated for the analysed data. Figure 4.33 gives the results obtained for the minimum amplitude values, for plate no. 5 under the sweep sine excitation. Results, obtained for the *static load* condition, show a fairly linear behaviour while for the *no load* condition a similar linear relationship can only be observed after a certain crack length. The results obtained with monitoring under the *dynamic load* are quite arbitrary. Performing, however, a linear regression over these data may possibly lead to a result similar to the data for the *no load* and *static load* conditions. The reason for the relatively good result when monitoring under the *static load* condition can be seen in a clear situation of the crack being fully open. For the other two load conditions the situation of the crack looks likely to be less clear.

The experimental data were also analysed using the classical Fourier transform approach. Examples of power spectra for the data from sensor no. 5 are given in Figure 4.34. Here, two dominant spectral components at about 1.5 and 2.7 kHz can be observed in Figure 4.34a for the data representing the undamaged plate. The amplitude of these two components decreases when the crack grows in the plate, as shown in Figure 4.34b and 4.34c, whereas the remaining parts of the spectrum are relatively unchanged. The results show that any direct comparison of spectra requires further signal processing analysis. A number of spectral statistical parameters, discussed in Chapter 5, were calculated for the sensor data under the

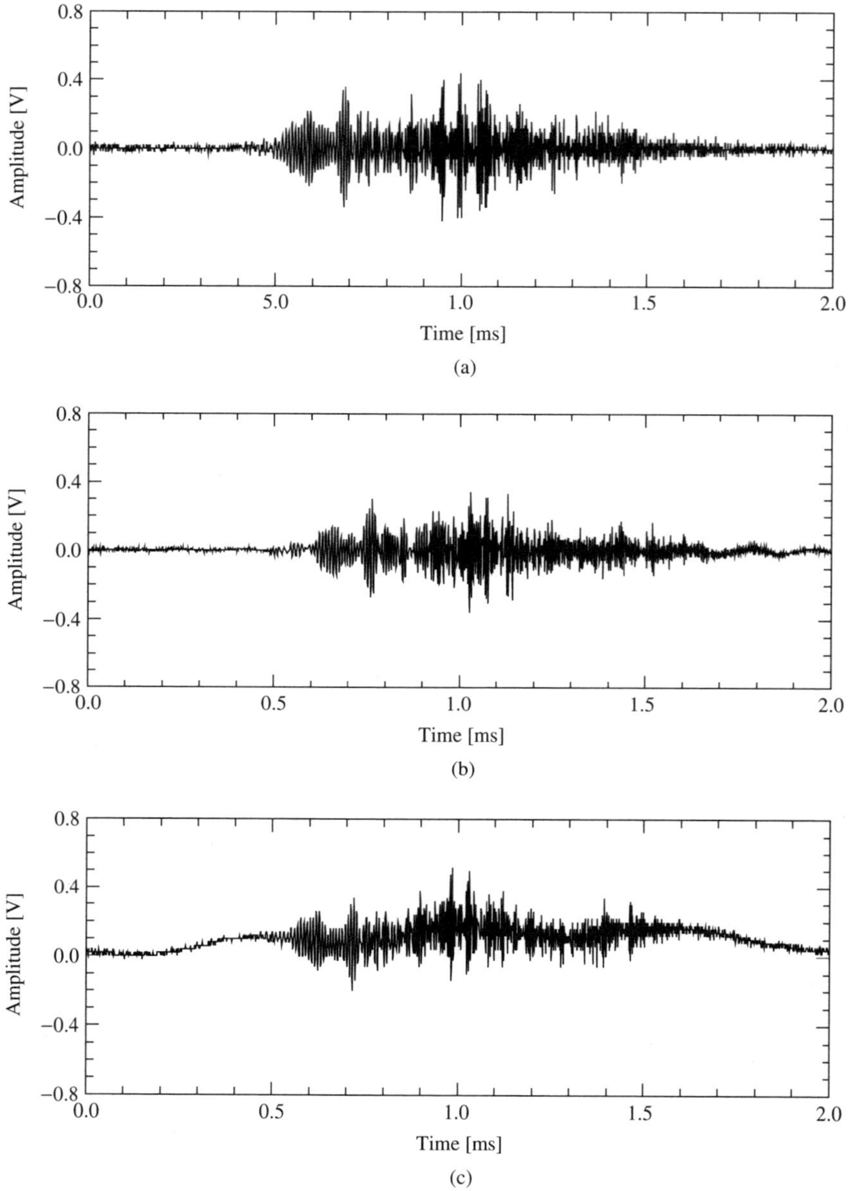

Figure 4.32 Acousto-ultrasonic responses to sine-sweep excitation for: (a) no load; (b) static load; (c) dynamic load (Biemans *et al.* 2001)

sine sweep excitation. Figure 4.35 shows an example of the Root Mean Square (RMS) of Spectral Difference characteristics for sensor no. 5. Again, results obtained for the *static load* condition show a monotonic behaviour while for the *no load* and *dynamic load* conditions the results are not satisfactory – the curves do not indicate any crack growing in a plate. More examples related to these investigations are given in (Biemans *et al.* 2001).

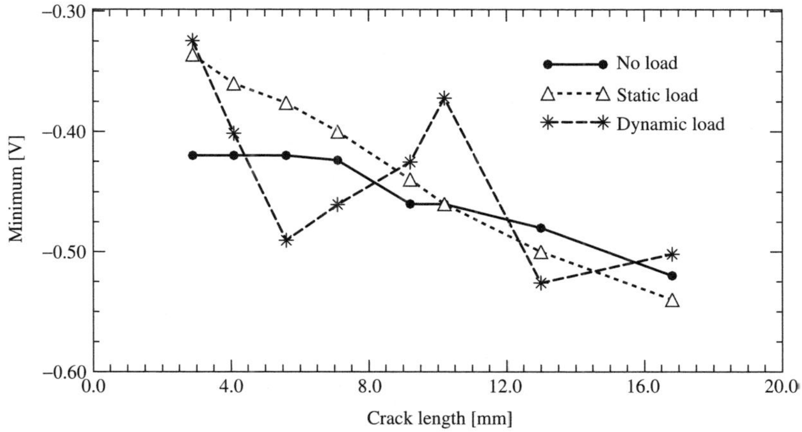

Figure 4.33 Minimum amplitude values for the acousto-ultrasonic responses (Biemans *et al.* 2001)

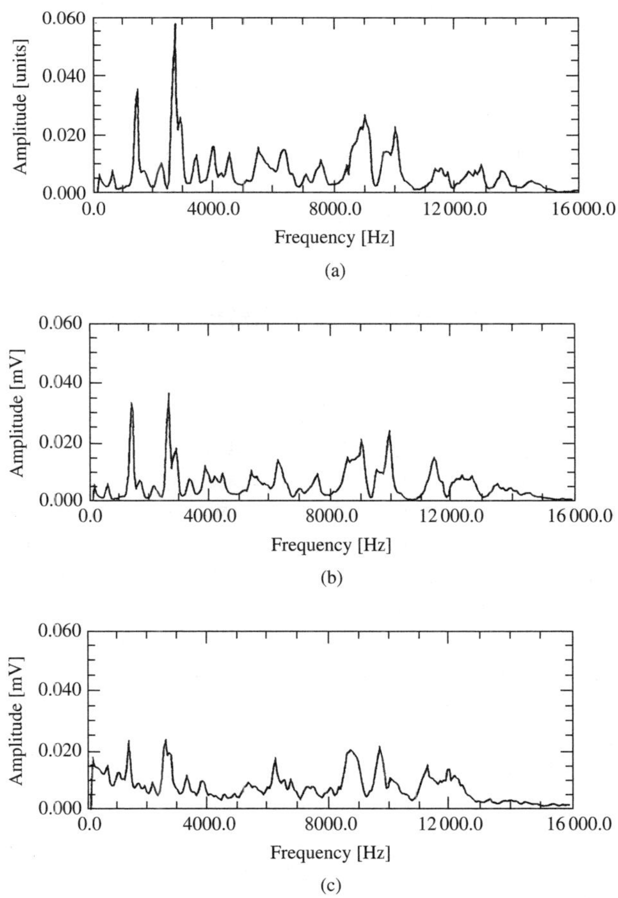

Figure 4.34 Power spectra for acousto-ultrasonic data acquired under the sine sweep excitation: (a) undamaged plate; (b) crack length – 12.5 mm; (c) crack length – 66.8 mm (Biemans *et al.* 2001)

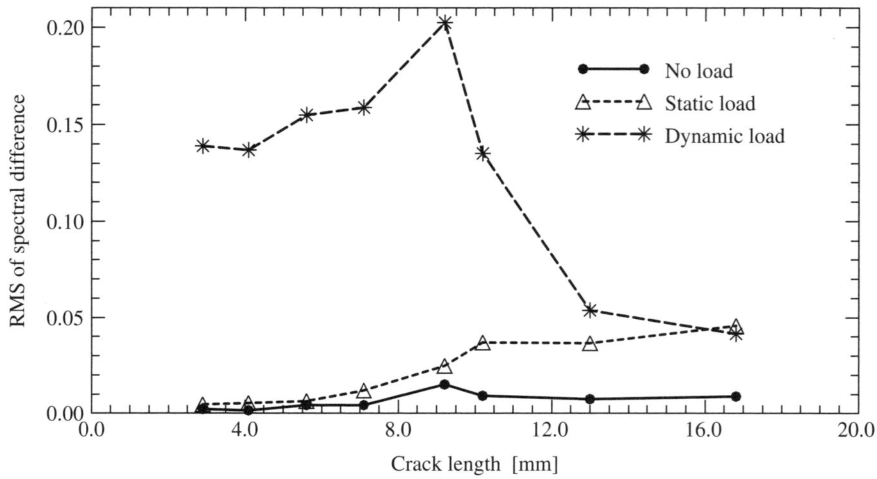

Figure 4.35 Root Mean Square for the Spectral Difference values for the sweep sine excitation (Biemans *et al.* 2001)

This preliminary study shows that satisfactory damage detection results can be obtained under all analysed load conditions. However, this requires the appropriate choice of excitation and signal processing. Also, the best results are mainly obtained under the *static load* when cracks tend to be clearly open or under the *dynamic load* when cracks open and close successively due to cycling loading.

4.8.2 Impact Damage Detection in Composite Structures Using Lamb Waves

This section presents damage detection analysis based on Lamb waves. The study involves the composite wing-box structure described in Section 4.7.2. Piezoceramic sensors have been used to introduce the excitation signals and to acquire the Lamb wave responses. The sensors chosen for these tests were $3 \times 4 \times 1$ mm piezoceramic P160 plates. A general rule of ultrasound excitation is to excite the actuator at its natural resonances rather than at any other frequency. This approach enables a very efficient conversion from the electrical to mechanical energy, as discussed in Section 4.6.3.

A two-dimensional numerical model was developed (Assaad *et al.* 1990) using the ATILA FE code in order to perform the modal and harmonic analysis of the piezoelectric transducers. The analysis resulted in four natural vibration modes for the frequency bandwidth between 100 kHz and 1 MHz. The second mode at 400 kHz was the best electromechanically coupled ($k_e = 43\,\%$). The computation of the displacement field, shown in Figure 4.20, shows that the analysed mode of vibration is a transverse mode.

The transducers have been positioned in various locations in order to detect different types of damage (e.g. impact damage, delaminations, substructural disbonds). Figure 4.36 shows sensor locations on the composite structure. A series of impacts, as described in Section 4.7.2, was performed in order to introduce various severities of damage to the composite skin. A pair of transducers, i.e. El (actuator) and R1 (sensor), was used to allow

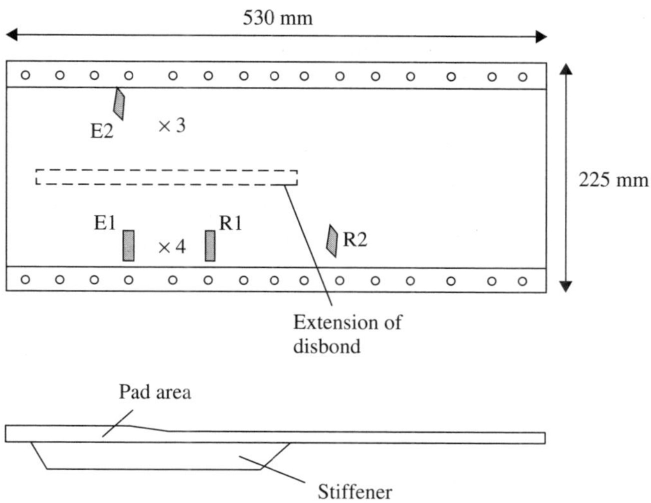

Figure 4.36 Impact positions and locations of Lamb wave transducers on the composite wing-box structure

for the inspection of the plate near impact location 4 (see Figure 4.36). One more pair of transducers, i.e. E2 (actuator) and R2 (receiver), were positioned in order to monitor the disbond growth between the stiffener and the composite plate under cyclical loading and detect possible delaminations at impact location 3 (see Figure 4.36).

A pulse 5-cycle sine burst signal at 400 kHz used to excite the actuators at their resonance frequencies. The Lamb waves excited by the actuators propagated along the plate and were received by the sensors. The responses were amplified, filtered and transferred to a digital oscilloscope. The data acquisition utilised 64 averages in order to improve the signal-to-noise ratio.

Preliminary tests have been performed on the structure in order to study the influence of external conditions, such as additional vibration and temperature, on Lamb wave testing procedure. This was followed by a sequence of testing steps:

- acquisition of Lamb wave signals before the tests (*no damage* condition);
- specimen loaded under 0.2 kN: acquisition of Lamb wave signals;
- vibration conditions: acquisition of Lamb wave signal during, the tests;
- impact tests performed with a mobile impactor using different energy levels (2 J, 4 J, 8 J, 12 J, and 24 J) in location 3 and 4: after each impact, inspections of the plate with Lamb waves were carried out;
- cyclic loading with different displacement levels (9 ± 7 mm, 14 ± 9 mm, 20 ± 12 mm, 22 ± 15 mm): acquisition of Lamb wave signals during the tests.

The first set of tests has demonstrated good reproducibility of the Lamb wave responses under external conditions (external vibration and loading). The performance of all the sensors was not affected by varying thermal conditions; Lamb wave responses remained unchanged after the thermal tests.

The second set of tests was performed in order to measure the sensitivity of Lamb wave signals to damage (delamination and disbond). Damage detection analysis utilised the presence of other modes, or modes conversion due to possible discontinuities in the structure.

Figure 4.37a shows the Lamb wave response from sensor R1 in the absence of defects. Here, the first wave packet was identified as the S_1 mode from a measurement of its group velocity using the time-of-flight analysis. When the specimen was loaded under 0.2 kN and submitted to an external vibration excitation, no influence on the shape of the Lamb

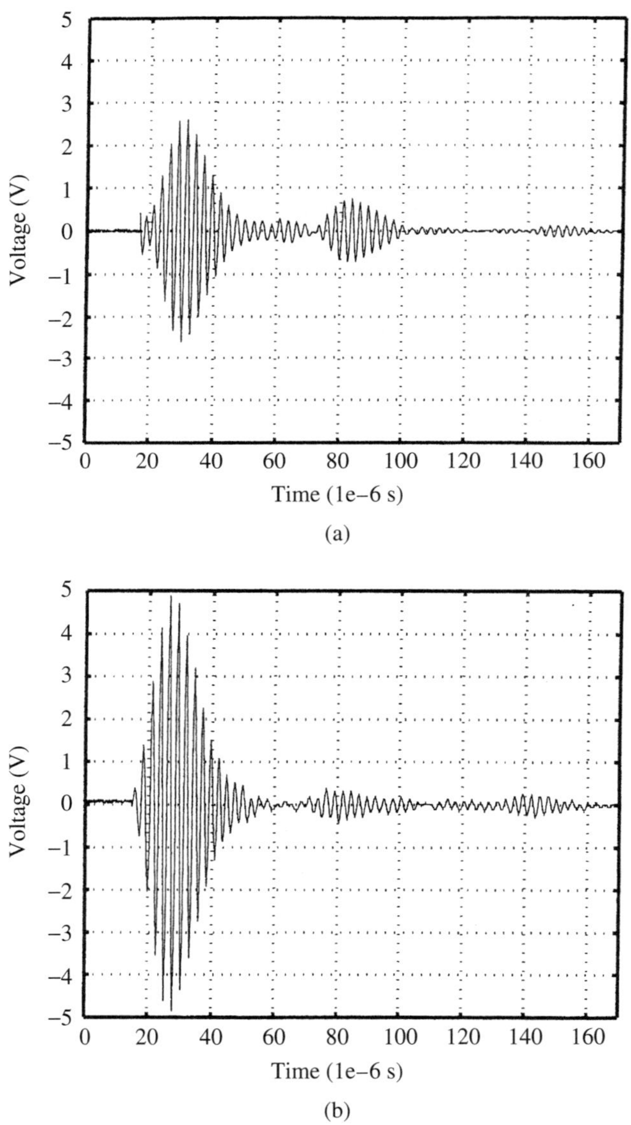

Figure 4.37 Lamb wave responses from sensor R1: (a) undamaged plate; (b) damaged plate

wave signal was observed, which proves that loading did not have any influence regarding the signal recorded. However, after a damaging impact at location 4, one of the modes is lost, as shown in Figure 4.37b. The presence of delamination was confirmed using a conventional ultrasonic A-scan. Figure 4.38 gives the local minima of the envelope function calculated using the Hilbert transform (Randall 1987). These minima, corresponding to mode changes due to damage, can be used for damage detection.

Figure 4.39 shows the time history of the Lamb wave response for the transmission test between the E2 and R2 transducers before the impact tests were performed (*no damage* condition). The signal exhibits more complex modes because of the geometry of the analysed plate (thickness variations and stiffener). Further testing reveals a clear

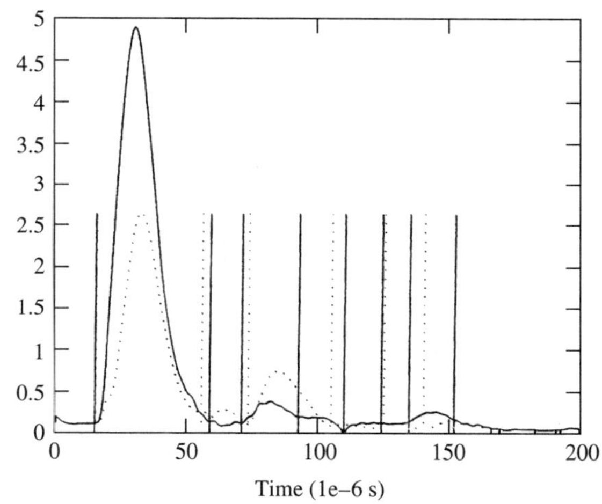

Figure 4.38 Local minima for the Lamb wave responses from sensor R1: dashed line – before testing, solid line – after damaging impact

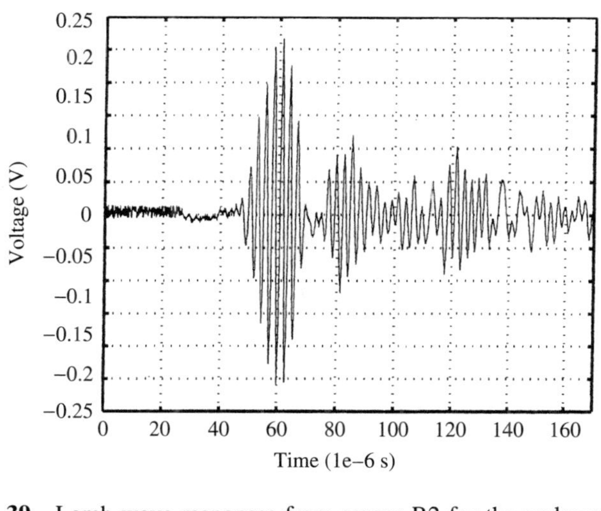

Figure 4.39 Lamb wave responses from sensor R2 for the undamaged plate

delamination in the composite plate due to the damaging impact at location 3, as shown in Figure 4.40.

Finally, the composite plate was submitted to cyclic loading. An example of local minima for the Lamb wave responses is presented in Figure 4.41. The result displays phase delays and mode conversions due to the growth (by 1 cm) of the disbond.

These simple experimental tests demonstrate the ability of the Lamb wave testing method to detect and monitor various types of defects in composite materials. The technology is capable to perform continuous, in-service health and usage monitoring of aircraft structures. Further examples are shown in Chapter 6.

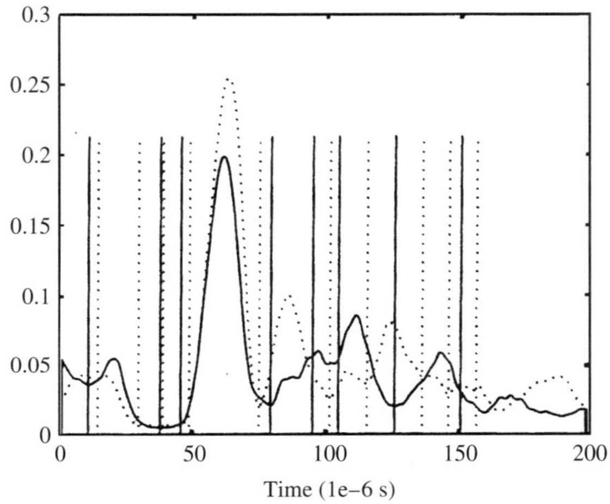

Figure 4.40 Local minima for the Lamb wave responses from sensor R2 dashed line – before testing (dashed line) and after damaging impact (solid line)

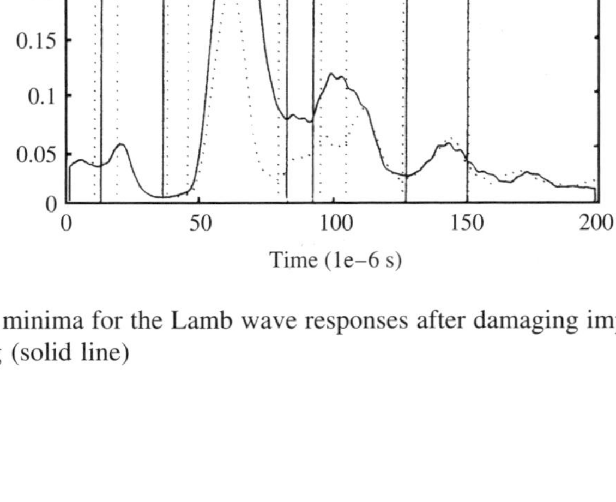

Figure 4.41 Local minima for the Lamb wave responses after damaging impact (dashed line) and after cycling loading (solid line)

4.9 SUMMARY

A brief introduction to damage detection methods based on sound and ultrasound has been given in this chapter. The presentation includes AE, Ultrasonics, Acousto-Ultrasonics and guided wave Ultrasonics. The material presented shows that there are various similarities between these techniques. All these techniques utilise piezoceramic transducers for damage monitoring and testing. Therefore a brief description of piezoceramic materials has been given for the sake of completeness. Simple application examples show how the methods work in practice. The potential of the methods for aerospace applications is presented in Chapter 6 where various large-scale and flight test examples are given.

REFERENCES

Acellent Technologies, web page: **www.acellent.com**

Achenbach, J.D. 1984. *Wave Propagation in Elastic Solids*, North–Holland, New York.

Agostini, V., Delsanto, P.P. and Zoccolan, D. 2000. Flaw detection in composite plates by means of Lamb waves, *Proceedings of the 15th World Conference on NDT*, Roma, Italy, 15–21 October.

Assaad, J., Bruneel, C., Decarpigny, J.N. and Nongaillard, B. 1990. The finite element code at ISEN (IEMN), Proceedings of the workshop held in Toulon, France, June 1990.

Betz, D., Thursby, G., Culshaw, B. and Staszewski, W.J. 2002. Acousto-ultrasonic sensing using fiber Bragg gratings, *Smart Materials and Structures*, Vol. 12, No. 1, pp. 122–128.

Biemans, C., Staszewski, W.J., Boller, C. and Tomlinson, G.R. 2001. Crack detection in metallic structures using broadband excitation of acousto-ultrasonics, *Journal of Intelligent Materials Systems and Structures*, Vol. 12, No. 8, pp. 589–597.

Buirks, A.S., Green Jr, R.E. and McIntyre, P. (eds). 1991. *Nondestructive Testing Handbook*, Vol. 7, *Ultrasonic testing*, American Society of Nondestructive Testing, Columbus, Ohio.

Chang, F.K. 1998. Smart layer built-in diagnostics for composite structures, *Proceedings of the 4th European Conference on Smart Materials and Structures and the 2nd MIMR Conference*, Harrogate, UK, 6–8 July, pp. 777–781.

Delsanto, P.P., Whitcombe, T., Chaskelis, H.H. and Mignogna, R.B. 1992. Connection machine simulation of ultrasonic wave propagation in materials – I: the one-dimensional case, *Wave Motion*, Vol. 16, pp. 65–80.

Delsanto, P.P., Schechter, R.S., Chaskelis, H.H., Mignogna, R.B. and Kline, R.B. 1994. *Connection machine simulation of ultrasonic wave propagation in materials – II: The two-dimensional case*, Vol. 20, pp. 295–314.

Delsanto, P.P., Schechter, R.S., and Mignogna, R.B. 1997. *Connection machine simulation of ultrasonic wave propagation in materials – III: The three-dimensional case*, Vol. 26, pp. 329–339.

Egusa, S. and Iwasawa, N. 1993. Polling characteristics of PZT/epoxy piezoelectric paints, *Ferroelectrics*, Vol. 145, pp. 45–60.

Envirocoustics S.A. web page: **www.pacndt.com**

Holroyd, T. 2001. *The Acoustic Emission and Ultrasonic Monitoring Handbook*, Coxmoor Publishing Company, Oxford, UK.

Khuri-Yakub, B.T., Cheng, C.H., Degertekin, F.L., Ergun, S., Hansen, S., Jin, X.C. and Orlakan, O. 2000. Silicon micromachined ultrasonic transducers, *Japanese Journal of Applied Physics*, Vol. 39, pp. 2882–2887.

Kiernanand, M.T. and Duke Jr, J.C. 1988. PC analysis of an acousto-ultrasonic signal, *Materials Evaluation*, Vol. 46, No. 9, pp. 1344–1352.

Kwon, O.Y., Kim, T.H. and Lee, K.J. 2000. Monitoring fatigue damage in adhesively bonded composite–metal joints by acoustic methods, *Proceedings of the 15th World Congress on NDT*, Roma, Italy, 15–21 October.

Lee, B.C. and Staszewski, W.J. 2002. Modeling of acousto-ultrasonic wave interactions with defects in metallic structures, *Proceedings of the International Conference on Noise and Vibration Engineering – ISMA* 2002, Leuven, Belgium, 16–18 September.

Lee, B.C. and Staszewski, W.J. 2003a. Lamb wave interactions with structural defects – modelling and simulations, *Proceedings of the SPIE's 10th International Symposium on Smart Structures and Materials, Conference on Modeling, Signal Processing and Control*, San Diego, California, 2–6 March, Paper No. 5049-22.

Lee, B.C. and Staszewski, W.J. 2003b. Modelling of Lamb waves for damage detection in metallic structures, Part I: wave propagation, *Smart Materials and Structures*, Vol. 12(5), pp. 804–814.

Lee, B.C. and Staszewski, W.J. 2003c. Modelling of Lamb waves for damage detection in metallic structures, Part II: wave interactions with damage, *Smart Materials and Structures*, Vol. 12(5), pp. 815–824.

Moulin, E., Assaad, J., Delebarre, C., Kaczmarek, H. and Balageas, D. 1997. Piezoelectric transducer embedded in a composite plate: application to Lamb wave generation, *Journal of Physics*, Vol. 82, No. 5, pp. 2049–2055.

Muravin, G. 2000. *Inspection, Diagnostics and Monitoring of Construction Materials and Structures by the Acoustic Emission Methods*, Minerva Press, London, UK.

Pedemonte, P., Staszewski, W.J., Aymerich, F., Found, M.S. and Priolo, P. 2001. Signal processing for passive impact damage detection in composite structures, *Proceedings of the SPIE's 8th International Symposium on Smart Structures and Materials, Conference on Modeling, Signal Processing and Control, Newport Beach*, California, 4–8 March, Paper No. 4326-19.

Randall, R.B. 1987. *Frequency Analysis*, 3rd edition, Brüel & Kjæl Publications, Nærum, Denmark.

Rose, J.L. 1999. *Ultrasonic Waves in Solid Media*, Cambridge University Press, Cambridge, UK.

Schmerr Jr, L.W. 1998. *Fundamentals of Ultrasonic Nondestructive Evaluation: A Modeling Approach*, Plenum Publishing Corporation, New York.

Scott, I.G. 1991. *Basic Acoustic Emission*, Gordon and Breach Science Publishers, New York, NY.

Staszewski, W.J. and Boller, C. 2002. Acoustic wave propagation phenomena modelling and damage mechanisms in ageing aircraft, *Proceedings of the Aircraft Integrated Monitoring Systems Conference – AIMS, Garmisch-Partenkirchen*, Germany, 27–30 May, pp. 169–184.

Staszewski, W.J. and Holford, K. 2001. Wavelet signal processing for acoustic emission data, *Proceedings of the 4th International Workshop on Damage Assessment Using Signal Processing Procedures – DAMAS-2001*, Cardiff, Wales, 25–27 June.

Staszewski, W.J., Biemans, C., Boller, C. and Tomlinson, G.R. 1999a. Crack propagation monitoring in metallic structures, *Proceedings of the International Conference on Smart Materials, Structures*, Bangalore, India, 7–10 July, pp. 532–541.

Staszewski, W.J., Biemans, C., Boller, C. and Tomlinson, G.R. 1999b. Impact damage detection in composite structures, *Proceedings of the 2nd International Workshop on Structural Health Monitoring*, Stanford, California, 8–10 September, pp. 754–763.

Vary, A. 1988. The acousto-ultrasonic approach. In: J.C. Duke Jr (ed.), *Acousto-Ultrasonics: Theory and Application*, Plenum Press, New York, pp. 1–21.

Viktorov, I.A. 1967. *Rayleigh and Lamb Waves*, Plenum Press, New York.

Wilcox, P.D., Dalton, R.P., Lowe, M.J.S. and Cawley, P. 1999. Mode transducer selection for long range Lamb wave inspection, *Proceedings of the 3rd International Workshop on Damage Assessment Using Advanced Signal Processing – DAMAS* 1999, Dublin, Ireland, 28–30 June, pp. 152–161.

Yoshikawa, S. Selvaraj, U., Moses, P., Witham, J., Meyer Jr, R. and Shrout, T. $Pb(Zr,Ti)O_3$ [PZT] fibers – fabrication and measurement methods, *Journal Intelligent Materials, Systems and Structures*, Vol. 6, pp. 152–158.

5

Signal Processing for Damage Detection

W.J. Staszewski and K. Worden

Department of Mechanical Engineering, Sheffield University, UK

5.1 INTRODUCTION

Signal processing and computation are crucial elements in the implementation and operation of any damage identification system. The generic system requires the availability of appropriate signal processing technology to extract features from different types of sensors and to translate this information into a diagnosis of location and severity of damage (Worden *et al.* 1997; Staszewski 2000a). The overall intelligent chain of processing for a multi-sensor architecture is summarised in Figure 5.1. The pre-processing, which in fact filters out all unwanted features, is the first element of this chain. This is followed by various procedures of feature extraction and selection. It is important to isolate data features which are sensitive to damage yet insensitive to environmental and operational conditions. This is in fact the most difficult task in damage identification. Since data features are often combined in patterns, the entire process of damage identification can be considered as a pattern recognition procedure.

Sensors are usually deployed in arrays. Multi-sensor architectures improve signal-to-noise ratio, offer better robustness and reliability and increase confidence in the results. Multi-sensor architectures require procedures which assess the informativeness of sensors. It is also important to establish where to locate sensors for optimal damage detection and how to detect sensor failures.

Information gathered from different types of sensors needs to be combined with linguistic and knowledge-based data. This part, called a data fusion procedure, leads to the final information about damage. Damage detection forms the primary objective of the overall problem of damage identification. A further analysis of this identification includes: severity and classification of damage, location of damage, and finally prediction of the remaining service life of the structure and possible breakdown (Cempel 1991; Rytter 1993). Altogether these objectives form four different levels of damage identification.

Health Monitoring of Aerospace Structures – Smart Sensor Technologies and Signal Processing.
Edited by W.J. Staszewski, C. Boller and G.R. Tomlinson ISBN: 0-470-84340-3

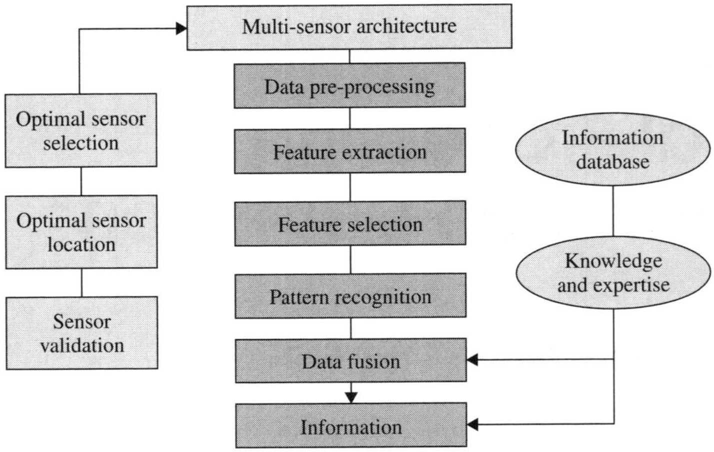

Figure 5.1 Signal processing for a multi-sensor architecture

The first three levels are mostly related to analysis, identification and modelling issues of engineering systems and signal processing. The last level of prediction falls into the field of fatigue analysis, fracture mechanics, design assessment, reliability and statistical analysis.

Recent advances in signal processing allow for automated damage detection systems. Sensor data can be processed either using software or hardware. Various methods and algorithms have been developed for damage identification. The last fifteen years have shown an explosion of new techniques such as Neural Networks, Genetic Algorithms or wavelets. Figure 5.2 shows examples of the most widely used software tools for damage identification. Different elements of advanced signal processing for a multi-sensor architecture are discussed in this chapter. This includes examples related to damage detection. More examples can be found in Chapters 4 and 6.

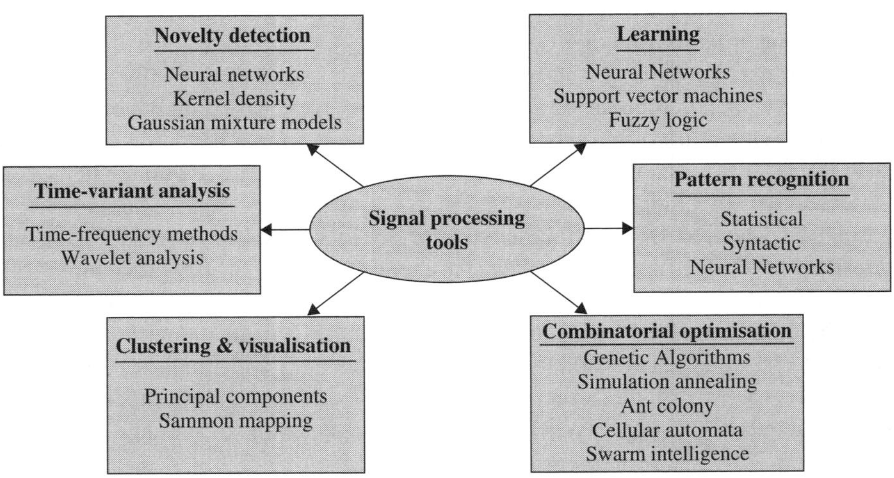

Figure 5.2 Examples of signal processing tools for damage identification

5.2 DATA PRE-PROCESSING

Data pre-processing is usually the first element of signal processing for damage detection. It involves normalisation, trend removal techniques, outlier analysis, averaging, smoothing and filtering. Following (Staszewski *et al.* 2000), some of these procedures are briefly described in this section. Normalisation identifies relationships between measurements and features. Trends show unwanted temporal relationships in the data. Outliers are feature patterns which are statistically far from the normal selection of patterns used for training. They can lead to poor generalisation of the learning process. Outliers can be eliminated using standard statistical analysis. The level of noise in the data can be reduced by local and/or global averaging, smoothing and denoising procedures. In what follows, the most commonly used signal smoothing procedures are briefly discussed. Recent procedures of denoising are based on orthogonal wavelet analysis, as discussed in Section 5.11.

5.2.1 Signal Smoothing

Data gathered from sensors can be corrupted by noise. The noisy signal $z(t)$ can be given by,

$$z(t) = x(t) + w(t) \tag{5.1}$$

where $x(t)$ is the signal and $w(t)$ is the noise. Often the level of noise can be reduced by local and/or global averaging. An alternative approach can be offered by smoothing and denoising procedures. Smoothing can be done using filtering or fitting. Fitting of the best-fit polynomial through a data set is a smoothing process in which the number of fitted coefficients is usually much less than the number of analysed data points.

5.2.2 Signal Smoothing Filters

The smoothness of a function indicates its differentiability. It corresponds to the decay of its Fourier transform and can be measured using the Sobolev norm obtained from the derivative norms,

$$\|g\|^2_{W^N} = \sum_k \|g^{(k)}\|^2 \tag{5.2}$$

There exist a number of low-pass filters which can be use to smooth or in other words remove the noise from data (Hamming 1989). However, only a few of them are suitable for optimal smoothing. One obvious choice includes the well-known Wiener filter based on Fourier analysis The other well-adapted low-pass filters for data smoothing include: Savitzky–Golay (Press *et al.* 1992), least-squares (Hamming 1989) and Digital Smoothing Polynomial (Press *et al.* 1992) filters. The simplest type of digital filter replaces the data value x_i by a linear combination of itself and some nearby neighbours, i.e.,

$$y_i = \sum_{n_a}^{n_b} c_n x_{i+n} \tag{5.3}$$

Savitzky-Golay filters derive directly from the time domain filter given by the above equation. They use the simplest possible so-called moving window averaging procedure.

Data points x_i are computed as the average of nearby neighbours from x_{i-n_a} to x_{i+n_b} using constant value coefficients, i.e. $c_n = 1/(n_a + n_b + 1)$.

5.3 SIGNAL FEATURES FOR DAMAGE IDENTIFICATION

5.3.1 Feature Extraction

The process of feature extraction in data analysis is necessary to enhance damage identification. Features are any parameters extracted from the measurements through signal processing. Feature extraction for damage detection includes signature analysis and advanced signature analysis (Staszewski 2000a). In general signature analysis employs simple feature extraction methods based on data reduction procedures in order to obtain scalar features. This includes for example statistical spectral moments, physical parameters of the analysed system or modal based criteria. Advanced signature analysis uses sets of features in the form of vectorial or pattern representations such as: spectra, envelope and phase characteristics. The past ten years have seen major developments in the area of advanced signature analysis. A number of algorithms based on time–frequency and time–scale procedures have been proposed. In general the choice of features involves a trade-off between the computational feasibility associated with low-level features and extensive pre-processing required for high-level features.

5.3.2 Feature Selection

It is important not only to find the best features which represent and explain differences between various damage classes but also to reduce the number of features for classification procedures. As explained above, feature extraction does not necessarily lead to reduced dimensionality. In order to achieve this reduction many feature selection procedures have been developed. Altogether these procedures either employ feature reduction techniques and combine existing features into new features or select subsets of features.

All methods of feature reduction fall into two categories. The first category includes methods which transform features into a new domain. The assumption is that the new domain not only reduces the dimensionality of feature space but also represents the data with minimum loss of information. This analysis includes linear and nonlinear transforms, factor analysis (Kim and Mueller 1978), projection pursuit (Friedman and Tukey 1974) and nonlinear mapping techniques. Examples of such methods are given in Section 5.9.

The second category of feature reduction involves the process of reducing redundances in the feature space. This can be done by selecting individual features from the initial feature set on the basis of theoretical and/or practical Engineering knowledge. Various optimal selection procedures are available for this task, as described in Section 5.16.

A number of measures have been developed to asses the informativeness of features. These measures are used as a criterion for feature selection. An example of mutual information is given in Section 5.16.1 Other methods in this area apply procedures based on transformations of the feature vector. This include the K–L decomposition, the Principle Component Analysis (PCA), Fisher discriminant analysis and methods based on classification binary trees. The K–L transform is a linear projection which arranges transformed features in order of their significance; the most significant features are those corresponding to the major eigenvalues of the correlation matrix. In particular the PCA, described in Section 5.9.1 evaluates features on the basis of the largest eigenvalues. The method based on Fisher's linear discriminant performs an optimal linear dimensionality reduction based on the magnitude of the discriminant vector. It leads to maximum class separation in the output space. The classification binary tree uses a binary decision tree based on Kolmogorov–Smirnov minimisation criterion of the Bayesian probability.

The 'goodness' of features for damage identification can also be estimated on the basis of the Fisher information matrix (Middleton 1960). This matrix is an inverse of the covariance matrix associated with the analysed features. Maximising the trace or determinant and minimising the condition number of the Fisher information matrix corresponds to the process of feature selection.

5.4 TIME-DOMAIN ANALYSIS

There exist a number of simple time domain measures in structural damage detection which are fault oriented. These include: $\overline{x}$, maximum x_{max}, minimum x_{min} and peak-to-peak X_{p-p} values. The peak-to-peak value X_{p-p} is defined as a difference between the maximum and minimum. Often the Root Mean Square (RMS), defined as

$$x_{RMS} = \sqrt{\frac{1}{T}\int_0^T x^2(t)dt} \tag{5.4}$$

is used to give an indication about the average energy.

Statistical moments are also used to describe the shape of the probability density function of the analysed data. The statistical moments of any type of distribution can be defined in the time domain by

$$m_i = \frac{1}{T}\int_0^T [x(t) - \overline{x}]^i dt \tag{5.5}$$

Similar definitions exist in the frequency domain. The mean and variance are the first and second central statistical moments. The variance describes the variability of a signal from the mean. The standard deviation from the mean is used widely in statistics to indicate the degree of dispersion. In general, higher moments are more sensitive to deviations in the data as a result of possible damage. Kurtosis, which is the normalised fourth moment,

$$K_r = \frac{\displaystyle\int_0^T [x(t) - \overline{x}]^4\, dt}{x_{RMS}^4} \tag{5.6}$$

is the most widely used fault indicator based on statistical moments. It gives an indication about *the spikiness* of the analysed data.

5.5 SPECTRAL ANALYSIS

Any observed signals, as acquired from measuring sensors in raw form, are in the time domain. It is possible to characterise signals in other domains by applying specific operations to them. Some of these may be viewed as transformations. Various linear and nonlinear transforms are used in practice. There may be various reasons for transforming to another domain, e.g. data reduction. Even if no data reduction is achieved, the new domain can lead to preferable statistical properties and parameters.

The most widely used transformation is the Fourier transform defined as

$$X(f) = \int_{-\infty}^{+\infty} x(t)e^{2\pi jft}\, dt \tag{5.7}$$

The power spectrum $S_{xx}(f)$, which is the squared modulus of the Fourier transform, gives an indication about the frequency content of the analysed data. Often, as a part of parametric selection, the moments of the power spectrum are used as

$$M_n = \int_0^{F_{max}} f^n S_{xx}(f)\, df \tag{5.8}$$

A number of statistical parameters can be obtained from power spectrum moments. The most widely used are the estimate of the positive zero crossing

$$E(0) = \sqrt{\frac{M_2}{M_0}} \tag{5.9}$$

and the expected number of peaks

$$E(p) = \sqrt{\frac{M_4}{M_2}} \tag{5.10}$$

The primary task of structural health monitoring is the detection of the damage condition. Trending indices are most successfully employed for this task. The trending indices are single number evaluation indices calculated from the time domain signal or the spectrum. The value of the index can be compared with a set of values which indicate a significant degree of damage, or plotted against time show a gradual deterioration. The indices based on spectra seem to provide particularly useful information. Some of these parameters are used to describe the difference in frequency spectra. These are known as the arithmetic and geometric mean. The arithmetic mean is defined as

$$A_m = 20 \log \left(\frac{1}{N} \sum_{i=1}^{N} \frac{A_i}{10^{-5}} \right) \tag{5.11}$$

whereas the geometric mean is given as

$$G_m = \frac{1}{N} \left(\sum_{i=1}^{N} 20 \log \left(\frac{A_i}{\sqrt{2}10^{-5}} \right)^2 \right) \tag{5.12}$$

where A_i is the amplitude of the ith Fourier coefficient of frequency spectrum.

The Matched Filter Root Mean Square is the parameter which involves direct comparison of the corresponding Fourier coefficients of the two spectra. It is defined as

$$M_{frms} = 10\log\left(\frac{1}{N}\sum_{i=1}^{N}\left(\frac{A_i}{A_i(ref)}\right)^2\right) \tag{5.13}$$

where $A_i(ref)$ is the amplitude of the ith Fourier coefficient in the reference spectrum and A_i is the amplitude of the ith Fourier coefficient in the current spectrum.

The other parameter is known as the Root Mean Square of Spectral Difference. This involves finding the root mean square of the difference between two spectra when the amplitude of the Fourier coefficients are given logarithmically. This parameter may be applied to either two successive sampled signal or one sampled and the original. This is obtained from the expression

$$R_{do} = \left(\frac{1}{N}\sum_{i=1}^{N}(L_{ci} - L_{oi})^2\right)^{1/2} \tag{5.14}$$

where L_{ci} is the dB level of the ith Fourier component of a spectrum and L_{oi} is the dB level of the ith Fourier component of a reference spectrum.

Examples of spectral analysis of Lamb wave responses are given in Section 4.8.1.

5.6 INSTANTANEOUS PHASE AND FREQUENCY

It is well known that the ultrasonic waves propagating in structures can interact with material boundaries or discontinuities such as damage. This behaviour can be studied using the instantaneous wave characteristics, namely envelope and phase. Several approaches exist for obtaining these characteristics. For the concept of the analytic signal

$$x_A(t) = x(t) + j\hat{x}(t) \tag{5.15}$$

where $\hat{x}(t)$ is the Hilbert transform of $x(t)$ being

$$\hat{x}(t) = \frac{1}{\pi}\int_{-\infty}^{+\infty} x(\tau)\frac{1}{t-\tau}\,d\tau, \tag{5.16}$$

the envelope and instantaneous frequency can be defined as

$$e(t) = \sqrt{x^2(t) + \hat{x}^2(t)}, \quad \phi(t) = \frac{1}{2\pi}\frac{d}{dt}\,arctan\,\frac{\hat{x}(t)}{x(t)} \tag{5.17}$$

respectively. The Hilbert transform based envelope can be easily implemented by using the Fourier transform. The reader is referred to (Randall 1987) for more details about the signal instantaneous characteristics.

In order to illustrate the application of the Hilbert transform for damage detection, a simple example can be used. A burst sine excitation is applied to monitor a crack growing in an aluminium component (Staszewski *et al.* 1999a). Figure 5.3 gives an example for the strain data captured by a piezoceramic sensor. Here, phase modulation can be clearly observed: two characteristics for different crack lengths are in and out of phase. The data were analysed using the Hilbert transform. Figure 5.4 shows the instantaneous frequency

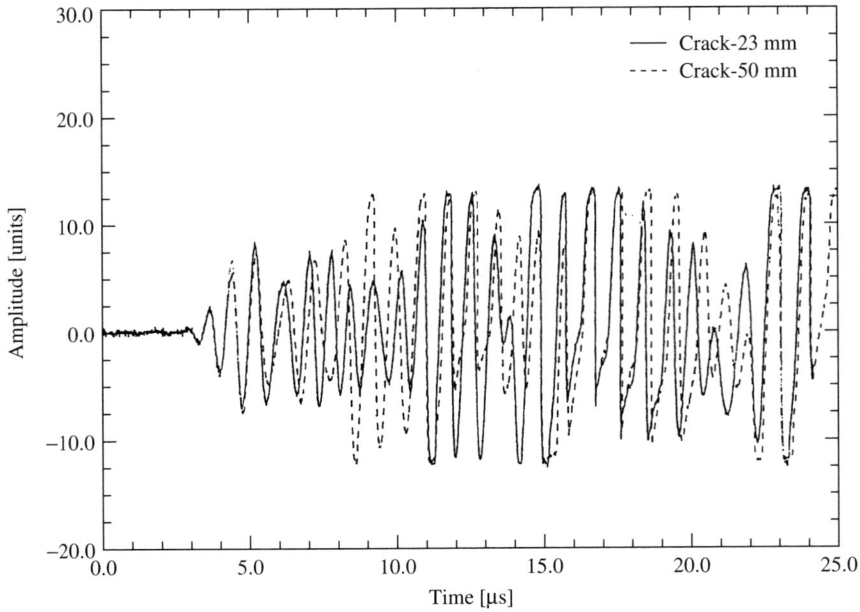

Figure 5.3 Crack detection in an aluminium plate – piezoceramic sensor response for a sine burst excitation (Staszewski *et al.* 1999a)

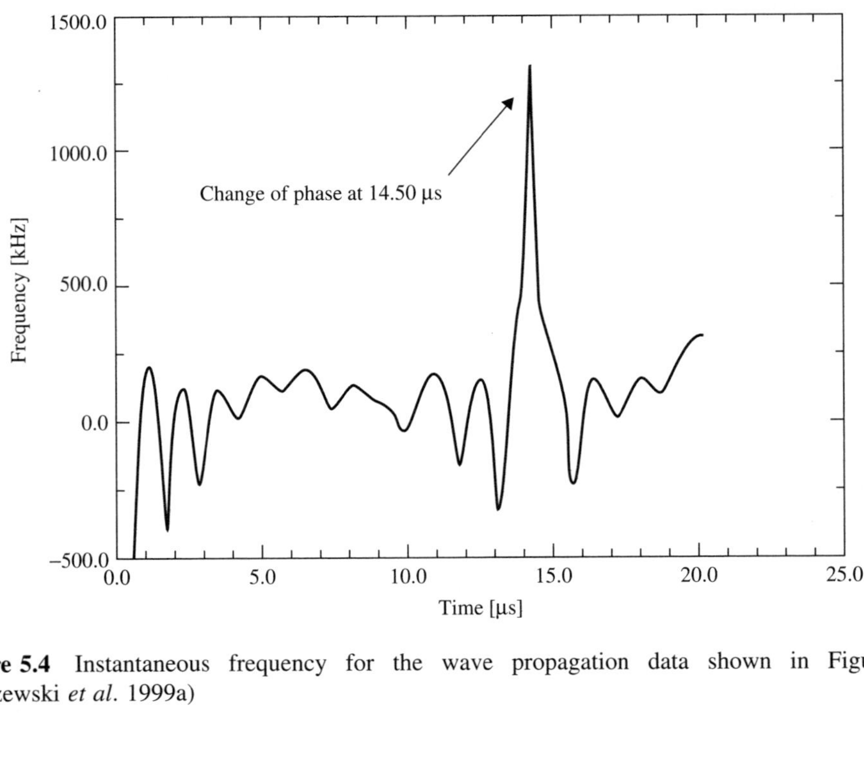

Figure 5.4 Instantaneous frequency for the wave propagation data shown in Figure 5.3 (Staszewski *et al.* 1999a)

for the 45.8 mm crack data. This characteristics displays a clear phase change at 14.5 μs. The analysis for the other two data sets representing the 18.8 and 68.8 mm cracks shows the same behaviour. It appears also that the time location of the phase change is related to the crack length and increases with the bigger crack.

5.7 TIME-FREQUENCY ANALYSIS

The majority of Fourier transform based methods assume the stationarity of the analysed data or the reduction to stationarity by a simple transformation. The analysis of nonstationary signals requires specific time-variant techniques which go beyond the classical Fourier approach. The past twenty years have seen major developments in the area of time-variant analysis. There exist many different methods: some are reviewed in (Priestley 1988; Cohen 1989; Flandrin 1989). These methods can be classified into three major groups: time-dependent models, time–frequency analysis and time–scale analysis, as shown in Figure 5.5.

Time–domain analysis is the study of a signal as a function of time. The evolution of a signal $x(t)$ can be described by a mathematical equation which allows one to analyse

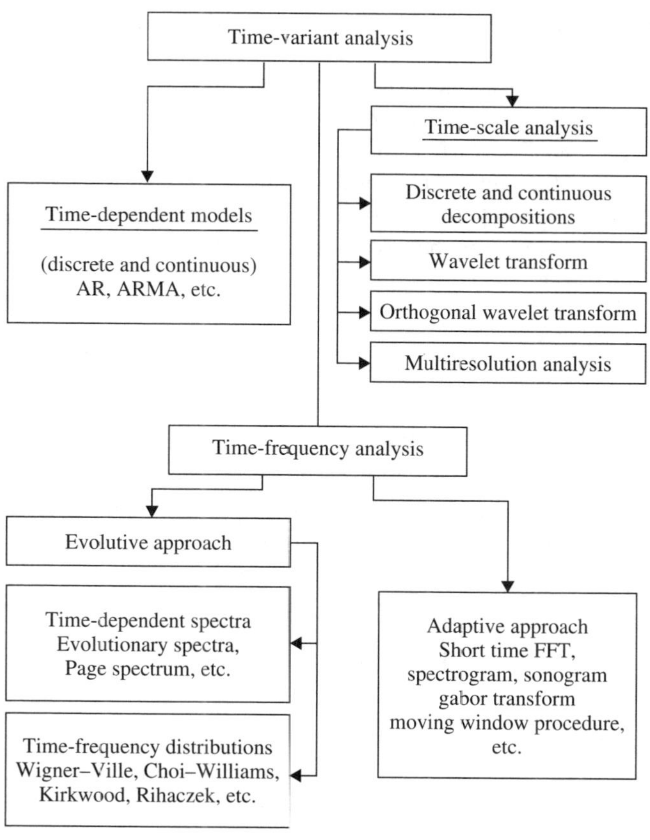

Figure 5.5 Time-variant signal processing methods (Staszewski 2000a)

different parameters (time of duration, maximum amplitude, statistical parameters, etc.) and physical properties (source of propagation, medium of propagation, etc.). Different properties and parameters can be obtained when a frequency analysis is performed. In many cases the signal description is simplified by the use of the Fourier transform, which gives information about the spectral contents of the analysed signal. Some physical properties which depend on frequency can also be better understood. When the spectral content of the signal changes in time, neither the time nor the frequency-domain is sufficient to accurately describe the signal properties. Many attempts have been made to overcome this drawback. The analysis in both time and frequency domains using the so-called windowed Fourier transform is the first important development. This analysis makes use of a window function. The idea of the windowed Fourier transform is to analyse the frequency content of a signal within a window which is fixed in size and moves with time along the signal. Effectively the signal is divided into segments before Fourier analysis is applied. The signal is assumed to be stationary within each segment. Different positions of the window cover the whole time domain. Thus the windowed signal $x_w(t)$ encodes the window position τ and time t. If the Fourier transform is applied to such a signal the result is a new signal representation $F(t, f)$ which in fact is a function of time and frequency. This analysis leads to the short-time Fourier transform. The biggest limitation of this approach is the comparison between the time and frequency resolution of the analysis. Good time resolution implies a small time window, which results in a poor frequency resolution and vice versa. The optimum, obtained for the Gaussian window, is called the Gabor transform.

The natural extension of this analysis can be offered by time–frequency distributions. In a physical interpretation, the spectrum of the signal is a power frequency distribution. By analogy the function $F(t, f)$ is also called a distribution. The values of the function $F(t, f)$ are defined in the time–frequency domain. The analysis of a signal using this joint time–frequency approach is called time–frequency analysis. The general class of time–frequency distributions for the given time signal $x(t)$ can be defined as the Cohen distribution (Cohen 1989),

$$F_c(t, f, \phi) = \int_{-\infty}^{+\infty} \int_{-\infty}^{+\infty} \int_{-\infty}^{+\infty} \phi(\theta, \tau) x^*(u - \frac{\tau}{2}) x(u + \frac{\tau}{2}) \times e^{-i2\pi(\theta t + f\tau - \theta u)}\, du\, d\tau\, d\theta, \tag{5.18}$$

where $\phi(\theta, \tau)$ is an arbitrary function called the kernel. This representation provides for different time–frequency distributions. Different kernels can be chosen to obtain desired properties. Some of the most widely used time–frequency distributions are given in (Cohen 1989). The best known – the Wigner distribution – can be obtained from Equation (5.18) by taking $\phi(\theta, \tau) = 1$, which results in,

$$W_x(t, f) = \int_{-\infty}^{+\infty} x^*(t - \frac{\tau}{2}) x(t + \frac{\tau}{2}) e^{-i2\pi f\tau}\, d\tau \tag{5.19}$$

The Wigner distribution has a number of properties useful for the analysis of nonstationary, transient data. These properties together with the theoretical background can be found in a number of publications (Claasen and Mecklenbrauker 1980a, 1980b; Boashash 1991, 1992). The 1990s has shown an increased interest in time–frequency analysis in fault detection. A number of references for machinery diagnostics are given in

(Wang 1993; Staszewski 1994; Hammond 1997). Gearboxes in particular attracted many investigations and have been analysed using the Wigner–Ville distribution (Forrester 1990, 1992; McFadden and Wang 1990; Staszewski and Tomlinson 1993) and the Gabor transform (Wang and McFadden 1993). The other important development in this area includes the Choi–Williams distribution (Choi and Williams 1989). The Choi–Williams distribution is a Wigner–Ville distribution with a smoothing kernel which attenuates interference terms.

5.8 WAVELET ANALYSIS

Time–frequency analysis studies time variations of the data's spectral characteristics. A different approach to nonstationary signals evolves if nonstationary signals are considered as a superposition of a number of components which are more or less localised in time. This can be done using signal decomposition based on *a priori* chosen scaled functions. Signal decomposition using scaled functions results in time-scale representations – it leads directly to the wavelet transform. Wavelet analysis has been one of the most important and fastest evolving mathematical and signal processing tools of the last twenty years.

Altogether the wavelet transform can be classified as continuous or discrete. In general, continuous wavelets are better for time–frequency analysis and discrete wavelets are more suitable for decomposition, compression and feature selection. However, the choice of wavelets is not always clear.

5.8.1 Continuous Wavelet Transform

The continuous wavelet transform can be defined as

$$W_{\psi}(a,b) = \frac{1}{\sqrt{a}} \int_{-\infty}^{+\infty} x(t)\, \psi^{*}\left(\frac{t-b}{a}\right) dt \tag{5.20}$$

where b is a translation indicating the time locality, a is a dilation or scale parameter, $\psi(t)$ is an analysing (basic) wavelet and $\psi^{*}(\cdot)$ is the complex conjugate of $\psi(\cdot)$. Each value of the wavelet transform $W_{\psi}(a,b)$ is normalised by the factor $1/\sqrt{a}$. This normalisation ensures the integrated energy given by each wavelet $\psi(t)$ is independent of the dilation a. By analogy with the Fourier transform, the wavelet transform is a linear transformation which decomposes an arbitrary function $x(t)$ into the elementary functions $\psi_{a,b}(t)$ which are obtained by translation and dilation from the analysing wavelet $\psi(t)$. Here dilation represents the harmonic or periodic nature in terms of harmonic/periodic decomposition. These two operations are sufficient to produce a basis which can represent all functions in the analysed space. There are a number of different complex and real-valued functions used as analysing wavelets; examples are given in (Chui 1992a, 1992b). In general all these functions satisfy the condition,

$$\int_{-\infty}^{+\infty} |\psi(t)|^2\, dt < \infty \tag{5.21}$$

Clearly, this implies that $\psi(t)$ decays to zero at $\pm\infty$.

The Morlet wavelet defined as

$$\psi(t) = e^{j\omega_0 t} e^{-|t|^2/2}. \tag{5.22}$$

is one of the most widely used functions in the continuous wavelet analysis. In the physical interpretation, the modulus of the transform shows how the energy of the data varies in time and frequency. The energy of a signal is mainly concentrated on the time–scale plane around the so called ridge of the wavelet transform. The ridge of the wavelet transform is a set of points (a, b) for which the transform behaves like the analytic signal. For linear systems, the ridge can be approximately given by the local maxima of the amplitude of the transform. The historical perspective and theoretical developments related to the wavelet transform are given in (Mallat 1999). Applications to structural health monitoring problems include (Staszewski 1998a, 2000b).

More precise definitions and theoretical developments related to ridges and skeletons of the wavelet transform can be found in (Carmona 1997; Staszewski 1998b).

The frequency modulation processes presented in Figure 5.4 can be studied using the continuous wavelet transform, as shown in Figure 5.6. Here, the modulus of the wavelet transform is given together with the original data and power spectrum. The time–scale analysis displays the varying frequency content of the sensor data. This phenomenon can be better observed in Figure 5.7, where the ridge of the wavelet transform is presented.

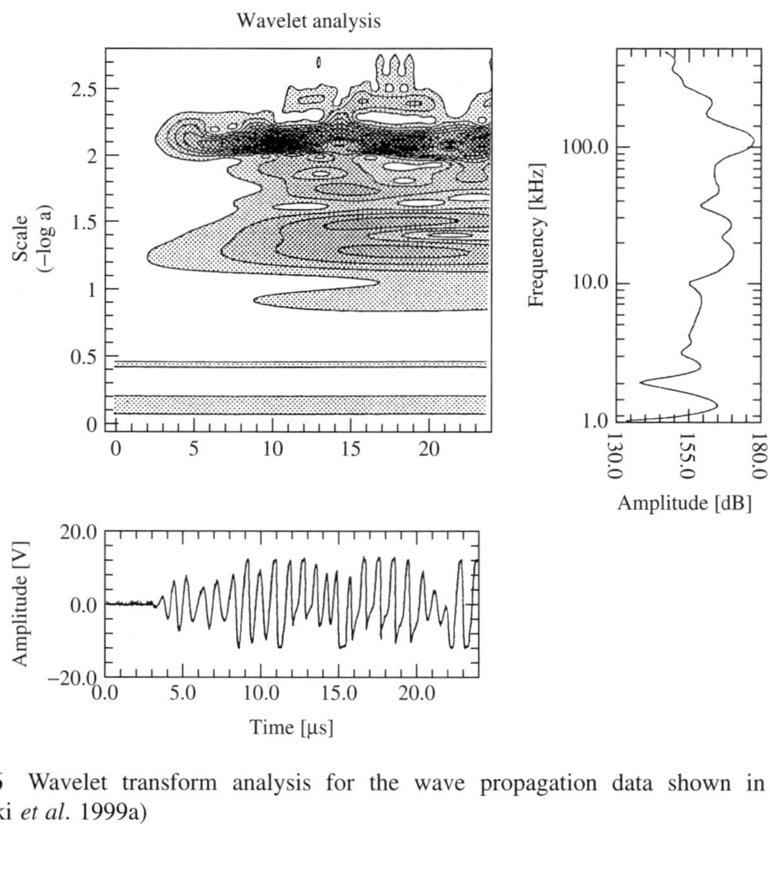

Figure 5.6 Wavelet transform analysis for the wave propagation data shown in Figure 5.3 (Staszewski *et al.* 1999a)

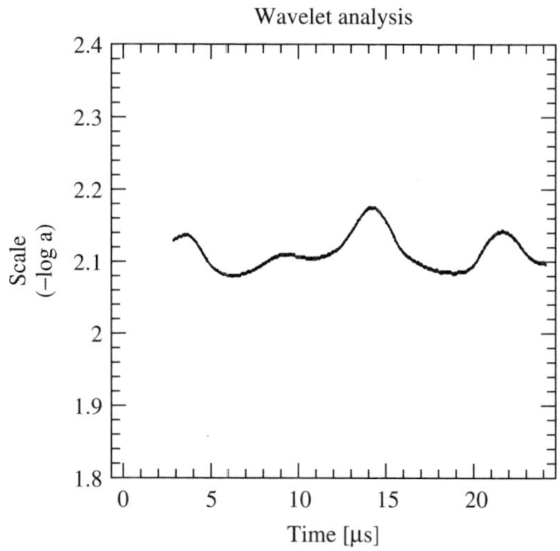

Figure 5.7 Wavelet transform ridge for the data shown in Figure 5.6 (Staszewski *et al.* 1999a)

5.8.2 Discrete Wavelet Transform

The representation given by the continuous wavelet transform is redundant. The analysed data can be completely characterised by discrete samples on the dyadic time–scale grid given by,

$$a = 2^{-m} \quad b = n2^{-m} \tag{5.23}$$

where m, n are integers. This concept leads to the discrete wavelet transform. Substituting Equation (5.23) into (5.20) yields,

$$w_n^m = \int_{-\infty}^{+\infty} x(t)\psi_{m,n}^*(t)dt \tag{5.24}$$

where $\psi_{m,n}(t)$ are the translated and dilated basic wavelets given by,

$$\psi_{m,n}(t) = 2^{m/2}\psi(2^m t - n) \tag{5.25}$$

The concept of the discrete wavelet transform leads to the orthogonal wavelet transform if the functions $\psi_{m,n}(t)$ are orthonormal. There exist many functions which can be used for this orthogonal decomposition. The simpliest basis can be constructed using the Haar function $h(t)$ that is equal to 1 on $[1, \frac{1}{2}]$, -1 on $[\frac{1}{2}, 1]$ and 0 outside the interval $[0, 1]$. More recently the orthogonal basis of Daubechies' wavelets have been proposed (Daubechies 1992). These wavelets cannot be represented in an explicit mathematical form. The so-called scaling functions, corresponding to a finite impulse response (FIR) filter, can be used to construct them.

The original data can be reconstructed from the wavelet synthesis formula,

$$x(t) = \sum_{m=-\infty}^{+\infty} \sum_{n=-\infty}^{+\infty} w_n^m \psi_{m,k}(t) \tag{5.26}$$

This clearly shows that the analysed signal can be represented as a sum of m so-called wavelet levels

$$x_m(t) = \sum_{n=-\infty}^{+\infty} a_k \psi_{m,n}(t) \tag{5.27}$$

where a_k are the amplitudes of the contributing wavelets.

Each of these levels represents the time behaviour of the signal within different frequency bands and gives the contribution to the whole signal energy.

High-frequency strain data were gathered from piezoceramic sensors bonded on a composite wing-box panel, as described in Section 4.7.2. The data can be decomposed using the orthogonal wavelet transform. Figure 5.8 gives the example result for 24 and 4 J

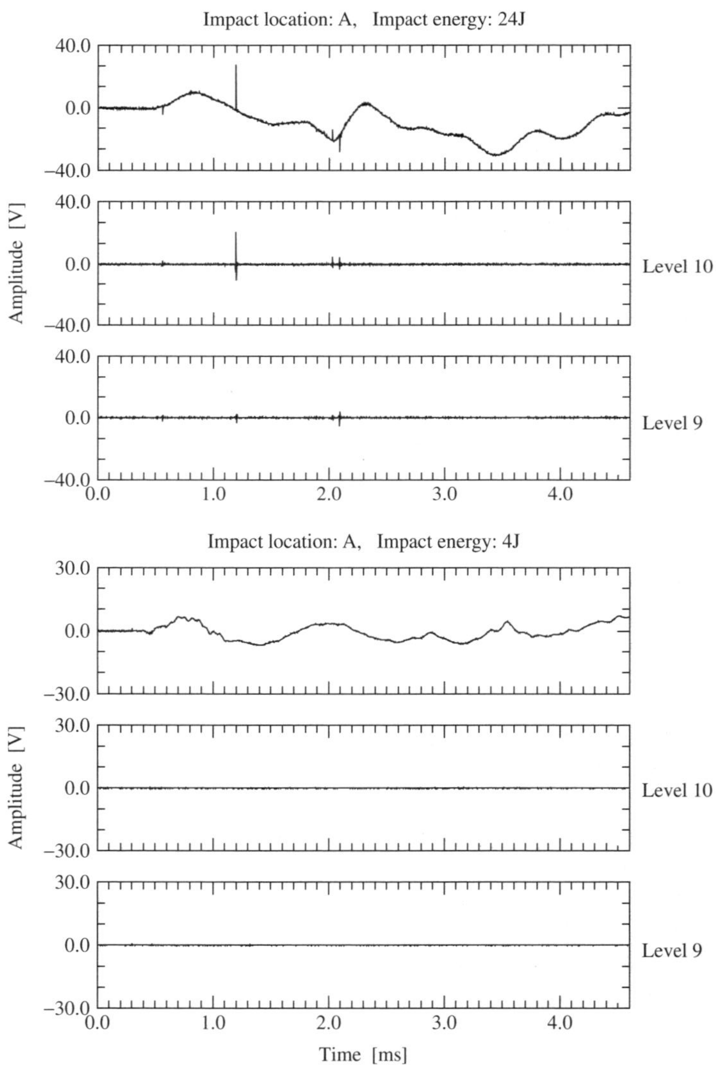

Figure 5.8 Orthogonal wavelet decomposition of the impact strain data (Staszewski *et al.* 1999b)

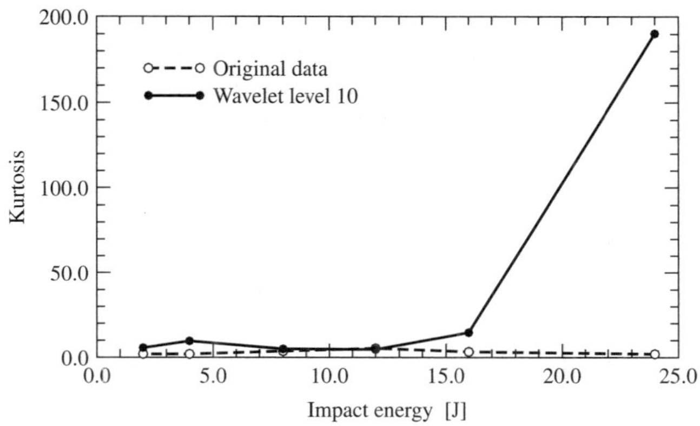

Figure 5.9 Kurtosis characteristic for the impact strain data (Staszewski *et al.* 1999b)

impacts. The top part of each characteristic shows the original data used for the wavelet decomposition. Here, a number of clear spikes at positions: 0.6, 1.2, 2.0 and 2.1 ms can be seen for the 24 J impact. These spikes were only observed when the impact caused a delamination in the composite panel. The highest two (9–10) wavelet levels presented in Figure 5.8 show that the spikes in the data are dominant in the 9th and 10th wavelet levels which represent the high frequency of the analysed signal. The orthogonal wavelet decomposition can separate the spikes from the low frequency impact characteristic. No spikes can be observed for the 4 J impact data.

The spikiness of the data can be analysed using kurtosis, described in Section 5.4. Figure 5.9 gives the values of kurtosis for the analysed impact strain data and different impact energy levels.

The results show that the values of kurtosis estimated from the 10th wavelet levels can distinguish between the impacts which did not introduce any damage to the composite panel and the 24 J impact which caused the delamination. The values of kurtosis were also estimated for the original sensor data; the characteristic remains flat for all energy levels. The results show that wavelet based filtering can extract features which could be related to damage in the structure.

5.9 DIMENSIONALITY REDUCTION USING LINEAR AND NONLINEAR TRANSFORMATION

The data extracted for damage detection is often inappropriate for analysis due to redundancy, high degree of correlation or a large feature space. There exist different procedures which are used to deal with this problem. Transformation is one of the standard techniques used for data compression and dimensionality reduction.

Transform coding projects a feature vector $\underline{X}$ to a vector $\underline{Y}$ using a transformation $\Re$

$$\underline{Y} = \Re(\underline{X}) \tag{5.28}$$

The assumption is that the new vector $\underline{Y}$ keeps most of the energy in only a few vector elements and thus concentrates the energy better than the initial vector $\underline{X}$. The compression is then achieved by setting some vector elements below a threshold to zero and discarding them. The Fourier transform, discussed in Section 5.5, is a simple example of a linear transform which can be used for compression or dimensionality reduction. Unfortunately, Fourier coefficients often do not concentrate the energy well enough to obtain satisfactory results.

5.9.1 Principal Component Analysis

Principal Component Analysis (PCA) is one of the most commonly used linear techniques for dimensionality reduction. PCA is a classical multivariate statistics technique which maps the original to a reduced data space using linear transformation. The procedure seeks to retain as much as possible of the variance in the analysed data.

For the random variable of measurements $x = (x_1, x_2, \ldots, x_n)^T$ the mean can be calculated as

$$\mu_x = Ex \tag{5.29}$$

whereas the covariance is defined as

$$C_x = E(x - \mu_x)(x - \mu_x)^T \tag{5.30}$$

where n is a number of samples, E is the expected value and T denotes the vector transpose. The covariance indicates the correlation between two different vectors of measurements. The k-th principal component of x is defined as the normalised eigenvector v_k corresponding to the eigenvalue λ_k of the covariance matrix C. The eigenvectors and eigenvalues can be found from the solution of the equation

$$C_x v_k = \lambda_k v_k \tag{5.31}$$

for $k = 1, 2, \ldots, n$. This is equivalent to finding the solutions of the characteristic equation

$$\begin{gathered} |C_x - \lambda I| = 0 \\ v_k = \lambda_k v_k \end{gathered} \tag{5.32}$$

where I is the unity matrix. The PCA procedure is thus a singular value decomposition of the covariance matrix. The procedure finds the directions in which the data set has the most significant amount of energy. These directions can be used as features for damage detection.

Figure 5.10 shows examples of Lamb wave responses acquired from piezoceramic sensors bonded on an aluminium plate (Lee *et al.* 2003). The responses, undamaged and damaged (1 mm hole in the plate) conditions, were taken for various temperature levels between 35 °C and 70 °C. The data was projected in to the PCA space. Figure 5.11 gives an example of the first two principal components plotted against each other. The analysis was used to explore various damage features which are independent of environmental conditions.

5.9.2 Sammon Mapping

There exist a number of mapping techniques which are capable of reducing the dimensionality of data. Sammon mapping is a nonlinear extension of the PCA technique. The

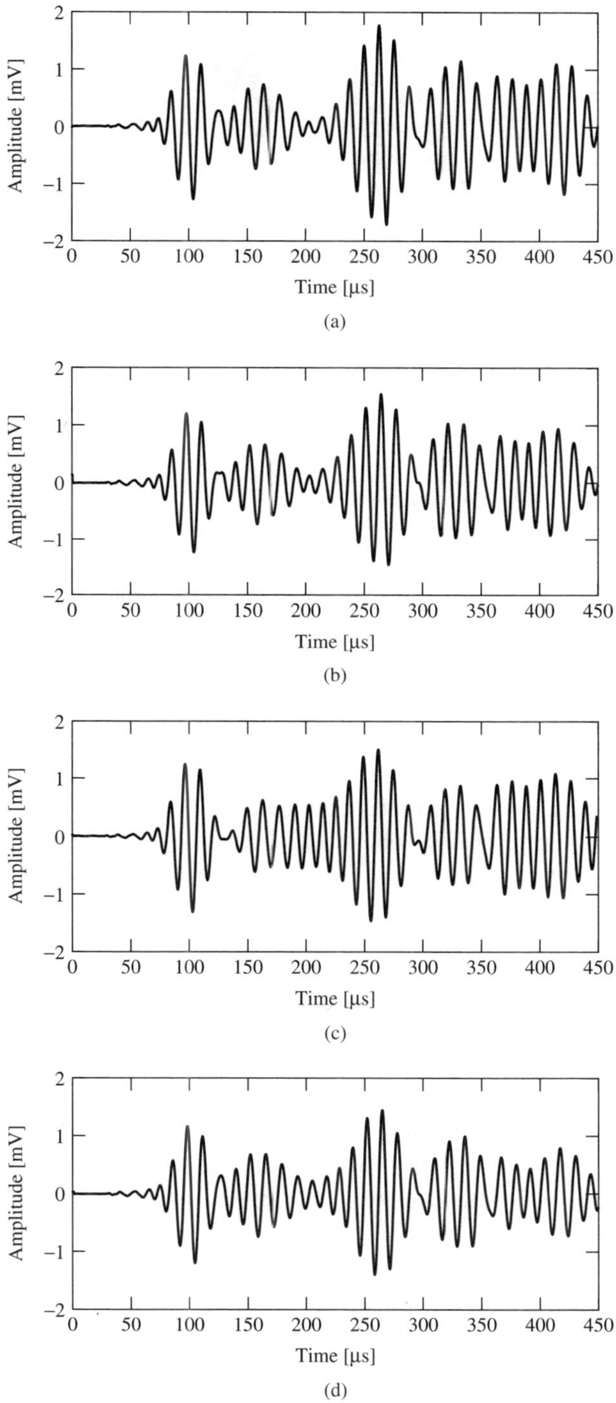

Figure 5.10 Lamb wave responses from the piezoceramic sensors bonded on the aluminium plate: (a) undamaged plate, temperature 35 °C; (b) undamaged plate, temperature 70 °C; (c) damaged plate, temperature 35 °C; (d) damaged plate, temperature 70 °C (Lee *et al.* 2003)

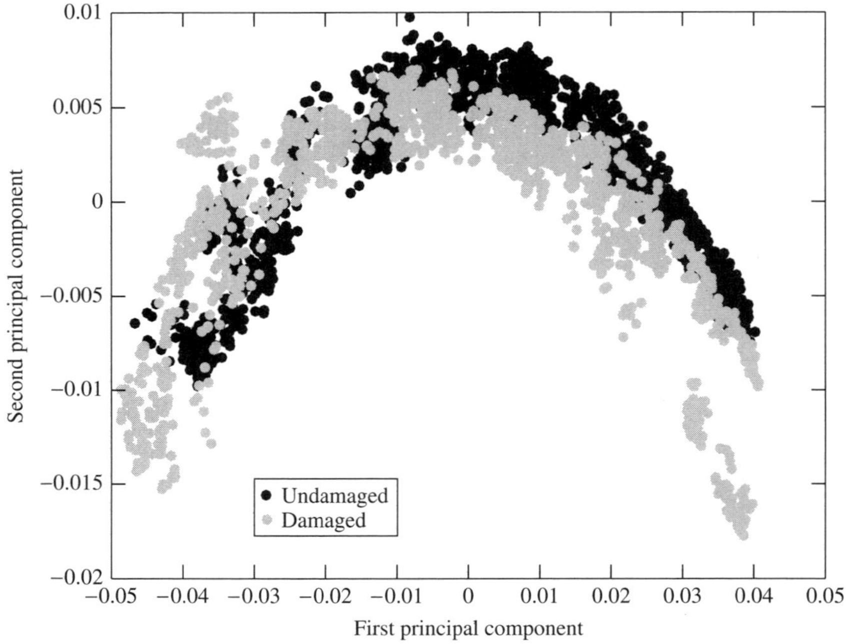

Figure 5.11 Principal component analysis for the data presented in Figure 10 (Lee *et al.* 2003)

method projects the N- dimensional data into a two-dimensional (2-D) space preserving all interpoint distances, i.e. two samples that are close to each other in the original space remain close in the new space. The objective is to reveal the underlaying structure of the data preserving the original topology. The algorithm of calculation typically starts with random coordinates in the 2-D space. An error (or stress) function is calculated in the next step as

$$e_s = \frac{1}{N} \sum_{i,j=1} \left(\sum_{i,j=1}^{1} \frac{(d_{ij}^o - d_{ij}^n)^2}{d_{ij}^o} \right) \tag{5.33}$$

where d_{ij}^o and d_{ij}^n are the distances between points i and j in the old and new spaces, respectively. A gradient descent can be used in order to move the points in the new space following the direction given by the error, i.e. minimising the error value. Although, this iterative procedure is very effective for data visualisation, the major drawback is that it is calculation intensive.

5.10 DATA COMPRESSION USING WAVELETS

Recent developments in the are of data compression include the application of wavelet analysis (Coifman *et al.* 1989; Wickerhauser 1992; Sweldens 1995). The wavelet transform can give a compression basis which is independent of the data set and can reveal local temporal correlations in the data. Also there exist fast algorithms for wavelet transform calculations. A wavelet algorithm of compression is based on a linear decomposition given by the orthogonal wavelet transform.

The orthonormal basis can be used for a decomposition of a given function $x(t)$ according to the wavelet synthesis formula given by Equation (5.26).

The main idea of wavelet based compression is to keep a small number of coefficients which represent the major energy of the signal. The remaining coefficients are set to zero according to the threshold function,

$$F_t(x_m^k) = \begin{cases} 0 & |x_m^k| < t \\ 1 & |x_m^k| \geq t \end{cases} \tag{5.34}$$

It is clear that the smaller the number of wavelet coefficients kept in this analysis, the higher the compression ratio which can be achieved. There exist many wavelet functions which can be used for compression. Two properties which are important for this analysis are: the smoothness and compact support of the wavelet function. Here, the smoothness of a function corresponds to the decay of its Fourier transform and the support of the function means the smallest closed set outside which the function vanishes identically. In general the smoother wavelets (e.g. higher-order Daubechies or Lemarié wavelets) are better for regular, stationary, periodic data and the compactly supported wavelets (e.g. lower-order Daubechies wavelets) are more suitable for nonstationary, transient data. This problem is discussed in (Staszewski 1998c). Often time–scale–frequency wavelets, e.g. Malvar wavelets, offer a good solution to the problem.

The thresholding of wavelet coefficients is not always an effective procedure of feature selection for damage detection. Often the coefficients can be truncated according to their position in the vector, not according to their amplitude. This can be done using combinatorial optimisation procedures (Staszewski 1997). Another solution can be offered by *a priori* knowledge about a damage (Staszewski 1998c).

To improve the ratio of compression, the amplitudes and positions of wavelet coefficients can be quantised and encoded. Quantisation is the procedure which restricts the values of chosen wavelet coefficients to a limited number of levels. Encoding is an operation in which the whole scale of possible levels is divided into intervals represented by coded symbols. There exist various methods of quantisation and encoding; a good summary can be found in (Rabbani 1991).

Figure 5.12 shows examples of Lamb wave responses from optical fibre sensors bonded on a composite plate. The data were acquired for the defect free and delaminated parts of the carbon fibre composite plate (Staszewski *et al.* 1997a). and decomposed using the orthogonal wavelet transform. The original data were reconstructed using 16 wavelet coefficients; the remaining coefficients were set to zero. Figure 5.13 shows the reconstructed vectors. A clear feature between the original and reflected Lamb wave modes results from the defect in the composite plate.

More application examples in the area of damage detection can be found in (Staszewski 1997a, 1998c; Staszewski *et al.* 1997b). An example of the effect of data compression on a fault detection neural network is given in (Staszewski *et al.* 1997b).

5.11 WAVELET-BASED DENOISING

The thresholding procedure of wavelet coefficients used for compression is applied to all wavelet coefficients. The choice of specific coefficients to eliminate can remove some

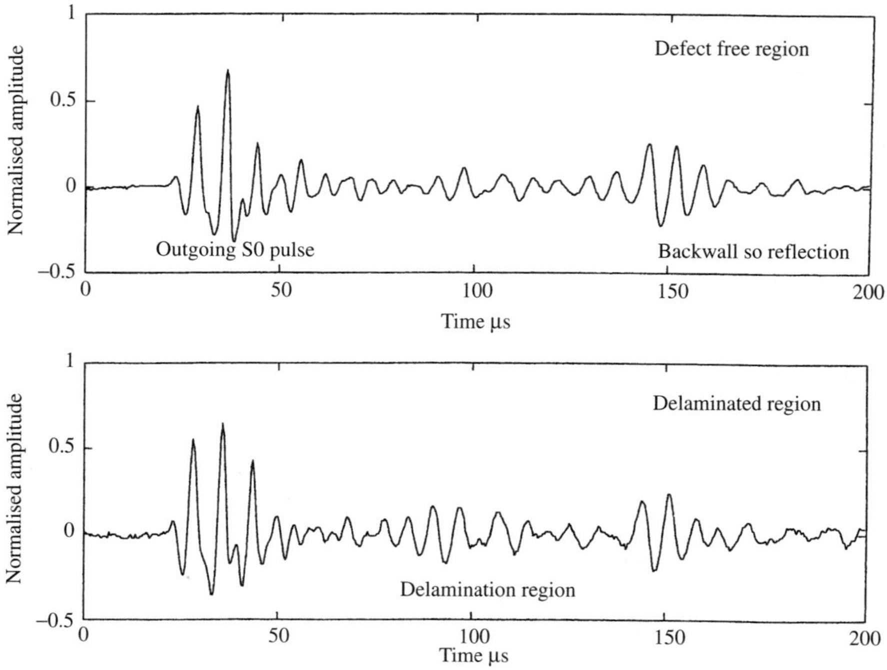

Figure 5.12 Lamb wave responses from optical fibres sensors bonded on the composite plate (Staszewski *et al.* 1997a)

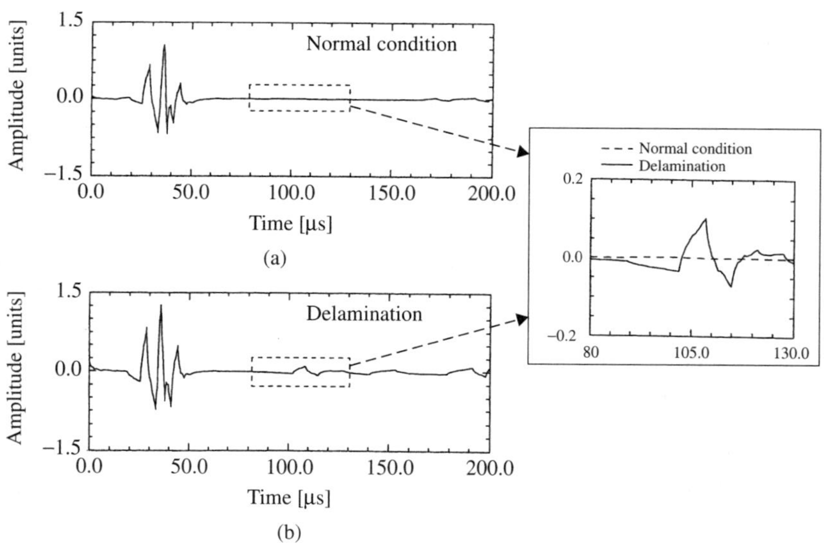

Figure 5.13 Reconstructed Lamb wave responses shown in Figure 5.12 after wavelet-based compression (Staszewski *et al.* 1997a)

undesirable features from the signal. One particular application is the procedure of denoising. The noise $w(t)$ in the data is usually related to high frequencies which are represented by high level wavelet coefficients. If one sets to zero all wavelet coefficients at the highest level, the high frequency part of the signal will be removed.

Alternatively thresholding procedure can be applied within each level separately to track specific frequency bands. In (Donoho and Johnstone 1994a, 1994b) a threshold at two standard deviations, i.e. $c_t = 2\sigma_{c_{mn}}$, has been used for denoising. Alternatively, the so-called optimal threshold value c_t is chosen as (Mallat 1999),

$$c_t = \sigma\sqrt{\ln N} \tag{5.35}$$

where N is a number of samples.

Often the attenuation of wavelet coefficients yields better denoising than coefficient selection. This requires the amplitude of all noisy coefficients that are above c_t to be decreased by c_t. The procedure, called a soft thresholding, uses the limit (Mallat 1999),

$$c_t = \sigma\sqrt{\log_2 N} \tag{5.36}$$

The question remains how to estimate the noise standard deviation σ of the noisy data. It is clear that one needs to suppress the influence of $x(t)$ in Equation (5.1). A robust estimator of σ^2 can be obtained from a median measurement of the highest level wavelet coefficients as (Mallat 1999),

$$c_t = \frac{1}{0.6745}\, Med\ \left(|\langle z, \psi_{m,n}\rangle|\right) \tag{5.37}$$

where $m = \log_2 N$ and $0 \leq n < N/2$. The median is used here rather than the averaged value since it can isolate outliers of potentially high amplitude.

Altogether, the procedure of denoising falls into three steps: wavelet transform, thresholding of wavelet coefficients and inverse wavelet transform.

Figure 5.14 shows the results of the denoising procedure based on wavelets. Here, outputs from fibre optic sensors are used in the analysis. The measured data (Figure 5.14a) is the strain which results from the 5.88 J impact on the composite specimen [5]. The orthogonal wavelet decomposition, using the 10th order Daubechies wavelet function was applied to the original sensor data. Figure 5.14b gives the same sensor data with the noise removed using the thresholding procedure with the median filter.

5.12 PATTERN RECOGNITION FOR DAMAGE IDENTIFICATION

Damage detection can be regarded as a problem of pattern recognition. A pattern can be a set of features. These features are given by continuous, discrete or discrete-binary variables which can be formed in vector or matrix notation. Patterns represent different conditions and indicate whether the analysed structure is undamaged or damaged. The severity of damage can also be established. The problem which describes how to distinguish between different classes of patterns is known as pattern recognition.

There exist two classical approaches to pattern recognition; these are (Schalkoff 1992): statistical and syntactic methods. Syntactic pattern recognition classifies data according to

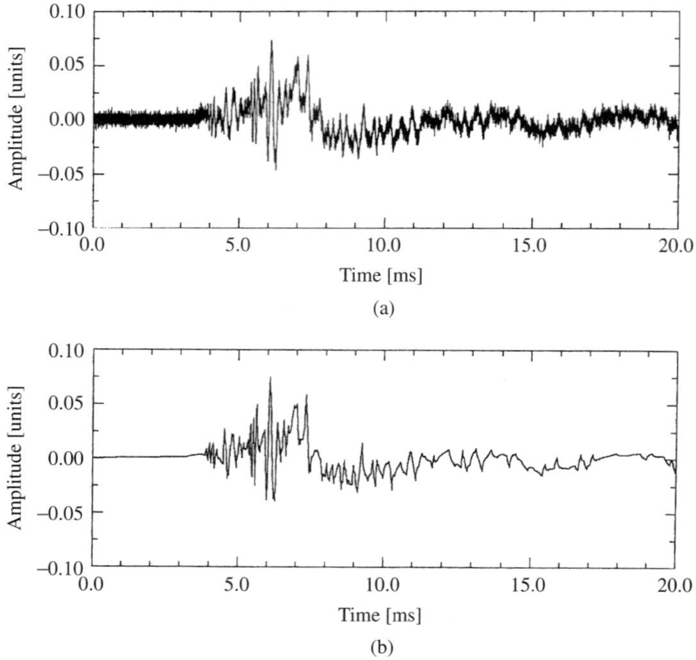

Figure 5.14 Wavelet denoising for the optical fibre strain data: (a) original data; (b) denoised data (Staszewski *et al.* 2000)

its structural description and is not commonly used for damage detection. Statistical pattern recognition assigns features to different classes using statistical density functions. Recent advancements in pattern recognition include applications of Artificial Neural Networks, discussed in Section 5.13, and methods of novelty detection (Tarassenko *et al.* 1995; Worden 1997a, 1997b). Novelty detection establishes a description of normality using features representing undamaged conditions and then tests for abnormality or novelty. These methods provide only a damage detection level, however they do not require any *a priori* knowledge about damage. In what follows an example of statistical pattern recognition for damage detection is given.

The piezoceramic data representing the crack growing in an aluminium plate (see Section 4.8.1) is used in this example (Staszewski *et al.* 1999). The data were decomposed using the orthogonal wavelet transform. The variance was then calculated for each wavelet level of decomposition. Figure 5.15 shows an example of three variance characteristics, plotted in a semilogarithmic scale, as a function of wavelet levels. These characteristics were used as patterns in similarity analysis.

The mean vector $\overline{\mu}$, of wavelet variance characteristics $var_i x_m^n$, given by

$$\mu(m) = \frac{1}{N}\sum_{i=1}^{N} var_i x_m^n \tag{5.38}$$

was calculated for the data representing the initial crack length. This vector formed the template for the similarity analysis. The damage index was calculated as a Euclidean

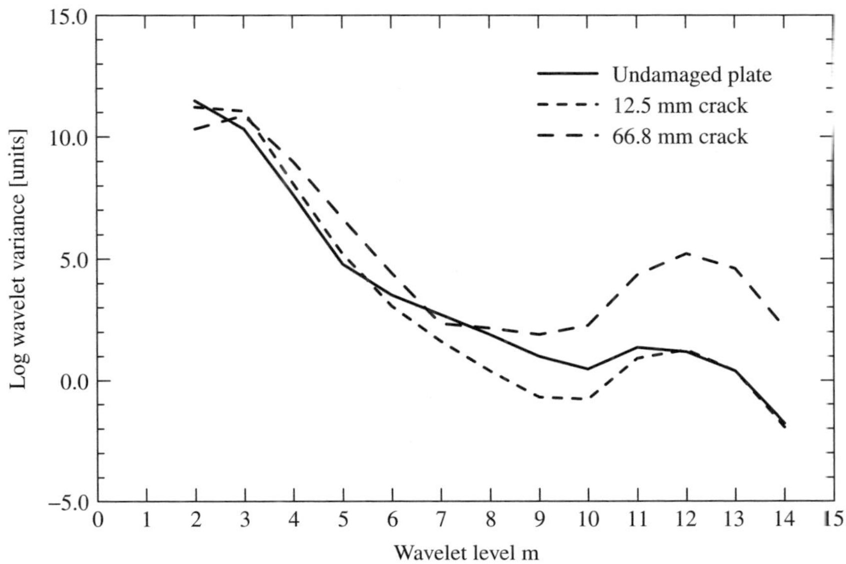

Figure 5.15 Wavelet variance characteristics for the piezoceramic sensors data (Gaussian noise excitation) (Staszewski *et al.* 1999a)

distance between the template $\overline{\mu}$ and the wavelet variance characteristics $\overline{x} = varx_m^n$, from which the damage is to be evaluated,

$$d_{x,\mu}^2 = (\overline{x} - \overline{\mu})^T (\overline{x} - \overline{\mu}) \tag{5.39}$$

where, T denotes the vector transpose.

The values of damage index from the *no fault* condition of the plate were used to establish statistics for the expected variations in damage indices. The resulted 95 % statistical confidence levels were calculated as $\nu \pm 1.96\sigma$, where ν and σ are respectively the mean and standard deviations of the damage indices. The alarm or warning level was also established. It represents the index value above which it can be considered that the damage exists in the plate. The alarm value was taken as $d^2 + 4\sigma_{d^2}$, where d^2 and σ_{d^2} are the mean and the standard deviation of the damage index representing the *no fault* condition.

Finally, Figure 5.16 shows the damage index as a function of crack length. Here, the confidence and alarm levels are indicated by dotted and dashed lines respectively. It can be seen that the method can detect a 6–7 mm crack in the aluminium plate.

5.13 ARTIFICIAL NEURAL NETWORKS

Artificial Neural Networks are one of the most significant developments of advanced signal processing in recent years. There exists a vast literature on the subject including a comprehensive introduction given in (Haykin 1994). In the following only a brief summary of the topic will be given.

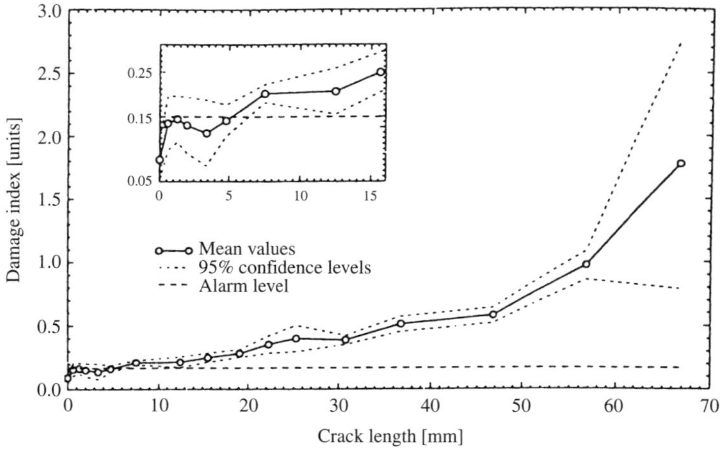

Figure 5.16 Damage index based on wavelet analysis (Staszewski *et al.* 1999a)

5.13.1 Parallel Processing Paradigm

Artificial Neural Networks (ANN) are biologically inspired; their origin goes back to the pioneering work on neurons as structural constituents of the brain in the 1910s [54], and the mathematical modelling of single neurons in the 1940s (McCulloch and Pitts 1943). Neurons are structural cells of the brain which process information. Although they are relatively slow, the overall processing power of the human brain is enormous due to massive parallel processing. ANNs attempt to simulate the perceptual power of the human brain using this parallel processing paradigm. The most important elements of this paradigm include (Haykin 1994):

- **Nonlinearity**

  Neurons are nonlinear devices and therefore neural networks are also nonlinear. Moreover, the nonlinearity can be distributed throughout the network according to its structure. ANNs are usually designed to be nonlinear

- **Adaptivity and Learning**

  ANNs can modify their behaviour in response to the environment. Once a set of inputs with desired outputs is presented to the network, the connections between the neurons self-adjust to produce consistent responses. This training is repeated for many examples until the network reaches a stable state.

- **Evidential Response**

  A neural network can be designed to provide information not only about the response but also about the confidence in the response; examples include pattern recognition analysis where patterns can be classified with a confidence level.

- **Fault Tolerance**

  If a neuron or its connections are damaged, processing of information is impaired. However, a neural network can exhibit a graceful degradation in performance rather than total failure.

- **VLSI Implementation**

  The parallel nature of neural networks make them ideally suited for implementation using microchip technology.

ANNs are suitable for sensor data processing problems that require parallelism and optimisation due to high-dimensionality of the problem space and complex interactions between the analysed variables.

5.13.2 The Artificial Neuron

Despite the diversity of network paradigms, nearly all consist of very similar building blocks – the artificial neurons. The structure of these neurons has changed very little since the first study in (McCulloch and Pitts 1943). The usual artificial neuron consists of two blocks: summation and activation, as shown in Figure 5.17.

The input values $x_i \in 0, 1$ are weighted by a factor w_i before they are passed to the body of the neuron. The weighted inputs are then summed to produce an *activation* signal z,

$$z = \sum_{i=1}^{n} w_i x_i \tag{5.40}$$

This signal is then passed through a nonlinear *activation* or *transfer function*. Any one of a number of functions can be used for processing. If the output signal y is described as,

$$y = kz \tag{5.41}$$

where k is a constant coefficient, the neuron is linear and consequently networks made up from these types of neurons will be linear networks. In the literature linear neurons and networks are called *Adeline* and *Madeline* (Widrow and Hoff 1960), respectively.

As stated above, there exist a number of different activation functions. The well-known McCulloch–Pitts (MCP) model uses a hard threshold function. The MCP neuron fires if the weighted sum z exceeds some predefined threshold β, i.e. if,

$$z > \beta \tag{5.42}$$

and does not fire if,

$$z \leq \beta \tag{5.43}$$

The weighted sum z can also be transformed by the sigmoidal function,

$$y = \frac{1}{1 + e^{-z}} \tag{5.44}$$

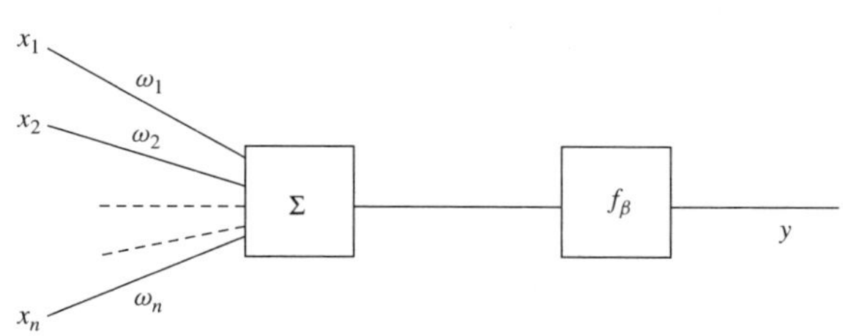

Figure 5.17 McCulloch–Pitts neuron (Worden and Staszewski 2000)

or hyperbolic tangent function,

$$y = tan[h(z)] \tag{5.45}$$

The nonlinear activation function helps to solve the noise-saturation dilemma of how the network can handle both small and large signals.

5.13.3 Multi-Layer Networks

Although a single neuron can model a number of simple logical problems, the real processing power is obtained when neurons form networks. A simple single-layer ANN comes from connecting neurons in a layer. However, multi-layer networks have been proven to have capabilities beyond those of a single layer. There have historically been two important developments in multi-layer network technology.

The first serious study of networks was carried out by Rosenblatt (Rosenblatt 1962), who proposed a three-layer structure – the *Perceptron*. The first layer is the input layer which simply distributes signals to the processing layers. The first of these is the *hidden* layer which is referred to as the *associative* layer while the second, which outputs signals to the outside world is termed the *decision* layer. Only the connections between the decision and associative nodes are adjustable in strength; those between the input and associative nodes are preset before training takes place.

A completely rigorous investigation into the capability of perceptrons is given in (Minsky and Papert 1988). Although perceptrons were initially received with enthusiasm, they were soon associated with many problems. In representing a function with N arguments, the generic perceptron was shown to need 2^N elements in the associative layer, i.e. the networks grow exponentially in complexity with the dimension of the problem. Although there are a number of ways of avoiding this problem, Rosenblatt's perceptrons are generally not used in applications.

The second important development was not actually a multi-layer network. After the publication of (Minsky and Papert 1988) which highlighted the problems with perceptrons, there was very little research carried out on ANNs for many years. The period of inactivity ended with the work of Hopfield (Hopfield 1984) in the 1980s. He considered the network from the point of view of dynamical systems theory. The outputs of the constituent neurons in Hopfield's networks were regarded as dynamical states which could evolve in time. The Hopfield network proved capable of solving a number of practical problems and reinvigorated ANN research. One immediate result of the resurgence in activity was the solution by various groups of the problems associated with Rosenblatt-type perceptrons.

5.13.4 Multi-Layer Perceptron Neural Networks and Others

Altogether ANNs can be classified as:

- **Feedforward networks**

  Signals are passed through the network in only one direction; there are always connections between the neurons in adjacent layers, there may or may not be more far-reaching connections. Feedforward networks have no memory, their output is solely determined by the current inputs and the values of the weights.

- **Recurrent networks**

  Feedback connections are possible between network elements. Some recurrent networks recirculate outputs back to the inputs. Since their output is determined both by their current input and their previous outputs, recurrent networks have internal dynamics and can exhibit properties similar to short term memory in the human brain.

- **Cellular networks**

  A basic cellular network has a two-dimensional structure. The processing units – called cells – are connected only to their neighbouring cells. Cells contain linear and nonlinear processing elements. Although only adjacent cells can interact directly with each other, the cells not directly connected can still affect each other due to a propagation effect. Cellular networks resemble structures found in cellular automata.

Feedforward networks are the most often-used paradigm. Such multi-layer networks can be organised in many different types of architecture, as described in (Freeman and Skapura 1991; Hertz *et al.* 1991; Haykin 1994); however, the most frequently-used by far is the Multi-Layer Perceptron (MLP) (Rumelhart and McClelland 1988).

The MLP is a set of neurons arranged together in layers (Figure 5.18). Feature vectors pass into the input layer nodes, progress forward through the network hidden layers and the result finally emerges from the output layer. The nodes are connected to each node

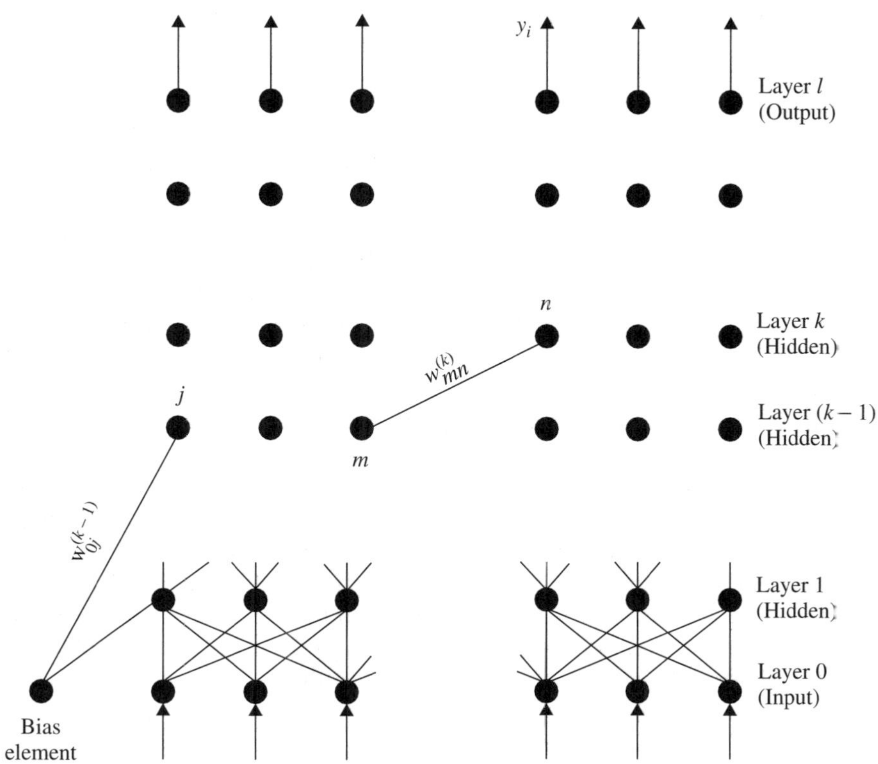

Figure 5.18 The Multi-Layer Perceptron (MLP) network (Worden and Staszewski 2000)

in the preceding layer and to nodes in the following layers. The connections are through a set of weights w_{ij}, where i stands for the i-th node of the previous layer and j for the j-th node of the current layer. Signals pass through the node as follows: a weighted sum is performed at i of all the signals x_j from the preceding layer, giving the excitation z_i of the node; this is then passed through a nonlinear activation function f to emerge as the output of the node x_i to the next layer. This can be modelled as,

$$x_i = f(z_i) = f\left(\sum_j w_{ij} x_j\right) \tag{5.46}$$

and as discussed previously, various choices for the activation function f are possible. Implementation of the algorithm and MLP software can be found in (Worden 1997c).

Radial-Basis Function (RBF) networks (Broomhead and Lowe 1988) form another important group of feedforward networks. An RBF is a multi-dimensional function that depends on the distance between the input vector and a number of cluster centres in the pattern space. RBFs are used as ANNs to find the surface in a multi-dimensional space that provides the best fit to the training data. The most basic form of the RBF network involves three different layers. The input layer consists of input (sensory) nodes; the hidden layer serves to carry out the nonlinear processing; the output layer supplies the response of the network to the activation patterns applied to the input layer. A number of different functions can be used as transfer functions of the hidden neurons in the RBF networks; the important point about the RBF is that these nonlinear functions are *local* as opposed to the global behaviour of the MLP. The most often used transfer function is the Gaussian,

$$f(r) = e^{-r^2/\sigma^2} \tag{5.47}$$

Other examples include (Luo and Kay 1992): multiquadratic, inverse multiquadratic, thin-plate-spline, piece-wise linear and cubic approximation functions.

The *Boltzmann Machine* (Haykin 1994), represents an example of a recurrent network. This network is based on the *Simulated Annealing* algorithm of combinatorial optimisation. It shares many common features with the Hopfield network. However, in contrast with the Hopfield network, it permits hidden layers and uses stochastic neurons with a probabilistic firing mechanism. Other examples of recurrent networks include (Hertz *et al.* 1991): *Bidirectional Associative Memory* and Grossberg's network.

The Kohonen map (Haykin 1994), briefly discussed in the next section, is a typical example of a cellular network.

Often a simple unitary network is not sufficient for complex tasks. ANNs may need to be combined to enhance their performance. An excellent review on combining neural networks is given in (Sharkey 1996). There exist two main techniques: an ensemble-based and a modular approach. The ensemble-based approach combines a set of nets which are general function approximators and essentially accomplish the same tasks. The modular approach is much wider and is based on the *divide-and-conquer* rule. It includes hybrid systems of network architectures which are joined together to accomplish disparate tasks.

Ability to learn is one of the most important features of ANNs. Learning algorithms are categorised as supervised and unsupervised.

Supervised learning requires pairing of each input vector with a target vector representing the desired output – a so called training pair. Networks learn from seeing training pairs applied sequentially. The algorithms usually parallel the action of learning in biological neurons, as broadly encoded in *Hebb's rule* (Haykin 1994). This algorithm increases the synaptic strength or weight if both the source and destination neuron are activated and thus reinforces connections which are activated often. Examples of supervised learning algorithms include the steepest descent and error correction algorithms. Steepest descent algorithms adjust the weights of a network during each training pair presentation; the objective is to reduce the Mean Square Error (MSE) averaged over the learning process. Error correction rules alter the weights of a network to reduce the error in the output response to the current training pattern. Algorithms use similar learning procedures and can be additionally separated into layered network and single element categories, and finally into nonlinear and linear rules. The best known first-order optimisation learning procedures are: the μ–LMS steepest descent algorithm for a linear single element, the α–LMS error correction rule for a linear single element and the backpropagation algorithm. The backpropagation algorithm is a stochastic steepest descent learning rule used to train single or multi-layer nonlinear networks. This algorithm is often used to train MLP networks. Better performance can usually be obtained by using higher-order optimisation methods. Examples of second-order methods used to train MLPs are conjugate gradient methods and the Levenburg–Marquardt algorithm.

Unsupervised or self-organising learning schemes represent a completely different philosophy of training which in a way provides a more plausible model of the biological learning mechanism. The idea of these schemes is to discover significant features in the analysed data and to do the discovery without a *teacher*. The algorithms are provided with a set of rules of a local nature. The Hebb algorithm is also one of the best-known unsupervised learning schemes. Self-organising feature maps form a special class of networks that are based on unsupervised competitive learning; i.e. the output neurons of these networks compete among themselves to be fired. As a result of this competition only one neuron is active at any one time. Example of these include Kohonen maps and topologically ordered maps discussed in more details in, (Haykin 1994; Bishop 1995).

5.13.5 Applications

Broadly speaking, ANNs can offer solutions to four different problems:

- **Autoassociation**

  A signal is reconstructed from noisy or incomplete data.

- **Regression/Heteroassociation**

  Input-output mapping, i.e. for a given input data produce a required output characteristic.

- **Classification**

  Assign input data to given classes.

- **Detection**

  Detect statistical abnormalities in the input data.

The first two tasks are often associated with modelling applications using neural networks. The last category includes the problem of *novelty detection*. Novelty detection establishes a description of normality using features representing initial or normal conditions and then tests for abnormality or novelty. The application to condition monitoring or structural health monitoring is manifest and leads to damage detection. The higher levels of damage identification are associated with the classification function of neural networks.

5.14 IMPACT DETECTION IN STRUCTURES USING PATTERN RECOGNITION

This section describes an example of impact damage detection using neural networks. The strain data obtained from a drop test on a composite component is used in the analysis (Staszewski 1999).

The strain data were obtained from Bragg grating optical fibre sensors which were embedded in a composite skin. The skin design together with the layout of the sensors is presented in Figure 5.19. In an unloaded condition, the skin was impacted at positions $B1$ to $B7$ (Figure 5.20), while taking data from sensors 1.1 and 7.1. Impacts were also made at the sensor positions 1.1 and 7.1. For each impact location, two impacts were made. This allowed two different time traces to be recorded with different time resolutions and time durations of the events. The analysis involved two different energy levels of impacts: 2 and 3 J. Figures 5.21a and 5.21b show an example of the low and high time resolution strain outputs from sensor 1.1 for the impact position B6. The power spectrum of this strain data is presented in Figure 5.21c. The strain data obtained from the tests were employed as features for neural network analysis, which was the basis of the sensor data processing procedures used.

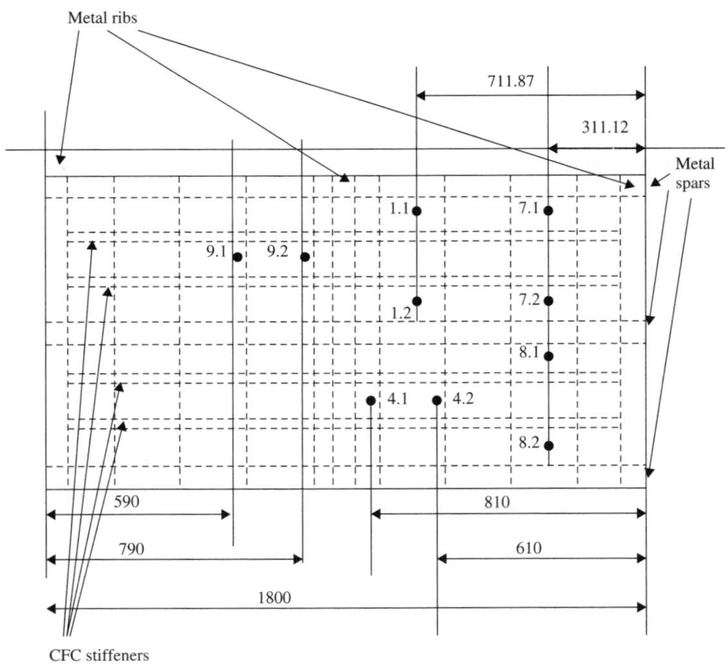

Figure 5.19 Composite skin design with locations of Bragg grating sensors (Staszewski 1999)

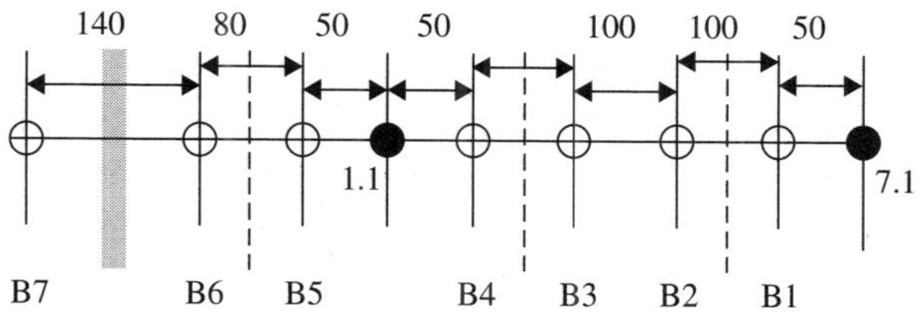

Figure 5.20 Impact locations for the composite skin shown in Figure 5.19 (Staszewski 1999)

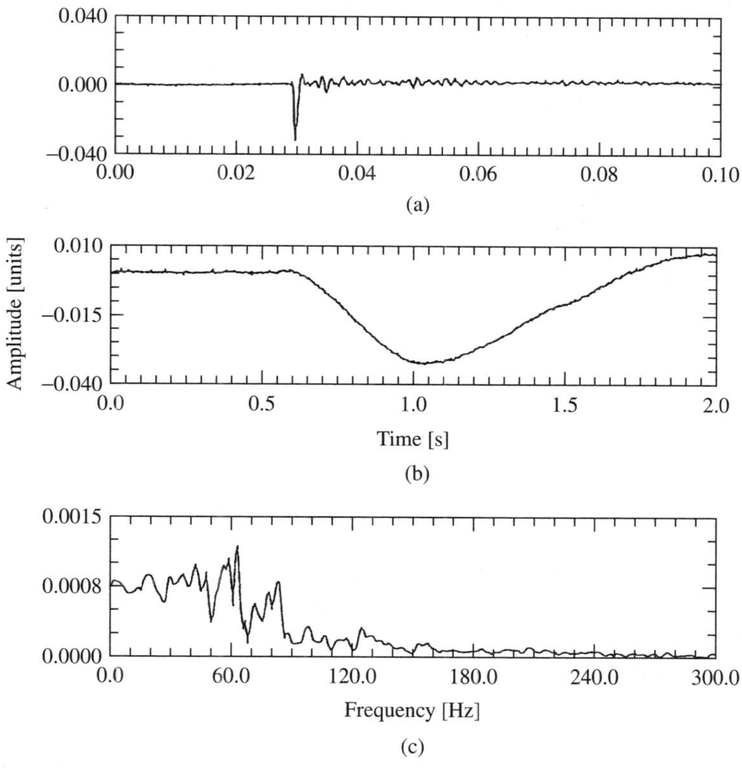

Figure 5.21 Low (a) and high (b) resolution strain outputs from the Bragg grating sensors. The spectrum of impact data is shown in (c). (Staszewski 1999)

It is obvious that for each sensor location, the amount of strain data available for the network training is unrealistically large. Thus a feature extraction was performed. The analysis involved three sets of time and frequency domain features:

- *Feature A*: RMS values of strain data from long duration time records;
- *Feature B*: Magnitude of the real and imaginary parts of the Fourier transform of the time data integrated over frequency; long duration time records of strain data were used to obtain these features;
- *Feature C*: Impact event time determined from short duration time records of strain data.

Features A and B proved to be successful in previous applications (Staszewski 1996). The integration in the frequency domain was performed over the first 250 frequency lines which covered the main spectral strain components. The relative timing of each event was established from short duration time records of the strain data. This involved thresholding with different amplitude levels while comparing with the absolute signal amplitude. For equal signal and threshold conditions, the time value was recorded. This process was repeated iteratively to generate a characteristic thresholding function. The major point of inflection was taken as the event time.

5.14.1 Detection of Impact Positions

The time and frequency domain sets of features A, B and C were used to train the neural network in order to predict the impact position. Two different network structures were employed for the feature sets. In the first network the input data were sets of four floating-point real numbers: RMS values (A) and impact event times (C) for two sensors used in the analysis. The second network involved the input data in the form of six floating real numbers: integrated absolute values of real and imaginary parts of the Fourier transform (B) and impact event times (C); all values calculated for two sensors (1.1 and 7.1) used in the analysis. The data were corrupted by zero-mean Gaussian white noise 5 % of the signal amplitude. The input data were then scaled by mapping it onto the interval $[-1,1]$ before randomly presenting it to the input layer of the network. The outputs of the networks were required to signal the position of the impact. Thus the network included nine output nodes representing nine different impact positions given in Figure 5.20 in the following order: node 1–7.7, node 2–$B1$, node 3–$B2$, node 4–$B3$, etc. A simple training strategy was allowed; the network was trained to produce a value 1.0 at output 1 and 0.0 at other outputs if the training data set was associated with impact at location 1. Outputs 2, 3 and 4 etc., were similarly associated with impacts at locations 2, 3 and 4 etc. This is referred to 1 of M strategy. A trial and error approach was adopted to obtain the appropriate dimension for the network. One hidden layer with twelve nodes was used in both networks.

The neural network used here was the MLP described in (Worden 1997). The MLP used the hyperbolic tangent activation function and the backpropagation learning algorithm. The number of presentations used to train the network was equal to 100 000; this was sufficient to obtain convergence. One cycle per epoch was used in the analysis. The learning algorithm used additionally a learning recall schedule which applied time varying values of learning and momentum coefficients.

The networks were tested using 1000 different sets of data for nine impact positions and two levels of impact energy: 3 and 4 J. All testing sets were corrupted by Gaussian white noise (5 % of the signal).

The Bayesian decision rule was adopted for the diagnostic network (Rytter 1993). Counting up the number of misclassifications when the converged network is tested on its training set was used to give an estimate of the probability of misclassification.

Figure 5.22 shows an example of the classification matrix from a network testing procedure for B and C features (4 inputs network). Here each row i corresponds to the class of the training data associated with a specific impact location. The value in column j represents the number of times the data in the class was associated with location j during testing. Thus any nonzero values off the diagonal represents misclassification. It can be seen that the probability of misclassification is equal to 0.0012.

```
* Classification matrix :

              1     2     3     4     5     6     7     8     9

Class 1 :  1000     0     0     0     0     0     0     0     0
Class 2 :     0  1000     0     0     0     0     0     0     0
Class 3 :     0     0  1000     0     0     0     0     0     0
Class 4 :     0     0     0  1000     0     0     0     0     0
Class 5 :     0     0     0     0   989     0    11     0     0
Class 6 :     0     0     0     0     0  1000     0     0     0
Class 7 :     0     0     0     0     0     0  1000     0     0
Class 8 :     0     0     0     0     0     0     0  1000     0
Class 9 :     0     0     0     0     0     0     0     0  1000

* Probability of error : 0.00122222
```

Figure 5.22 Example of a classification matrix based on neural network analysis (Staszewski 1999)

5.14.2 Detection of Impact Energy

Neural network analysis was also used to predict the energy of the impacts. Again two neural network structures were tested. The analysis involved the same A, B and C feature sets and the same number of input nodes as in the previous section. The outputs of the networks were required to predict the energy level of the impact. One hidden layer with two nodes was used in both networks. Thus the networks used here were 6 : 2 : 1 and 4 : 2 : 1. The analysis employed the same training parameters as in the previous section. The networks were tested using 200 different sets of data corrupted by Gaussian white noise (5 % of the signal).

In order to validate the results, the percentage error was introduced; the definition being

$$PE(\hat{y}) = 100 \sum_{i=1}^{N} \frac{y_i - \hat{y}_i}{y_i} [\%] \tag{5.48}$$

where y_i and $\hat{y}_i$ denote the desired and actual inputs, and N is a number of testing sets.

Figure 5.23 shows an example of the testing results for B and C features used. Here the dashed line represents the desired velocity and the solid line gives the actual network response. The averaged and maximum prediction errors are equal to 0.6 % and 10.1 % respectively.

5.15 DATA FUSION

Data fusion integrates data from a multitude of sensors with the objective of making a more robust and confident decision than is possible with any one sensor alone. The reasons why multi-sensor systems are desirable are (Esteban and Starr 1999):

- higher signal-to-noise ratio;
- robustness and reliability;
- better information regarding independent features;

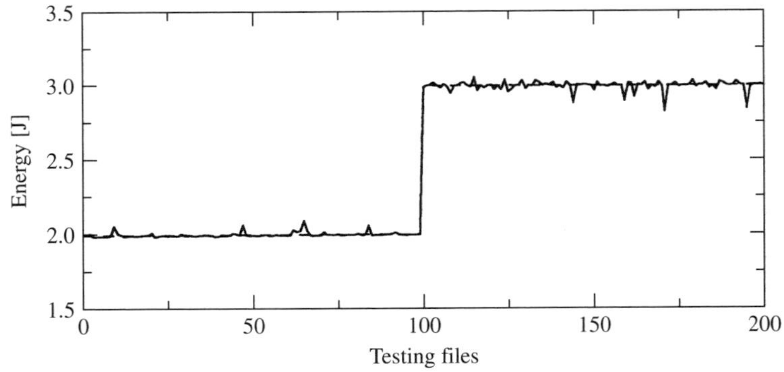

Figure 5.23 Comparison between the desired (dashed line) and actual network response (solid line) impact locations (Staszewski 1999)

- more complete picture of the monitored system;
- improved resolution;
- increased hypothesis discrimination;
- reduced measurement times.

Many different strategies are available for data fusion. The most common approach is the single sensor processing chain of Figure 5.1 replicated a number of times and fused together. An example strategy, which is usually called *central-level* or *centralised* fusion (Klein 1999) is to fuse the information provided from the data pre-processing step in the feature extraction step. Another approach could be a pattern-level fusion architecture where the feature extraction and selection are carried out for each sensor independently and the data are fused at the pattern recognition level. These strategies are illustrated in Figure 5.24. Various combinations of these strategies are possible in practice for different number of sensors. Various models of data fusion are available in practice. This includes the JDL, Boyd, Waterfall and Omnibus models, as described in (Klein 1999; Worden and Staszewski 2000). Probabilistic fusion is one of the most commonly used approaches. Usually the Bayesian probability is utilised. A damage detection example is presented in (Gros 1997). The problem concerns the detection of impact damage in composite materials. A sample was subjected to four methods of inspection following an impact: visual inspection, infrared inspection, radiographic inspection and eddy current inspection.

Dempster–Shafer Theory is a means of decision-fusion which is formulated in terms of probabilities but extends probability theory in a number of important respects. The method is based on the three fundamental propositions: *belief*, *doubt* and *uncertainty*. The basic idea of *belief* was introduced by Dempster in (Dempster 1967) and extended in Shafer's treatise (Shafer 1976). The basic model is formulated in similar terms to probability. Sensor evidence (e.g. measurements) can describe various events for observation (e.g. damage). Each event or union of events is assigned a degree of probability mass on the basis of sensor evidence. The difference between this *evidential* theory and probability theory is that the total probability mass need not be exhausted in the assignments to individual events. There is allowed to be a degree of *uncertainty*. The *belief* in an event B is the total probability which is committed to the support of the proposition that B has occurred. The *doubt* in the proposition B is the total support for the negation of the

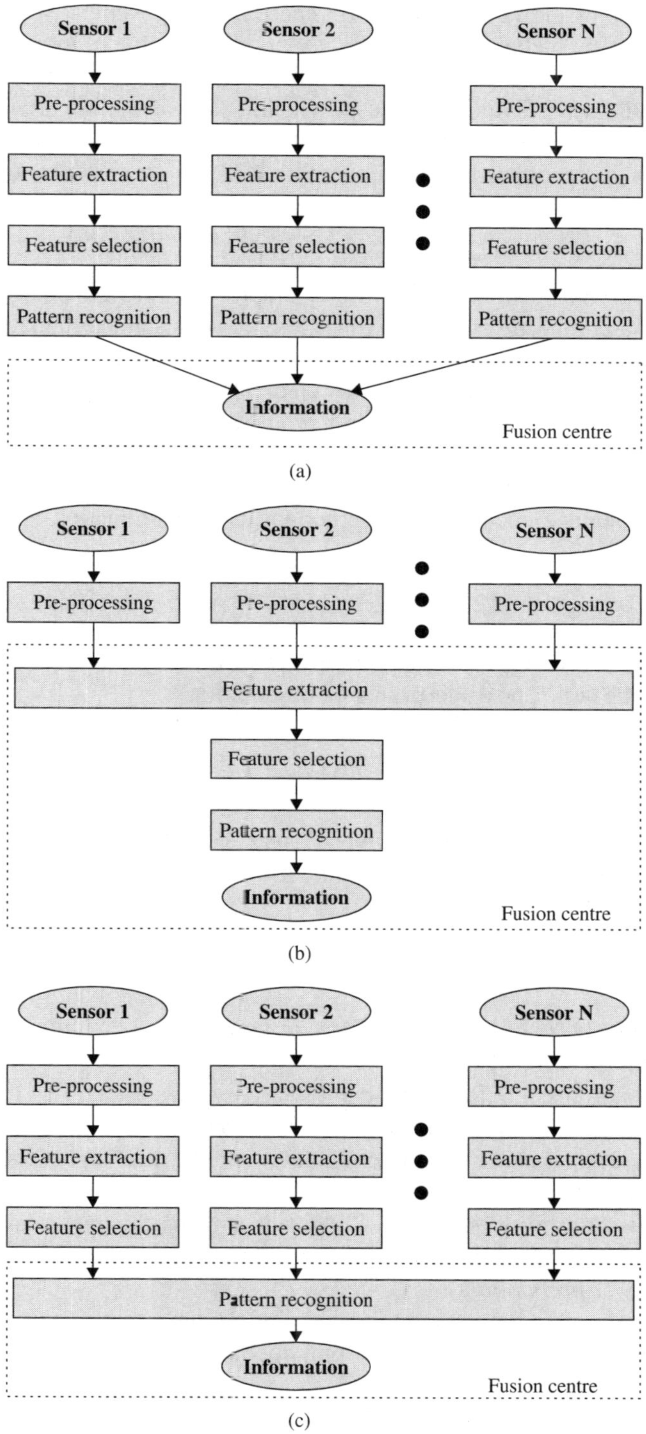

Figure 5.24 Data fusion strategies: (a) single sensor processing chain; (b) central-level processing; (c) pattern level processing

proposition B. One of the fundamental differences between Dempster-Shafer and probability theory is that the belief and doubt *do not necessarily sum to unity*. The uncertainty in the proposition B is that portion of the probability mass which does not support B or its negation. If further evidence were provided, some of the uncertainty could move in support of B but the mass assigned to the doubt cannot move. This means that the possible belief in B is bounded by the quantity called the *plausability*. In other words

$$\mathit{belief} + \mathit{uncertainty} = 1 - \mathit{doubt} = \mathit{plausability} \tag{5.49}$$

The entire idea can be illustrated using a simple example (Worden and Staszewski 2000). Consider a composite structure which may have sustained damage at one of two internal sites A and B which are indistinguishable. It is known that the only possible damage mechanism at site A is delamination (denoted D), but site B may fail by delamination or fibre fracture (denoted F) and the relative probabilities of the damage mechanisms are unknown. It is further known that failure at A is twice as likely as failure at B. What can one say about the likely damage type if a fault is found?

First of all, if damage occurs at A it is certainly by delamination and this forces the mass assignment,

$$m(D) = \frac{2}{3} \tag{5.50}$$

the remaining mass cannot be assigned with certainty, so it is assigned to the frame of discernment Θ (the set of possible events of observation),

$$m(\Theta) = m(D \cup F) = \frac{1}{3} \tag{5.51}$$

The belief in the delamination is simply *belief*$(D) = \frac{2}{3}$ as this is the only basic mass assignment to B. There is no such assignment to F so the belief *belief*$(F) = 0$. The plausibility in D is given by,

$$\mathit{plausability}(D) = m(D) + m(D \cup F) = 1 \tag{5.52}$$

and the plausibility of F is similarly calculated as $\frac{1}{3}$. The uncertainty interval for D is $[\frac{2}{3}, 1]$ and that for F is $[0, \frac{1}{3}]$.

Note that it is not possible to use probability theory here directly as the relative probabilities at site B are not known. It is possible to construct bounds on the probabilities though. Suppose delamination were impossible at site B, then the overall probability of delamination would be $\frac{2}{3}$ and this would be a lower bound. If delamination were certain at B, the overall probability would be 1. Note that these quantities are the belief and plausibility respectively. For this reason, the belief and plausibility are sometimes termed the *lower* and *upper probabilities*.

The interpretation of some common instances of the uncertainty interval for a proposition B is as follows (negation is denoted by a double underline)

- $[0, 0]$ B is impossible;
- $[1, 1]$ B is certain;
- $[0.75, 0.75]$ there is no uncertainty, B has a true probability of 0.75;
- $[0, 1]$ there is total ignorance regarding B;

- [0.25, 1] B is plausible, there is no support for $\underline{\underline{B}}$;
- [0, 0.75] $\underline{\underline{B}}$ is plausible, there is no support for B;
- [0.25, 0.75]tex both B and $\underline{\underline{B}}$ are plausible.

All this suffices to establish terminology, to explain how to compute belief functions and how to interpret the results. It does not provide a means of data fusion – that requires the use of *Dempster's combination rule*, as described in (Worden and Staszewski 2000).

5.16 OPTIMISED SENSOR DISTRIBUTIONS

Many types of sensor are available for damage detection. Some are localised, some are distributed, with their own advantages and disadvantages. Distributed sensing (see for example optical fibre sensors in Chapter 3) is often required for the location and assessment of structural damage. An optimised sensor distribution is one of the fundamental requirements in health and usage monitoring sensors. The problem is not only to determine optimal sensors but also to locate these sensors in the best possible positions. The problem of determining sensor locations is very much related to optimisation procedures. Recent years have shown considerable progress in this area; a number of methodologies have been proposed as summarised in (Staszewski and Worden 2001).

5.16.1 Informativeness of Sensors

A number of measures have been developed to assess the informativeness of features. These measures can be used as a criterion for both optimal selection and location of sensors. The methods based on information measures can reduce the dimensionality of the space by eliminating sensors (or locations of sensors) with low information content or with high redundancy with respect to other sensors. The mutual information derived from Information Theory (Shannon and Weaver 1959) assesses the information content of random variables and can be used as a criterion for feature selection (Battiti 1994).

The mutual information between a set C (severities of damage, e.g. crack lengths) and a set F (sensor parameters) can be defined as

$$I(C, F) = H(C) - H(C|F) \tag{5.53}$$

The entropy $H(C)$ and conditional entropy $H(C|F)$ in the above equation are given by

$$H(C) = -\sum_{C} P(C) log P(C) \tag{5.54}$$

and,

$$H(C|F) = -\sum_{F} P(F) \left(\sum_{C} P(C|F) log P(C|F) \right) \tag{5.55}$$

respectively. The entropy is a measure of randomness. The measure $P(C)$ is the probability of classes C and $P(C|F)$ is the conditional probability of classes C given knowledge of the feature input vector F.

This measure estimates arbitrary dependences between random variables and is capable of evaluating even nonlinear relationships between features and different damage classes.

It is always positive or zero; the later indicates that variables are statistically independent. The method is independent of the learning process. It has been successfully applied to optimal location of sensors (Wong and Staszewski 1998; Said and Staszewski 2000) and feature selection for a neural network based classification problem (Battiti 1994).

5.16.2 Optimal Sensor Location

There are two major problems associated with the optimal sensor location problems. The first problem is to find the best positions of sensors whereas the second is to establish the required number of sensors. Engineering expertise, knowledge and experience always come first before any signal processing attempt in the first problem. Designers and end-users of structures know better what are the critical areas which need to be analysed, controlled or monitored. Only then can intelligent signal processing help with the final adjustment of sensor positions. In contrast, the problem of optimal number of sensors very much relies on advanced signal processing techniques. Although, an exhaustive search of all possible combination of sensors can estimate relationships between sensors and show up any dominant (outlier) sensor, the approach is not practical. From the signal processing point of view, optimal sensor location is a problem of optimisation. A number of unconstrained (Newton methods) and constrained (linear and nonlinear programming) deterministic optimisation methods can be used for optimal sensor location, as reported in (Staszewski and Worden 2001). This include countless methods based on the availability of gradients and Hessians. Simple deterministic techniques are sufficient for local search; constrained optimisation has a great degree of complexity. Recent years have seen the development of combinatorial optimisation methods based on biological and physical analogies. These include Genetic Algorithms (GAs) (Goldberg 1999), Simulated Annealing (SA) (Otten and Van Ginneken 1989), Tabu Search (TS) (Glover 1989, 1990; De Werra and Hertz 1989) and a number of hybrids of these techniques.

The GA is the most widely used algorithm for optimal sensor location in damage detection problems. GAs use random selection algorithms to do a highly exploitative search through a parameter space. This space consists of individuals which are coded as finite-length strings over some type of alphabet based on genes. The procedure involves various types of coding. Often binary or integer numbers are used as genes. An initial population is selected randomly in the first step of the procedure. The number of individuals in a population depends on several parameters. This includes the size of each individual gene and the size of the solution space. In the next step individuals representing best solutions are chosen. The selection is based on a fitness function which operates on encoded individuals. A simple GA algorithm involves the following genetic operations:

- *reproduction* – an artificial version of natural selection; a process in which individual fittest individuals are copied to the next generation with probability proportional to their fitness;
- *crossover* – a method of combination between pairs of individuals in which the randomly chosen substrings from each individuals are switched;
- *mutation* – an operation involving random switching of genes in an individual;
- *new blood* – new entirely random individuals which form perturbations into the populations in order to prevent the population from stagnating;

- *elite* – a process in which the best solutions in a populations are copied automatically into the next population to prevent the loss of the fittest genetic material.

All these operations form an iterative process in which new generations are created until the population is dominated by a few relatively fit individuals.

An example, demonstrating a GA-based algorithm for optimal sensor location, can be found in (Staszewski *et al.* 1997d). The problem is to find an optimal location of sensors for a structure consisting of a rectangular 530 × 300 mm composite plate riveted to top flanges of four aluminium panels (Figure 5.25). The composite plate was instrumented with 17 piezoceramics (*PZT Sonox P5*, 15 mm × 15 mm) fixed on the lower surface of the plate, as shown in Figure 5.26. A series of impacts were applied on a composite panel. The strain data were recorded using a *DIFA SCADAS II* 24 channel measuring system running the *LMS 3.4.04* data acquisition software. For each impact, 8192 samples were recorded at a frequency of 25 kHz. Figure 5.27 shows an example of the strain data recorded. An MLP neural network was used for impact damage location. The input patterns to the network therefore constituted either 34 features (two features for each sensor). The maximum amplitude and the time-of-flight to the maximum amplitude were

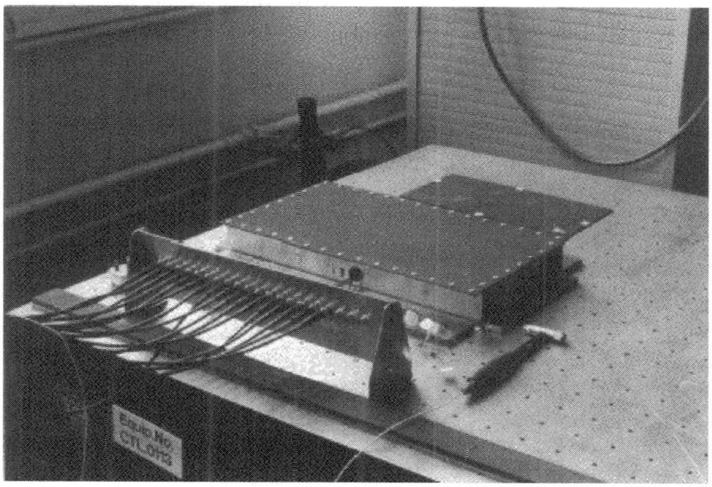

Figure 5.25 Composite structure used for optimal sensor location studies (Staszewski *et al.* 1997c)

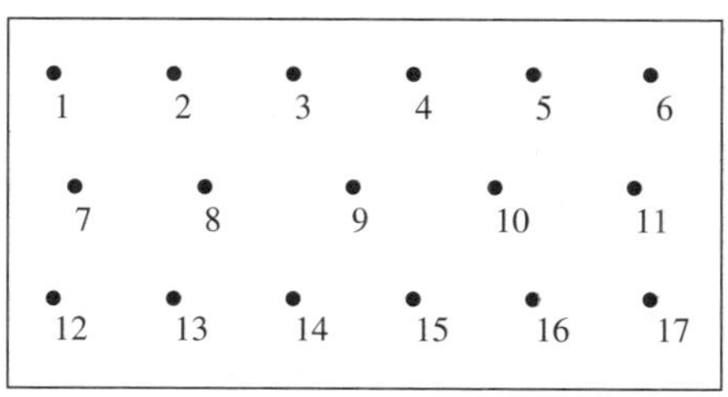

Figure 5.26 Sensor locations for the composite structure presented in Figure 5.25 (Staszewski *et al.* 1997c)

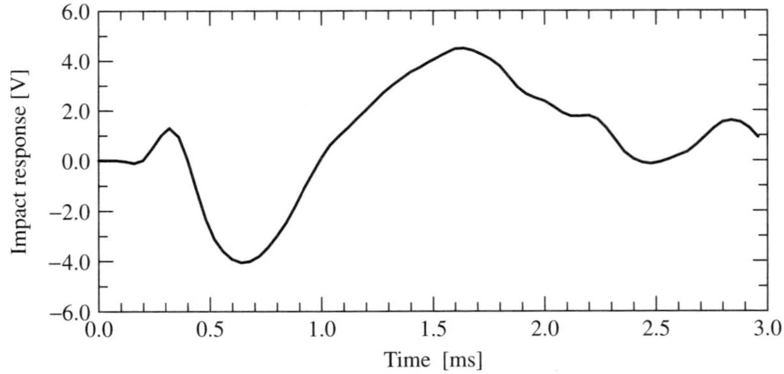

Figure 5.27 Example of impact strain data from piezoceramic sensors (Staszewski *et al.* 1997c)

used as the features. The network was required to return the x and y coordinates of the impact position. A GA algorithm was used to establish the best four sensors for the impact location problem. The sensor positions on the lower face of the panel were coded using strings of integers $1, 2, \ldots, 17$. With this coding, the individual (3, 15, 10 and 2) represents a distribution in which sensors are placed at positions 3, 15, 10 and 2. An initial population of 40 individuals was generated randomly. The inverse of the percentage location error from the network was used as the fitness measure for the GA. An elite of three individuals was used to prevent the loss of the fittest genetic material. The individuals of the initial populations were reproduced according to their fitness values to form the next generation. The simple GA used two basic genetic operations: crossover with probability 0.8 and mutation with probability 0.05. To prevent the population from stagnating, five entirely new random individuals were introduced at each generation. The GA was iterated until the population was dominated by a few relatively fit individuals. Up to 500 generations were allowed. Figure 5.28 shows the maximum and average fitness values obtained over a typical run. The best six-sensor distribution obtained was (1, 4, 9, 14, 15, 16). Figure 5.29 shows the graphical representation of the distribution.

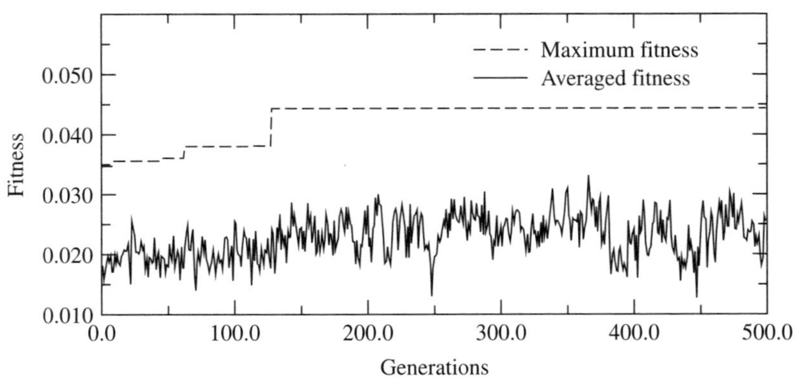

Figure 5.28 Maximum and averaged fitness values for the optimal sensor location procedure based on GAs (Staszewski *et al.* 1997c)

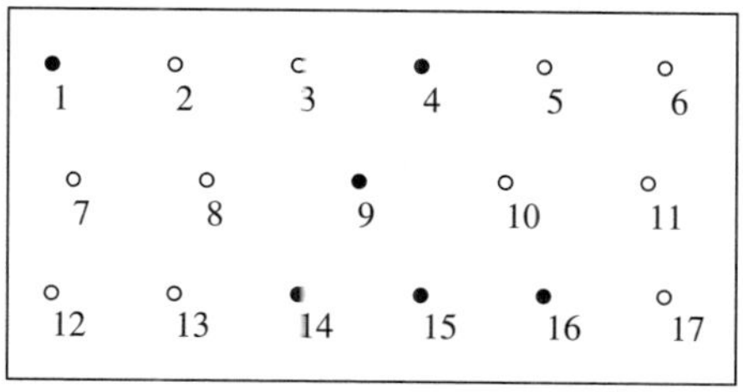

Figure 5.29 Optimal six-sensor distribution for the composite structure shown in Figure 5.25 (Staszewski *et al.* 1997c)

5.17 SENSOR VALIDATION

Multi-sensor architectures need to incorporate validation procedures which are important to detect sensor failures. There are two approaches available in practice, namely passive and active. In the active approach a signal (e.g. light in optical fibres or electric signal in piezoceramics) is sent through the sensor. The transmission characteristic of the sensor can then be compared to the expected response using a novelty index. In the passive approach the response probability distribution of the sensor is computed. Subsequent measurements are then taken to detect outliers which can indicate sensor failures. There exist a number of algorithms based on statistical analysis and neural networks. The neural network approach is similar to the novelty detection method described in Section 5.13.5. The original Kramer's approach (Kramer 1992) assembles the sensor values into a vector y. The auto-associative neural network is used to construct the nonlinear PCA mapping, i.e. decomposing the data into nonlinear PCA components and then reconstructing. Suppose – as a worst case – that one of the sensors fails completely and the corresponding component of the vector y is therefore zero, the de-mapping function will still be able to reconstruct part of the missing signals, the part which is correlated with the other sensor signals. However, the input value will not be close to the output value for that channel allowing for effective sensor validation. This strategy has been used for sensor validation in a ten degree-of-freedom dynamic system (Worden 2003).

5.18 CONCLUSIONS

Signal processing and computation are crucial elements in the implementation and operation of any damage identification system. A number of signal processing methodologies for damage detection have been discussed in this chapter. The material presented shows that various techniques are available for a multi-sensor architecture used in health and usage monitoring systems in aerospace structures. It is important for successful and reliable damage detection that such systems are designed and operated with signal processing techniques in mind at all times.

REFERENCES

Battiti, R. 1994. Using mutual information for selecting features in supervised neural network learning, *IEEE Transactions on Neural Networks*, Vol. 5, pp. 537–550.

Bishop, C.M. 1995. *Neural Networks for Pattern Recognition*, Oxford University Press.

Boashash, B. 1991. Time–frequency signal analysis. In: S. Haykin (ed.), *Advances in Spectrum Analysis and Array Processing I*, Prentice–Hall Advanced Reference Series in Engineering, Englewood Cliffs, New Jersey.

Boashash, B. 1992. *Time–Frequency Signal Analysis: Methods and Applications*, Longman Cheshire, Melbourne.

Broomhead, D.S. and Lowe, D. 1988. Multivariable functional interpolation and adaptive networks, *Complex Systems*, Vol. 2, pp. 321–355.

Carmona, R.A., Hwang, W.L. and Torresani, B. 1997. Characterization of signals by the ridges of their wavelet transforms, *IEEE Transactions on Signal Processing*, Vol. 45, No. 10, pp. 2586–2592.

Cempel, C. 1991. *Vibroacoustic Condition Monitoring*, Ellis Horwood, Chichester.

Choi, H.I. and Williams, W.J. 1989. Improved time–frequency representation of multicomponent signals using exponential kernals, *IEEE Transactions on Acoustics, Speech and Signal Processing*, Vol. ASSP 37, pp. 862–871.

Chui, C.H.K. 1992a. *Wavelet Analysis and its Applications – I: An Introduction to Wavelets*, Academic Press, Boston.

Chui, C.H.K. 1992b. *Wavelet Analysis and its Applications – II: A Tutorial in Theory and Applications*, Academic Press, Boston.

Claasen, T.A.C.M. and Mecklenbrauker, W.F.G. 1980a. The Wigner distribution – tool for time–frequency analysis – Part 1: continuous time signals, *Philips Journal of Research*, Vol. 35, pp. 217–250.

Claasen, T.A.C.M. and Mecklenbrauker, W.F.G. 1980b. The Wigner distribution – tool for time–frequency analysis – Part 2: discrete time signals, *Philips Journal of Research*, Vol. 35, pp. 276–300.

Cohen, L. 1989. Time–frequency distributions – a review, *Proceedings of the IEEE*, Vol. 77, pp. 941–981.

Coifman, R.R., Meyer, Y., Quake, S. and Wickerhauser, W.V. 1989. Signal processing and compression with wave packets, *Proceedings of the International Conference on Wavelets*, Masson Paris.

Daubechies, I. 1992. *Ten Lectures on Wavelets*, SIAM, Philadelphia.

De Werra, D. and Hertz, A. 1989. Tabu search techniques: A tutorial and application to neural networks, *OR Spectrum*, Vol. 11, pp. 131–140.

Dempster, A.P. 1967. Upper and lower probabilities induced by a multi-valued mapping, *Annals of Mathematical Statistics*, Vol. 38, pp. 325–339.

Donoho, D. and Johnstone, I. 1994a. Ideal denoising in an orthonormal basis chosen from a library of bases, *C.R. Acad. Sci. Paris, Série I*, Vol. 319, pp. 1317–1322.

Donoho, D. and Johnstone, I. 1994b. Ideal spatial adaptation via wavelet shrinkage, *Biometrika*, Vol. 81, pp. 425–455.

Esteban, J. and Starr, A.G. 1999. Building a data fusion model, *Proceedings of Eurofusion99 International Conference on Data Fusion, Stratford-upon-Avon*, pp. 187–196.

Flandrin, P. 1989. Some aspects of non-stationary signal processing with emphasis on time–frequency and time–scale methods. In: J.M. Combes, A. Grossman and Ph. Tchamitchian (eds), *Wavelets – Time–Frequency Methods and Phase Space*, Springer-Verlag, Berlin, Germany.

Forrester, B.D. 1990. Analysis of gear vibration in the time–frequency domain, *Proceedings of the 44th Meeting of the Mechanical Failures Group of the Vibration Institute, Virgina Beach, 3–5 April*, pp. 225–239.

Forrester, B.D. 1992. Time–frequency analysis in machine fault detection. In: B. Boashash (ed.), *Time–Frequency Signal Analysis*, Longman Cheshire, Melbourne.

Freeman, J.A. and Skapura, D.M. 1991. *Neural Netwroks: Algorithms, Applications and Programming Techniques*, Addison-Wesley.

Friedman, J.H. and Tukey, J.W. 1974. A projection pursuit algorithm for exploratory data analysis, *IEEE Transactions on Computing*, Vol. 23, pp. 881–889.

Glover, F. 1989. Tabu search – Part 1, *ORSA Journal of Computing*, Vol. 1, pp. 190–216.

Glover, F. 1990. Tabu search – Part 2, *ORSA Journal of Computing*, Vol. 2, pp. 4–32.

Goldberg, D.E. 1999. *Genetic Algorithms in Search, Optimization and Machine Learning*, Addison-Wesley Publishing Co. Inc., Reading, MA.

Gros, X.E. 1997. *NDT Data Fusion*, Arnold.

Hamming, R.W. 1989. *Digital Filters*, 3rd edition, Prentice–Hall, Englewood Cliffs, New Jersey.

Hammond, J.K. 1997. *Mechanical Systems and Signal Processing – Special Issue on Time–Frequency Analysis*, Vol. 11.

Haykin, S. 1994. *Neural Networks: A Comprehensive Foundation*, Macmillan College Publishing Company, New York.

Hertz, J., Krogh, A. and Palmer, R.G. 1991. *Introduction to the Theory of Neural Computation*, Addison-Wesley.

Hopfield, J.J. 1984. Neural networks and physical systems emergent collective computational abilities, *Proceedings of the National Academy of Sciences*, Vol. 52, USA, pp. 2554–2558.

Kim, J. and Mueller, C.W. 1978. *Factor Analysis: Statistical Methods and Practical Issues*, Sage Publications, Beverly Hills, CA.

Klein, L.A. 1999. *Sensor and Data Fusion: Concepts and Applications*, 2nd edition, TT14, SPIE Press.

Kramer, M.A. 1992. Autoassociative neural networks, *Computers in Chemical Engineering*, Vol 16, pp. 313–328.

Lee, B.C., Manson, G. and Staszewski, W.J. 2003. Environmental effects on Lamb wave responses from piezoceramic sensors, *Proceedings of the 5th International Conference on Modern Practice in Stress and Vibration Analysis*, Glasgow, Scotland, 9–11 September.

Luo, M.A. and Kay, M.G. 1992. Data fusion and sensor integration: state of the art 1990s. In: M.A. Abidi and R.C. Gonzalez (eds), *Data Fusion in Robotics and Machine Intelligence*, Academic Press.

Mallat, S. 1999. *A Wavelet Tour of Signal Processing*, Academic Press, San Diego.

McCulloch, W.S. and Pitts, W. 1943. A logical calculus of the ideas immanent in nervous activity, *Bulletin of Mathematical Biophysics*, Vol. 5, pp. 115–133.

McFadden, P.D. and Wang, W. 1990. Time–frequency domain analysis of vibration signals for machinery diagnostics: *1 – Introduction to the Wigner–Ville distribution, Research Report OUEL 1859/90*, Department of Engineering, Oxford University.

Middleton, D. 1960. *An Introduction to Statistical Communication Theory*, McGraw–Hill, New York.

Minsky, M.L. and Papert, S.A. 1988. *Perceptrons*, MIT Press, Cambridge, MS.

Otten, R.H.J.M. and Van Ginneken, L.P.P.P. 1989. *The Annealing Algorithm*, Kluwer, Boston.

Press, W.H., Teukolsky, S.A., Vetterling, W.T. and Flannery, B.P. 1992. *Numerical Recipes in C*, Cambridge University Press, Cambridge.

Priestley, M.B. 1988. *Non-Linear and Non-Stationary Time Series Analysis*, Academic Press, London.

Rabbani, M. 1991. *Digital Image Compression*, SPIE Optical Engineering Press, Bellingham, WA.

Ramón y Cajál, S. 1911. *Histologie du Systéme Nerveux de l'Homme et des Vertébrés*, Maloine, Paris.

Randall, R.B. 1987. *Frequency Analysis*, 3rd edition, Brüel & Kjæl Publications, Nærum, Denmark.

Rosenblatt, F. 1962. *Principles of Neurodynamics*, Spartan, New York.

Rumelhart, D.E. and McClelland, J.L. 1998. *Parallel Distributed Processing: Explorations in the Microstructure of Cognition*, MIT Press, Cambridge, MS.

Rytter, A. 1993. Vibration based inspection of civil engineering structures, PhD thesis, Department of Building Technology and Structural Engineering, University of Aalborg, Denmark.

Said, W.M.I. and Staszewski, W.J. 2000. Optimal sensor location for damage detection using mutual information, *Proceedings of the 11th International Conference on Adaptive Structures and Technologies (ICAST), Nagoya, Japan*, 23–26 October, pp. 428–435.

Schalkoff, R. 1992. *Pattern Recognition: Statistical, Structural and Neural Approaches*, John Wiley & Sons, Inc., New York.

Shafer, G. 1976. *A Mathematical Theory of Evidence*, Princeton University Press.

Shannon, C.E. and Weaver, W. 1959. *The Mathematical Theory of Communication*, University of Illinois Press.

Sharkey, A.J.C. 1996. On combining artificial neural networks, *Connection Science*, Vol. 8, pp. 314–337.

Staszewski, W.J. 1994. The application of time–variant analysis to gearbox fault detection, PhD thesis, Department of Mechanical Engineering, University of Manchester.

Staszewski, W.J. 1996. Optimal sensor distribution for fault diagnostics – Part V: detection of impacts on a FE composite panel, Report No. DRG-BAe-2/96, Dynamics Research Group, Department of Mechanical Engineering, University of Sheffield.

Staszewski, W.J. 1997. Data compression with optimal wavelet coefficients, *Proceedings of the 2nd IEE/IEEE International Conference on Genetic Algorithms in Engineering Systems: Innovations and Applications, Glasgow, Scotland, 2–4 September.*

Staszewski, W.J. 1998a. Structural and mechanical damage detection using wavelets, *The Shock and Vibration Digest*, Vol. 30, No. 6, pp. 457–472.

Staszewski, W.J. 1998b. Identification of non-linear systems using multi-scale ridges and skeletons of the wavelet transform, *Journal of Sound and Vibration*, Vol. 214, No. 4, pp. 639–658.

Staszewski, W.J. 1998c. Wavelet based compression and feature selection for vibration analysis, *Journal of Sound and Vibration*, Vol. 211, pp. 735–760.

Staszewski, W.J. 1999. Damage trend prediction, Brite-Euram EC Report, MONITOR (Monitoring On-Line Integrated Technologies for Operational Reliabilities), Project No. BE95-1524, Report NO. MON-UoS-WP3-47, Department of Engineering, Sheffield University.

Staszewski, W.J. 2000a. Advanced data pre-processing for damage identification based on pattern recognition, *International Journal of Systems Science*, Vol. 31, No. 11, pp. 1381–1396.

Staszewski, W.J. 2000b. *Wavelets for Mechanical and Structural Damage Indentification. In series: Studia i Materialy*, published by the Polish Academy of Sciences Press, Gdansk, Monograph No. 510/1469/2000.

Staszewski, W.J. and Tomlinson, G.R. 1993. Time–variant methods in machinery diagnostics. In: H.G. Natke, G.R. Tomlinson and J.T.P. Yao (eds), *Safety Evaluation Based on Identification Approaches Related to Time–Variant and Nonlinear Structures*, Vieweg, Braunschweig/Wisbaden.

Staszewski, W.J. and Worden, K. 2001. An overview of optimal sensor location methods for damage detection, *Proceedings of the SPIE's 8th International Symposium on Smart Structures and Materials, Conference on Modeling, Signal Processing and Control in Smart Structures*, Newport Beach, California, 4–8 March, Vol. 4326, No. 21.

Staszewski, W.J., Pierce, S.G., Worden, K., Philp, W.R, Tomlinson, G.R. and Culshaw, B. 1997a. Wavelet signal processing for enhanced Lamb wave defect detection in composite plates using optical fibre detection, *Optical Engineering*, Vol. 36, No. 7, pp. 1877–1888.

Staszewski, W.J., Worden, K. and Jacob, P.J. 1997b. Effect of data compression on fault detection networks, *Proceedings of International Conference on Engineering Applications of Neural Networks – EANN '97, Stockholm, Sweden*, 16–18 *September*.

Staszewski, W.J., Worden, K., Tomlinson, G.R. and Ball, A. 1997c. Optimal sensor locations for impact detection on a composite panel, *Proceedings of the International Workshop on Damage Assessment Using Advanced Signal Processing Procedures – DAMAS 97, Sheffield, UK*, 30 *June*–2 *July*.

Staszewski, W.J., Biemans, C., Boller, C. and Tomlinson, G.R. 1999a. Crack propagation monitoring in metallic structures, *International Conference on Smart Materials, Structures and Systems, Bangalore, India*, 7–10 *July*, pp. 532–541.

Staszewski, W.J., Biemans, C., Boller, C. and Tomlinson, G.R. 1999b. Impact damage detection in composite structures – recent advances, *Proceedings of the 2nd International Workshop on Structural Health Monitoring, Stanford, California*, 8–10 *September*, pp. 754–763.

Staszewski, W.J., Read, I. and Foote, P. 2000. Damage detection in composite materials using optical fibres – recent advances in signal processing, *Proceedings of the SPIE's 7th International Symposium on Smart Structures and Materials, Conference on Smart Structures and Integrated Systems, Newport Beach, California*, 5–9 *March*.

Sweldens, W. 1995. Wavelets, signal compression and image processing. In: A. Fournier (ed.), *Wavelets and Their Applications in Computer Graphics*, SIGGRAPH '95 course notes, University of British Columbia.

Tarassenko, L., Hayton, P., Cerneaz, Z. and Brady, M. 1995. Novelty detection for the identification of masses in mammograms, *Proceedings of the 4th International Conference on Artificial Neural Networks, Cambridge*, pp. 442–447.

Wang, W.J. 1993. Gearbox condition monitoring and early damage diagnosis by two and three dimensional vibration signal analysis, PhD thesis, Department of Engineering Science, Oxford University.

Wang, W.J. and McFadden, P.D. 1993. Early detection of gear failure by time–frequency analysis – Part 1: calculation of the time–frequency distribution, *Mechanical Systems and Signal Processing*, Vol. 7, pp. 193–204.

Wickerhauser, M.V. 1992. Acoustic signal compression with wavelet packets. In: Ch.K. Chui (ed.), *Wavelets: A Tutorial in Theory and Applications*, Academic Press, Boston.

Widrow, B. and Hoff, M.E. 1960. Adaptive switching circuits, *IRE Wescon Conv. Rec.*, Part 4, pp. 96–104.

Wong, C.S. and Staszewski, W.J. 1998. Mutual information approach to sensor location for impact damage detection of composite structures, *Proceedings of the International Conference on Noise and Vibration Engineering, ISMA* 23, *Leuven, Belgium*, 16–18 September, pp. 1417–1422.

Worden, K. 1997a. Structural fault detection using a novelty measure, *Journal of Sound and Vibration*, Vol. 201, pp. 85–101.

Worden, K. 1997b. Damage detection using a novelty measure, *Proceedings of the 15th International Modal Analysis Conference, Orlando*, pp. 631–637.

Worden, K. 1997c. *MLP – Multi-Layer Perceptron*, Version 3.4 – A Users Manual.

Worden, K. 2003. Sensor validation and correction using autoassociative neural networks and principle component analysis, *Proceedings of the IMAC XXI Conference, Kissimmee, Florida*.

Worden, K. and Staszewski, W.J. 2000. *Data fusion for smart sensors, technical report for DERA*, Department of Mechanical Engineering, Sheffield University.

Worden, K., Staszewski, W.J. and Tomlinson, G.R. 1997. Smart systems – the role of signal processing, *Proceedings of CEAS (Confederation of European Aerospace Societies) International Forum on Aeroelasticity and Structural Dynamics, Rome, Italy*, 17–20 June.

6

Structural Health Monitoring Evaluation Tests

P.A. Lloyd[1], R. Pressland[2], J. McFeat[3], I. Read[4], P. Foote[4], J.P. Dupuis[5], E. O'Brien[2], L. Reithler[5], S. Grondel[6], C. Delebarre[6], K. Levin[7], C. Boller[8], C. Biemans[9] and W.J. Staszewski[8]

[1] *DSTL, Farnborough, UK*
[2] *Airbus UK, Filton, UK*
[3] *BAE SYSTEMS, Military Aircraft, Warton, UK*
[4] *BAE SYSTEMS, Sowerby Research Centre, Filton, UK*
[5] *European Aeronautic Defence and Space (EADS) Company, Corporate Research Centre, Suresnes, France*
[6] *IEMN, Valenciennes, France*
[7] *Aeronautical Research Institute of Sweden (FFA), Bromma, Sweden*
[8] *Department of Mechanical Engineering, Sheffield University, Sheffield, UK*
[9] *DaimlerChrysler, Berlin, Germany*

6.1 INTRODUCTION

Large structural evaluators provide the opportunity for the direct assessment of the performance of several damage detection techniques under identical conditions. They also provide a mechanism for assessing the enhancement in maturity and capability of damage detection and Operational Loads Monitoring (OLM) techniques.

This chapter demonstrates the performance of the damage detection and OLM technologies, described in Chapters 3 and 4 respectively, in large complex structures and within

Health Monitoring of Aerospace Structures – Smart Sensor Technologies and Signal Processing.
Edited by W.J. Staszewski, C. Boller and G.R. Tomlinson ISBN: 0-470-84340-3

an operational environment. The tests are done as a mixture of ground-based evaluators and flight tests. The damage detection technologies use the ground-based evaluators, since the introduction of damage can compromise structural integrity, which is significant issue for flight tests.

The metallic evaluator was a riveted aluminium specimen and the composite evaluator was a wing-box structure. Flight tests were also performed using a civil aircraft as a testbed platform. Various monitoring technologies were used to detect and locate the damage. These include: embedded fibre optic sensors, Bragg grating fibre optic system, piezoceramic sensors and Acoustic Emission system.

An important feature of the assessment strategy was that in different cases a common test bench was provided on which all structural health monitoring techniques could be assessed simultaneously. This was provided in a way that all partners performed the test on the same type of specimen, manufactured at one location only, and under the same prescribed testing conditions. In some cases, and specifically where the test bench was very costly, such as the composite wing box or the test aircraft itself, all structural health monitoring techniques involved had to be implemented on the test bench on a back to back basis, while this principle was partially also used on the evaluator for the metallic specimens.

6.2 LARGE-SCALE METALLIC EVALUATOR

6.2.1 Lamb Wave Results from Riveted Metallic Specimens

The specimens used were three rectangular ($750 \times 300 \times 2$ mm) aluminium multi-rivet butt strapped metallic panels shown schematically in Figure 6.1. The plates were instrumented with two different types of low profile, thin piezoceramic transducers. The first set of transducers was bonded in a symmetrical configuration 120 mm below and above the middle rivet, as shown in Figure 6.1a. The second set of transducers was located between the rivet lines as shown in Figure 6.1b. Here, the metallic specimen is fully broken after completed fatigue test. The constant amplitude fatigue loading and a truncated FALSTAFF random loading were used in the tests to introduce and grow cracks in the specimen. FALSTAFF is a standardised fatigue test loading spectrum representing the lower wing root loading of a fighter aircraft. The truncation of the load sequence was necessary to avoid buckling of the specimens. Figure 6.2 shows an example of the crack propagation curve for the analysed specimens.

As described in Chapter 4, one of the general rules of ultrasound emission is to excite the actuating element at its natural resonances rather than at any frequency. This method enables a very efficient conversion from electrical to mechanical energy. A two-dimensional numerical model has been developed using the finite element code, ATILA (Hamonic *et al.* 1990) to allow modal and harmonic analysis of the piezoelectric transducer. From this analysis, four natural vibration modes were found ranging from 100 kHz to 1 MHz. The second mode at 400 kHz appears to be the best electromechanically coupled ($ke = 43\,\%$) and the computation of the displacement fields (Figure 6.3) has enabled its identification as a transverse mode.

The instrumentation used is shown schematically in Figure 6.4. A pulse from the pulse generator was used to simultaneously trigger the oscilloscope (HP 54522A 500 MHz) and

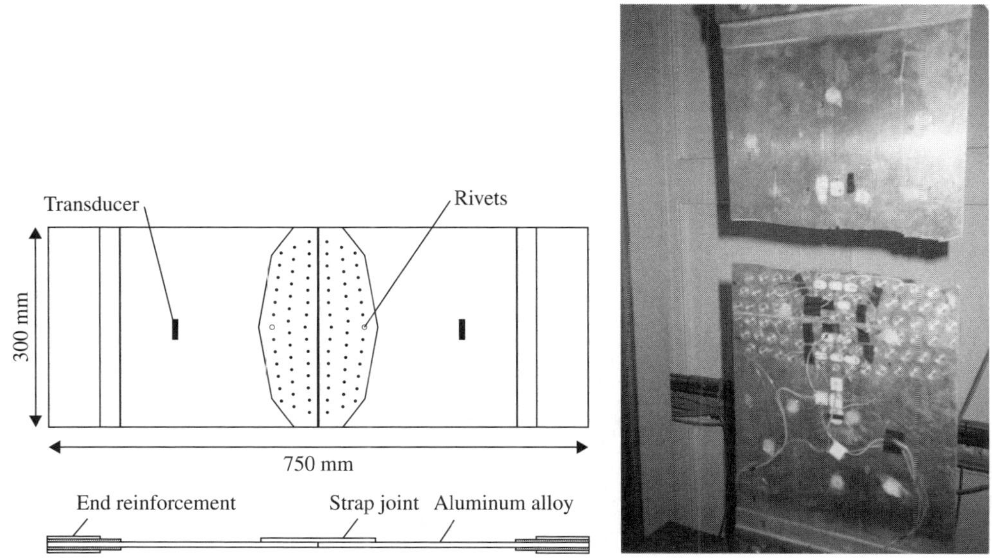

Figure 6.1 Aluminium multi-rivet butt strapped metallic panel: (a) schematic diagram; (b) fully broken after completed fatigue test

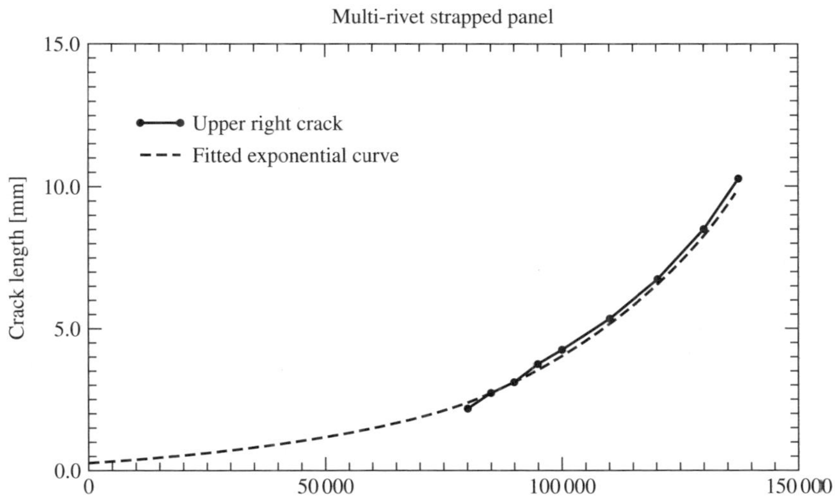

Figure 6.2 Example of crack propagation curve for the specimen shown in Figure 6.1

an arbitrary function generator (HP 33120A) that delivered a 5-cycle sinusoidal tone burst signal at 400 kHz (transducer transverse resonance).

Lamb waves excited by the transmitting transducer propagated along the plate towards the riveted area. Then the strap joint ensured transmission of a part of the incident-wave energy to the receiving transducer. Finally, the received signal was amplified, filtered and

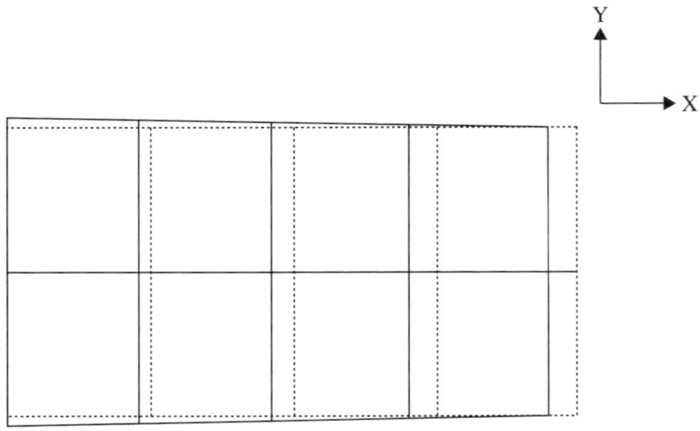

Figure 6.3 Transverse vibration mode of the piezoceramic transducer

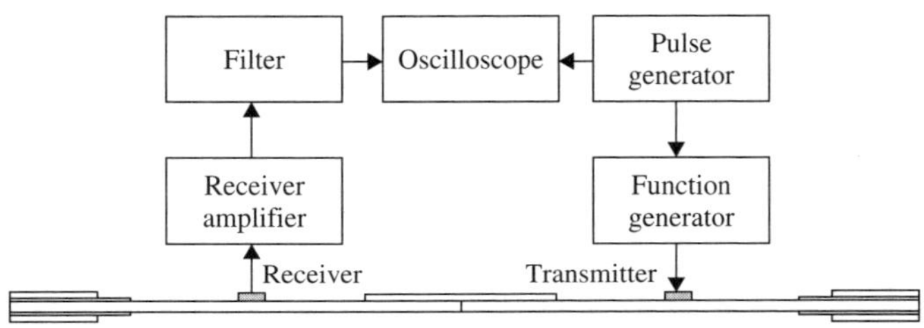

Figure 6.4 Instrumentation used for the crack detection technique based on Lamb waves

transferred to a digital oscilloscope, which averaged 64 signals to improve the signal-to-noise ratio.

In order to confirm that waves propagating in the aluminium riveted specimens were in fact Lamb wave modes, analytical curves of dispersion, which illustrate the plate-mode phase velocity as a function of the frequency-thickness, were computed. Figure 6.5 shows an example of the dispersion curve obtained from this analytical model in the case of an aluminium plate with a 2 mm thickness; it allows concluding that only the A_0 (antisymmetric) and S_0 (symmetric) modes can propagate at this frequency. In addition to that, experimental measurements of group velocity (time of flight) have confirmed this result. One specimen was tested back-to-back with Acoustic Emission equipment. This testing concluded that Acoustic Emission and Lamb waves detection technologies were compatible under prescribed protocols. The protocol adopted was to use the Lamb wave system at discrete intervals in order that the Acoustic Emission transducers do not 'pick-up' Lamb wave transmissions. Concerning the crack detection, visual inspection of the specimen was carried out and recorded every 10 000 cycles. Moreover X-ray inspection was used after each macro crack was detected.

Lamb wave signals resulting from the tests were composed of multiple packets of the fundamental lamb wave S_0 and A_0 modes due to the complexity of the interrogated structure. It was impossible to separate and 'individually' identify these modes. The Lamb wave packets also exhibited variations of the amplitude for different cycle numbers at the crack propagation curve. Therefore various signal processing techniques, described in Chapter 5, were used to analyse this data.

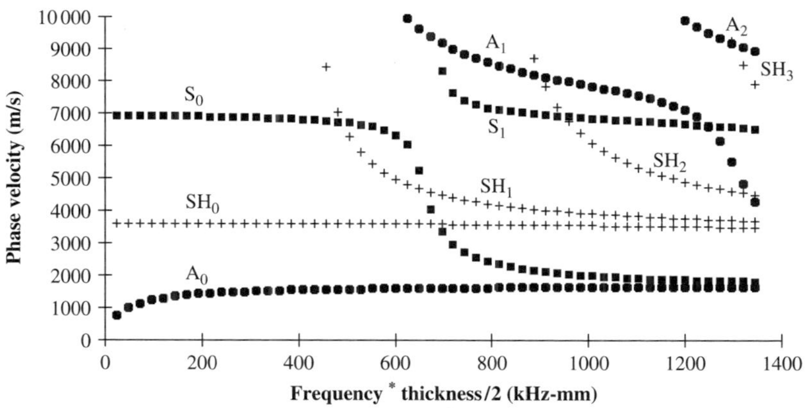

Figure 6.5 Example of the dispersion curve obtained from this analytical model in the case of an aluminium plate with a 2 mm thickness

A possible method for better understanding these waveform changes can utilise the instantaneous wave characteristics, namely envelope and phase functions. The idea developed here was to follow the delay variation in the time of flight of the Lamb wave packets, i.e. the change in the group velocity due to the interaction between the defects and the Lamb waves. Figure 6.6 presents the delay evolution of the maximum value of the envelope's first derivative as a function of the cycle number for the two different fatigue loadings respectively. Although each plate has been submitted to a different fatigue loading, it has turned out that some similar characteristics were obtained for both cyclical loadings. Indeed, delay variations of the order of 10 microseconds have been observed for the Lamb waves signal at the same time as macro cracks emergence (1 cm length) in both cases of fatigue loading. Furthermore, precise information on micro crack development has been obtained by using X-rays. The variations of the delay corresponding to these micro cracks were of the order of one microsecond. These changes were in agreement with the information obtained from the Lamb wave analysis.

The second signal processing approach was based on the wavelet analysis described in Chapter 5. The wavelet damage index was calculated for the Lamb wave data (shown in Figure 6.7) representing a crack propagating in the riveted specimens. Figure 6.8 shows an example of the damage index plotted as a function of crack length.

The studies presented in this section show the feasibility of Lamb wave testing for crack detection and monitoring in complex metallic structures. The performance of piezoceramic sensors was satisfactory; no disbond or loss of coupling was observed.

6.2.2 Acoustic Emission Results from a Full-Scale Fatigue Test

The type of equipment developed and reported in this section should be more accurately regarded as modified Acoustic Emission. The traditional approach to Acoustic Emission usually results in extremely large amounts of data that requires complex post processing which generally leads to a preference for shorter periods of surveillance.

The modified Acoustic Emission system *BALRUE* (Figure 6.9) developed jointly by *Airbus UK* and *Lloyds Register of Shipping* and built by *Ultra Electronics* is radically

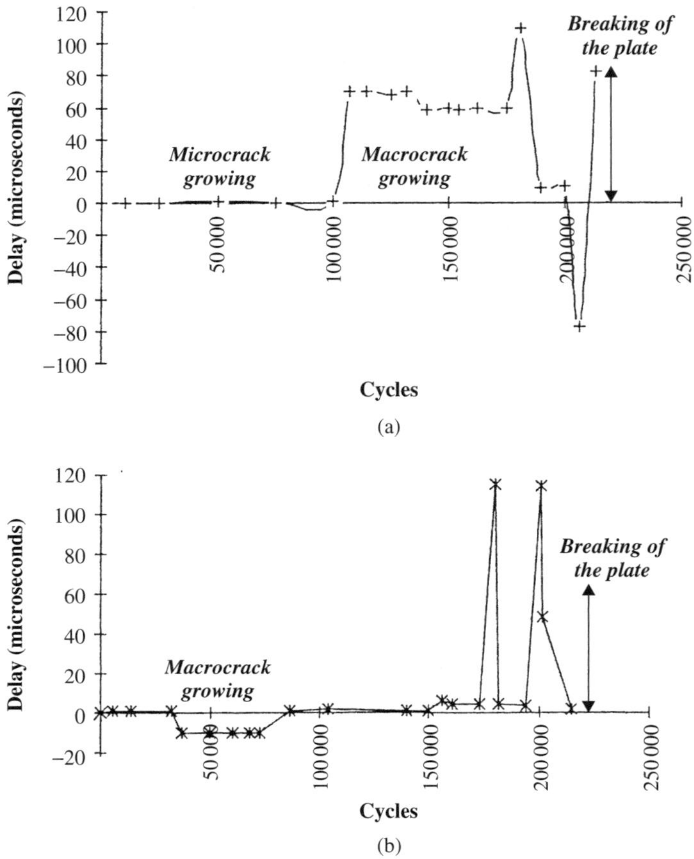

Figure 6.6 Lamb wave delay as a function of fatigue cycles: (a) constant amplitude loading; (b) Falstaff loading

different from conventional acoustic emission in that it demands continuous monitoring. The System contains phenomenological filters built into the hard- and firmware. These perform dramatic data reduction resulting in positive identification and location of crack growth in fatigue loaded structures. The method is appropriate to structures designed on damage tolerance principles. In composite materials the fretting at the periphery of the damage is detected and located although the sensors need to be closer than on metallic structures. Subsequent to the successful application of the System on the A340 Landing Gear Support Structure (Figure 6.10) the system was deployed on the full scale fatigue test of the A340-600 (Figure 6.11) with the specimen being located in Germany and the data remotely acquired in the UK. Miniature piezoceramic sensors tuned to 300 KHz were applied to the inner wing – 24 sensors were sufficient to survey the component which was under continuous fatigue loading representing flight conditions. Approximately 50 flight cycles after nucleation of a crack were required to unambiguously establish the presence of crack growth and location of the crack tip.

During a year of surveillance all the damages that were detected by conventional NDT were also detected by the *BALRUE* System. In addition a number of sites in the early

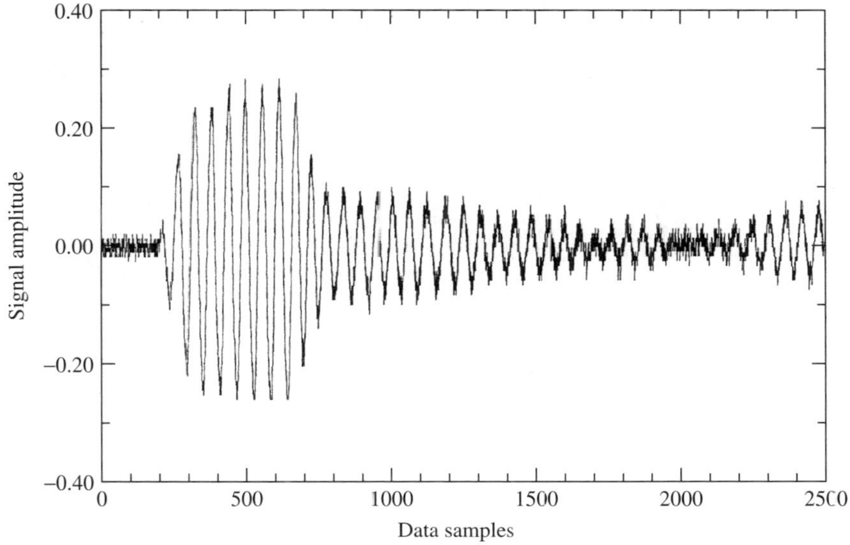

Figure 6.7 Example of Lamb wave data

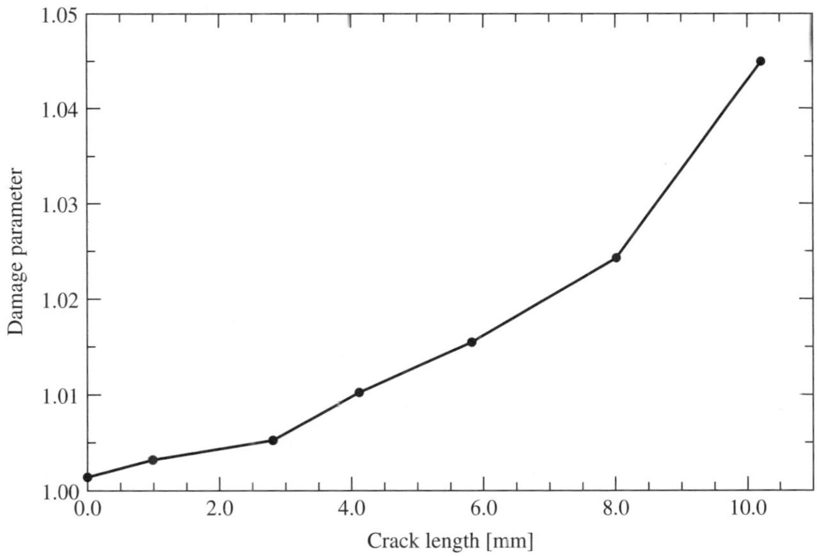

Figure 6.8 Damage index plotted as a function crack length

stages of fretting and crack growth were detected ahead of the NDT inspections. The surveillance area is established by the circle that encloses three adjacent sensors. This also enables location of the crack tip by triangulation using methods such as the Tobias algorithm. The system is ready for upgrading to flight-worthy status to enable in-service SHM should it be required.

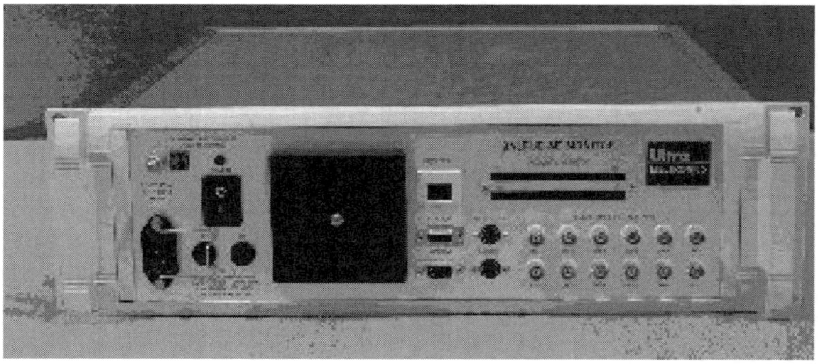

Figure 6.9 Modified Acoustic Emission system BALRUE

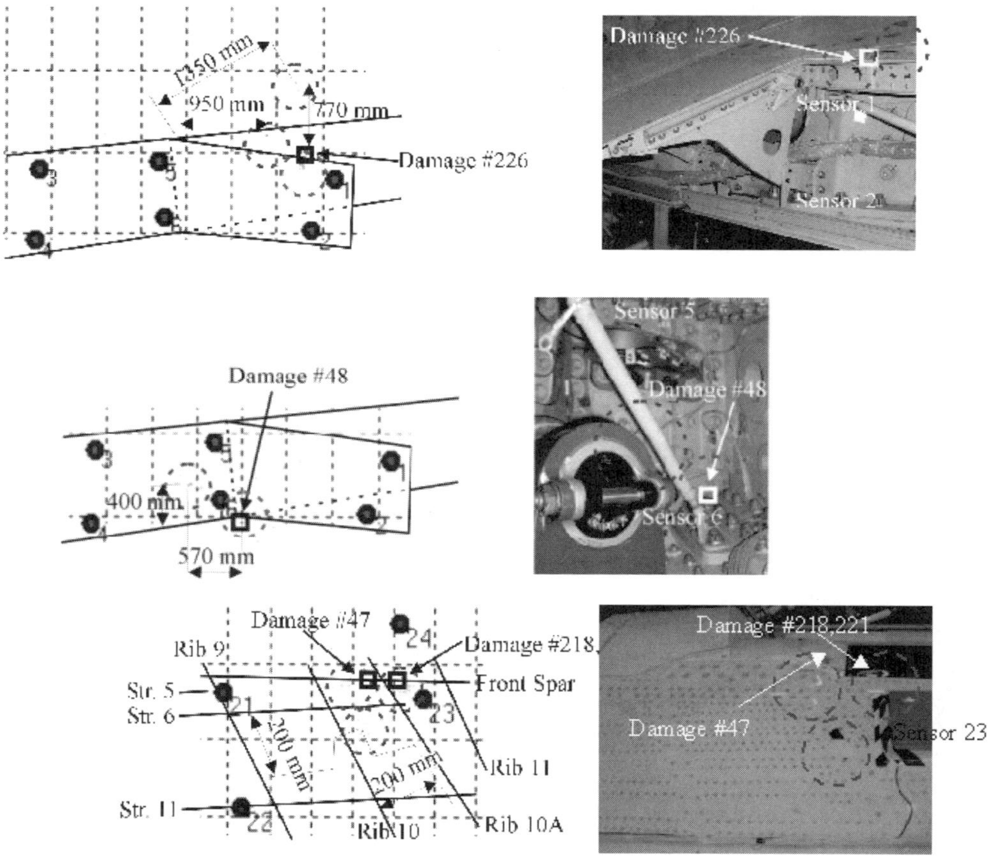

Figure 6.10 Application of the BALRUE system on the A340 Landing Gear Support Structure

Figure 6.11 Application of the BALRUE system on inner wing of the A340-600 aircraft

6.3 LARGE-SCALE COMPOSITE EVALUATOR

6.3.1 Test Article

The composite evaluator was a wing-box structure. It comprises carbon fibre composite skins with bonded stringers, attached to a metallic sub-structure. The skins were made from 913C-HTA composite material. The skins incorporate blocked pairs of 0°-plies into which optical fibres or other sensors can be embedded with minimal effect on structural performance. These 0°-plies also allow optical fibres or other connections to emerge from the ends of the wing-box. The skins have four bonded stringers. The stringer lay-up was $[+45, -45, 0_2]_s$. This gives a laminate thickness of 1 mm. The stringers were formed around foam cores. Stringer height was 30 mm; width 20 mm and the feet of the stringers were 10 mm wide. The stringers were bonded onto the skin using *REDUX 319A* structural adhesive.

The skins were mounted onto metallic substructure consisting of three metallic spars. The partially completed wing-box is shown in Figure 6.12, where the internal construction can be seen. The wing-box skin was bolted onto the substructure. This provided a simple

Figure 6.12 Partly completed wing-box showing internal structure

Figure 6.13 Wing-box skin bolted onto substructure

and economic method, which allows alternative skins to be fitted. Bolts were used rather than countersunk fasteners, because there was no requirement to have flush fasteners. The fitted skin is shown in Figure 6.13.

6.3.2 Sensor and Specimen Integration

The skins were instrumented with electrical strain gauges. The external and internal gauge positions are shown in Figure 6.14. The strain gauges used were *HBM 6/120LY11*.

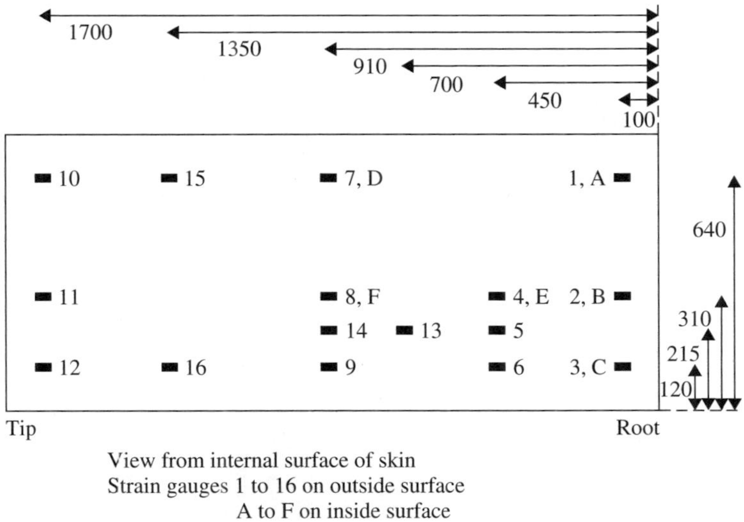

Figure 6.14 Position of strain gauges on wing-box skin

Two types of optical fibre sensors were also integrated into the wing-box skin during manufacture. Bragg grating sensors were embedded for the impact detection system, and lengths of Hi-Bi fibres for the damage detection system. The details of the sensor layouts for both systems will be described in Sections 6.3.4 and 6.3.5.

In Figure 6.15, the cured skin can be seen on the autoclave bedplate. The optical fibre sensors emerged from the free edges at the tip and root, where they were protected by small bore *PTFE* tubing. After cure, the skin was trimmed along its two long edges, and then drilled, using a template, for the bolts that would attach it to the wing-box substructure. The skin was then ready for the installation of conventional electrical strain gauges used for additional control of the test only and the other damage detection sensors used in the tests. The skin with these additional sensors installed is shown in Figure 6.16.

After completion, the wing-box skin was temporarily attached to its substructure and assembled into the test rig. Aluminium plates were bonded onto the root end of the skin to provide good load transfer into the test rig. A shimming adhesive was used to ensure a good fit between the skin and the rig root fittings. After cure, the rig fittings were removed and the aluminium plates drilled to receive the mounting bolts. After cleaning, the root fittings were re-fitted and the skin fitted to the substructure. The root fittings were designed to allow the embedded optical fibre sensors to be routed to the interrogation equipment. A view of the root fittings is shown in Figure 6.17.

At the tip end a loading bar was attached to the wing-box. This also allowed the optical fibre sensors to be connected to the interrogating equipment. The loading bar was the interface between the wing-box and the hydraulic jacks used to load the structure. This arrangement can be seen in Figure 6.18, which shows a view from the tip end of the structure.

The wing-box was installed in the test rig, fixed as a cantilever, and loaded by two hydraulic jacks at the tip. A number of additional displacement transducers used for control of the test were fitted as well. The hydraulic loading rams incorporated load cells

Figure 6.15 Wing-box skin on autoclave bedplate

Figure 6.16 Underside of skin with sensors installed

so that loading could be monitored during the test. The test instrumentation was connected to data logging equipment. The instrumentation consisted of the strain gauge fit (to top and bottom skins), load cells and displacement transducers (LVDTs). The arrangement of the LVDTs is shown in Figure 6.19. The displacement transducers fitted to the top skin, at the root, were attached to the loading frame. The displacement transducers fitted to the

Figure 6.17 Wing-box root-end fittings

Figure 6.18 View of wing-box loading arrangements

bottom skin were attached to an additional scaffolding framework. This can be seen in Figure 6.20. The displacement transducers and the load cells were calibrated before fitting.

The final part of the testing rig to be fitted was an instrumented impact tower. This enabled controlled impacts to be made at any location on the top skin of the wing-box. The drop weight was instrumented to enable force/time histories to be recorded via a

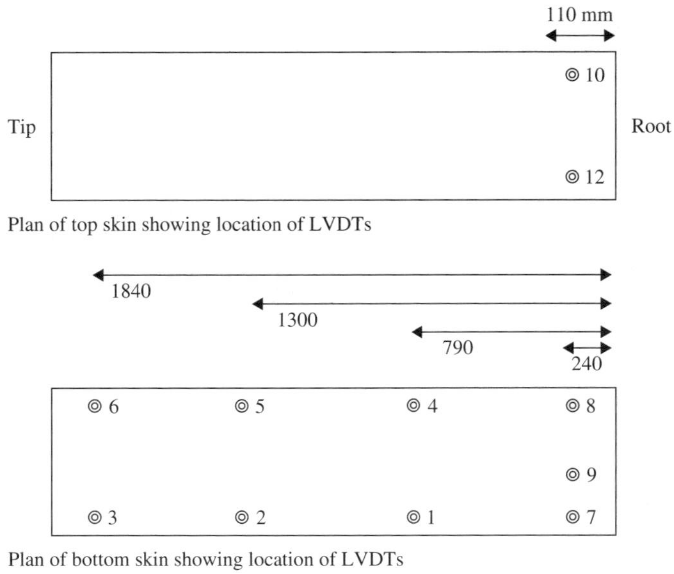

Figure 6.19 Locations of displacement transducers

Figure 6.20 Attachments for displacement transducers

digital oscilloscope. Figure 6.21 shows a view of the bottom of the impact tower close to the wing-box skin.

A general view of the test specimen installed in the testing frame is shown in Figure 6.22.

6.3.3 Impact Tests

The composite evaluator testing took place over a period of two weeks. In this period sensors were attached to the top skin of the wing-box, and the final assembly and fitting of

Figure 6.21 Instrumented impactor

Figure 6.22 Overall view of the composite evaluator test

the structure took place. After this the various damage detection systems were calibrated. For those techniques aimed at detecting impact events, a large number of low energy, non-damaging, impacts were introduced to allow a 'Velocity calibration' to be made to accommodate the non-isotropic properties of the composite skin. The schedule of events performed during the testing is shown in Table 6.1.

The loading cycles were imposed on the structure so that any initial loading effects on the composite or sensor systems could be assessed. Loading was taken to a maximum of approximately 2000 $\mu\varepsilon$, which corresponds to a high level of normal working strains in most carbon composite designs. The wing-box tip displacement as a function of the applied loading is presented in Figure 6.23. It can be seen that at the maximum load the tip displacement was 44 mm. The corresponding maximum strain as a function of load is presented in Figure 6.24. The maximum strain was recorded at the skin root. In Figure 6.22 data are plotted from strain gauge 3 (see Figure 6.19). Finally, Figure 6.25 shows the deflection of the wing-box at maximum load, as a function of distance along its length.

Damage was introduced into the wing-box skin using the calibrated impact tower. A total of six sites were used for the damage detection experiments. Their locations on the wing-box are indicated in Figure 6.26 and Table 6.2.

Table 6.1 Impact test programme

Task	Impact mass (kg)	Drop height (mm)
Initial calibration – low energy impacts	–	–
Loading to 2000 $\mu\varepsilon$ max	–	–
Loading to 500 $\mu\varepsilon$	–	–
Loading to 1000 $\mu\varepsilon$	–	–
Loading to 1500 $\mu\varepsilon$	–	–
Loading to 2000 $\mu\varepsilon$	–	–
Repeat calibration – low energy impacts	–	–
Impact position 1 – 5.88 J	1.2	500
10.0 J	1.2	850
20 J	1.2	1700
30 J	1.951	1570
30 J	1.951	1570
47 J	4	1200
Impact position 2 – 4.7 J	4	120
40 J	4	1020
Impact position 3 – 4.7 J	4	120
35 J	4	893
40 J	4	1020
Impact position 4 – 4.7 J	4	120
40 J	4	1020
Impact position 5 – 4.7 J	4	120
40 J	4	1020
50 J	4	1276
Impact position 6 - 4.7 J	4	120
40 J	4	1020

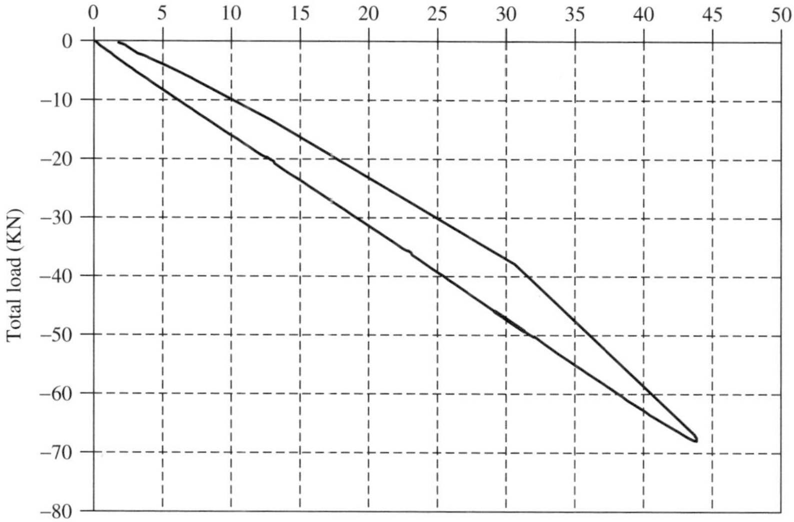

Figure 6.23 Wing-box load/displacement response

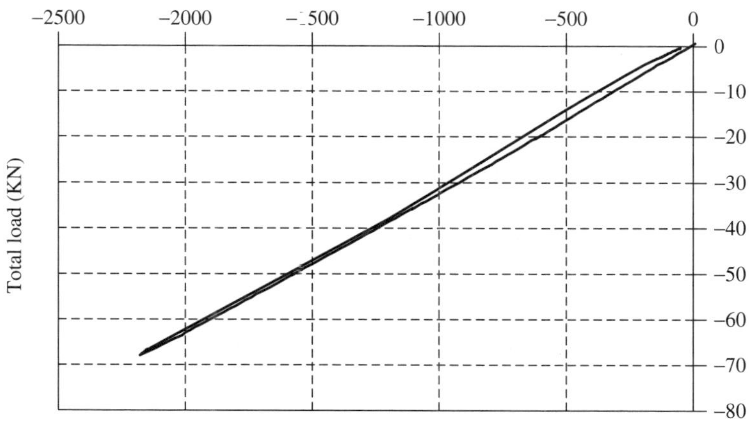

Figure 6.24 Wing-box load/strain response

After each impact, the skin was examined using an ultrasonic inspection system. This was manually scanned over the region of interest, and provided C-scan maps of damage. This technique was intended to provide a rapid assessment of damage state, and so the images produced were of lower quality than would be obtained from an immersion system, but sufficient for the purpose considered here, which was to estimate the size of the induced damage. C-scan examples are given in Figure 6.27 for different values of impact energy. It can be seen that no identifiable damage was introduced until the impact energy was increased to 47 J. In all of the scans shown, a horizontal band can be seen, which is due to the presence of the bonded stringer on the remote face of the skin.

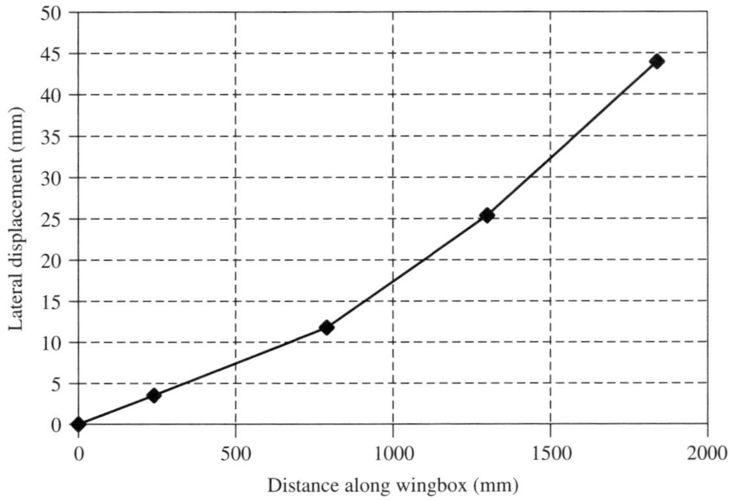

Figure 6.25 Displacement as a function of distance at maximum load

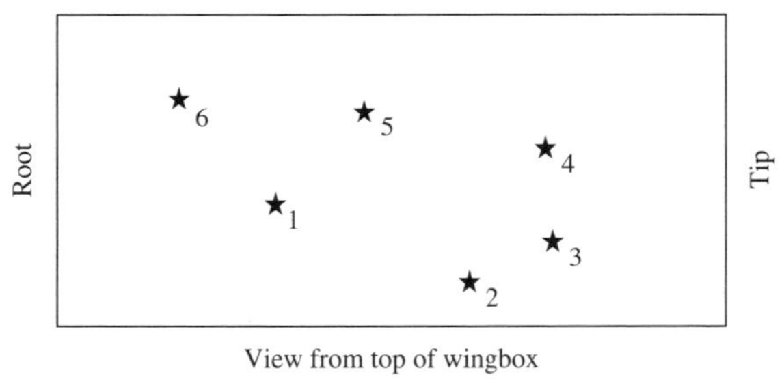

Figure 6.26 Locations of impact sites

Table 6.2 Impact locations on the wing-box structure

Impact number	Width (mm)	Length from root (mm)
1	312	584
2	111	1111
3	214	1335
4	449	1312
5	544	824
6	581	325

From the ultrasonic scans presented it is possible to estimate the area of damage induced by the impact. This is shown in Table 6.3. There is not a simple relationship between the impact energy since this depends upon the skin thickness, the local constraints and position on the structure. These effects are not considered in any more detail here. For the analysis presented in this chapter it is sufficient to know the location and severity of the damaged areas.

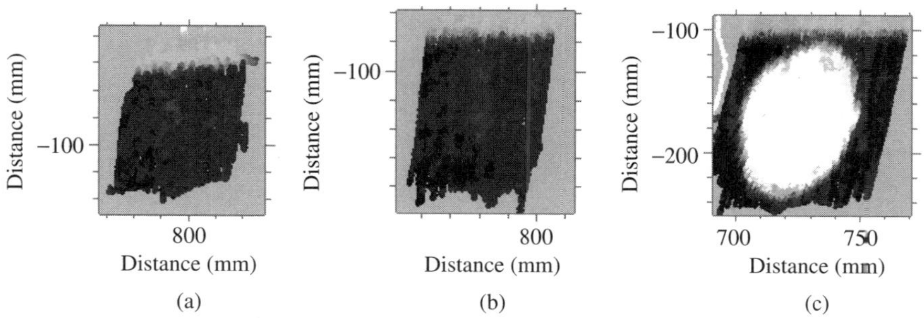

Figure 6.27 Ultrasonic scans for impact site 1: (a) impact energy 20 J; (b) impact energy 30 J; (c) impact energy 47 J

Table 6.3 Area of impact damage

Impact site	Impact energy (J)	Approx damage area (mm^2)
1	47	2640
2	40	2208
3	35	157
3	40	1731
4	40	1109
5	50	1141

When damage was introduced by using the impact tower, force/time histories were recorded from the force gauge fitted to the impactor tip. Examination of the data shows that it is possible to determine if damage was introduced into the structure. This can be seen from the following two examples. Figure 6.28 shows the force time history of a low level, nondamaging impact at position 5. The shape of the trace was seen to be very similar for all of the non-damaging impacts. In contrast, Figure 6.29 shows the trace recorded for a 50 J damaging impact, also at position 5. This shows that the shape of the curve is significantly different, with two peaks appearing and more high frequencies being present in the waveform. This suggests that the propagating stress waves, which are generated as a result of the impact, may contain characteristics that can be used to determine if damage has been induced. This becomes clearer in the analysis of the results from the impact event detection sensors deployed in the evaluator test.

6.3.4 Damage Detection Results – Distributed Optical Fibre Sensors

Optical sensors are small and lightweight and can, if necessary, be embedded into composite materials with little influence on mechanical properties. However, if discrete sensors are used it can mean that a large number of sensors are needed to ensure adequate coverage of a structure. This is not an attractive prospect since it would lead to great manufacturing complexity and require many connections between the sensors and the interrogation

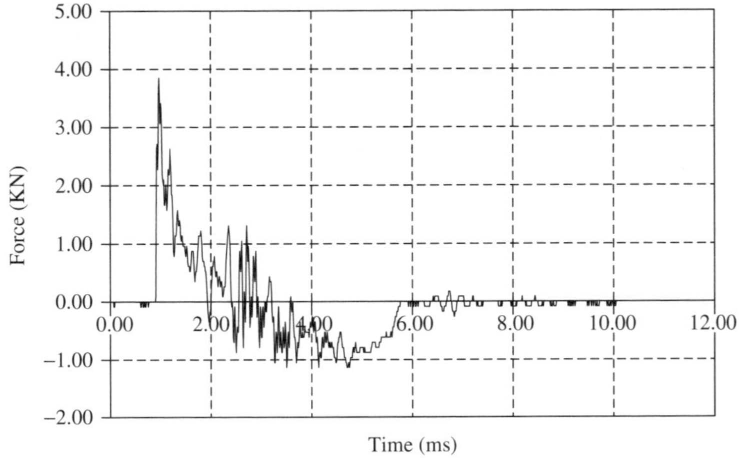

Figure 6.28 Force time history of nondamaging 4.5 J impact at position 5

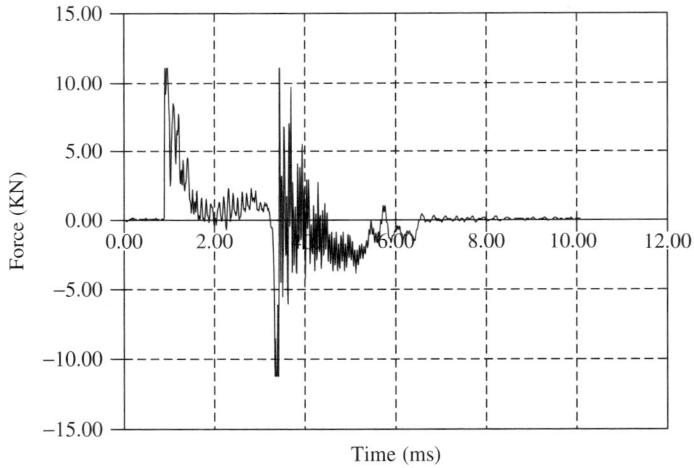

Figure 6.29 Force time history of damaging 50 J impact at position 5

system. An alternative is to employ an optical sensor in which the whole length of the fibre is employed as the sensor. A distributed sensor (Mondanos *et al.* 2000) has been evaluated within the tests described in Section 6.3.3.

The damage detection sensor consists of a length of single mode linear, stress induced (bow tie), birefringent fibre. In such a fibre, the degeneracy between the two mutually independent orthogonal polarisation modes has been lifted by introducing two regions of highly doped glass located on opposite sides of the core. The circular symmetry of the ideal fibre is broken thus producing an anisotropic refractive index distribution in the core region (internal asymmetrical stress). The refractive index difference between the two axes of the fibre produces a differential propagation velocity and a consequential time delay between the light travelling in the two axes. The difference between their effective

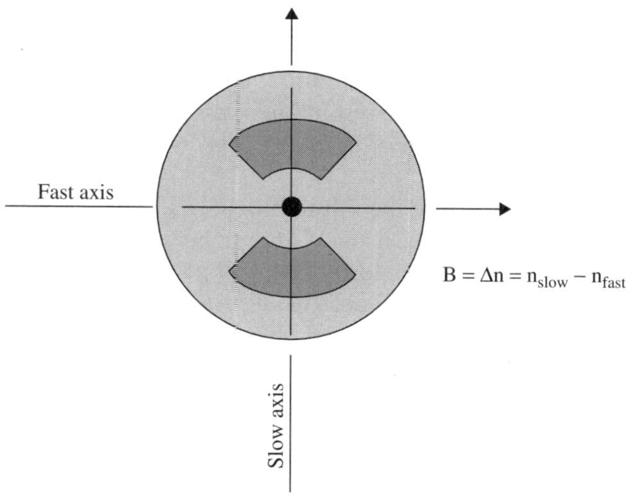

Figure 6.30 Cross-section view of a typical bow tie high birefringence fibre

refractive indexes determines the fibre *birefringence (B)*. The cross section of a bow-tie stress lobe fibre is shown in Figure 6.30. Any polarised optical signal launched onto an axis of the fibre will remain confined to that axis until it reaches the opposite end of the fibre. Localised modification of the birefringence, generally produced by a transversely applied force, causes light to couple between the polarisation modes.

For an optical fibre embedded in a composite structure, a damage event will create a perturbation of the birefringence and cause coupling of power between the two modes. Therefore, light initially launched onto just one polarisation mode will be partially coupled to the orthogonal mode by the damage effect. These two beams of light travel along the two modes with different velocities and emerge with a time delay relative to the position along the fibre where the perturbation occurs (Figure 6.31). If the delay time between two signals is greater than the source coherence time then the two outputs add in power, and there is no phase relationship. If the coherence time of the source is greater that the time delay then there is a phase relationship and the two outputs will add in amplitude with a cyclic response dependent on the phase relationship between the two. Therefore, the position of multiple perturbations along the fibre can be located by measuring the time delays. The magnitude and the angle of the applied force relative to the optical axes of the

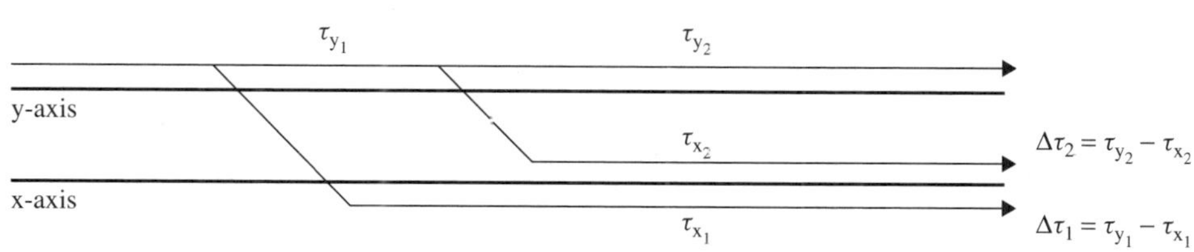

Figure 6.31 Schematic diagram of coupling between the polarisation axes of the fibre. Light travelling from each of the coupling points experiences a different time delay dependent upon the coupling position along the fibre

fibre determine the level of power coupled between the modes. The precision of the time delay measurement determines the spatial resolution of the perturbation position along the fibre. For a highly polarised preserving fibre the time delay is 1.686 ps/m. Therefore to achieve a resolution of 10 cm a time resolution better than 0.16 ps is required. The only technique to achieve time resolution of such precision is to use the time properties of a low coherence optical source.

Figure 6.32 shows a schematic of the fully distributed sensing system. The output light from the source is linearly polarised using a fibre polariser and aligned with one of the polarisation maintaining axes of the optical fibre. Any perturbation along the fibre couples a small amount of power to the orthogonal mode. At the output of the fibre the launch axis has a single beam, which has travelled along the full length of the fibre, and the orthogonal mode contains N-beams from N possible perturbations along the fibre. Each of these beams is mutually incoherent and they have no phase relationship therefore can be considered as independent beams.

At the end of the fibre the output is collimated and directed to the interferometer through a linear polariser at 45° to the fibre axes. The polariser superimposes the uncoupled and the N cross-coupled modes. The signals are then input to a Michelson interferometer in which one path can be scanned. The output from the interferometer is focused onto a detector.

One path of the interferometer is scanned to vary the time difference between the two paths. When the time delay in the interferometer matches any of the delays through the fibre then the relevant beams are brought back into coherence producing an interferogram. Its frequency is directly proportional to the velocity of the scanning mirror. The position of perturbations along the fibre can be located using the interferogram of a monochromatic source to calibrate the system. The monochromatic source is directly launched into the

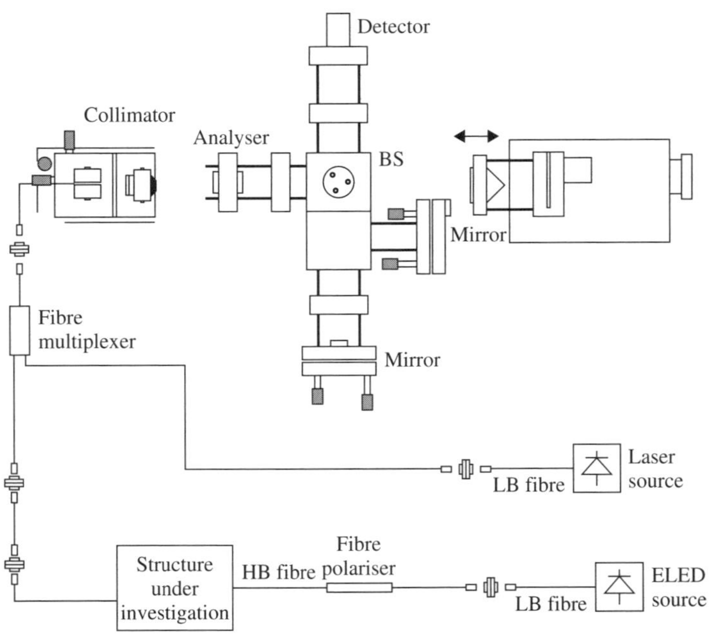

Figure 6.32 Block diagram of the distributed damage sensor

interferometer and can account for very accurate measurement of the scanning mirror displacement.

The basic system as described was evaluated at laboratory scale. This led to a re-evaluation of the system, and some aspects of the design were modified. The system for calibrating the movement of the scanning interferometer was modified to eliminate some nonlinear signal processing which could be a source of errors. In addition, significantly more care was taken with controlling the polarisation of the sensor light. Fibre polarisers were spliced onto the sensor fibre to provide stable polarisation conditions. An additional polarisation controller was added to provide minor corrections to avoid polarisation fading. The resulting system showed very little sensitivity to sensors being disconnected and reconnected, or to other changes such as temperature.

The optical fibre sensors were incorporated into the skin of the composite evaluator wing-box described in Section 6.3.1. Their arrangement through the thickness of the skin can be seen in Figure 6.33, which is a view from the root end of the skin (the scale of the figure is distorted for clarity). Sensors were placed in only a small number of interfaces in the structure. The ply numbers are given in Figure 6.33, and refer to the lay-up definition described in the section on the specimen design and manufacture. Plies 7, 8, 49 and 50 are 0° plies, which are aligned with the longest dimension of the skin. Plies 9 and 48 are 90° plies. The interface between two 0° plies was chosen because at this interface the presence of the fibre has the minimum impact on the mechanical properties of the composite. However, it is not an interface that is sensitive to delamination, which tends to form first at the interface between plies with the greatest change in ply direction. The greatest change is seen at the 0/90 interface, and that is why this was also chosen for comparison. Sensors were placed near the top and bottom of the skin to see if there was any effect on detection performance. The distribution of the sensor fibres in plan view is shown in Figure 6.34. Some locations were unavailable because rig fittings would impede fibres emerging from the edge of the skin.

During lay-up the positions for the fibre were marked out on the lay-up table. At the appropriate interface the cleaned fibre was introduced and aligned by hand. Each fibre was gently pressed onto the underlying ply to ensure that it would maintain its position during cure. At each end of the skin, the fibres were protected where they emerge with small-bore PTFE tubing.

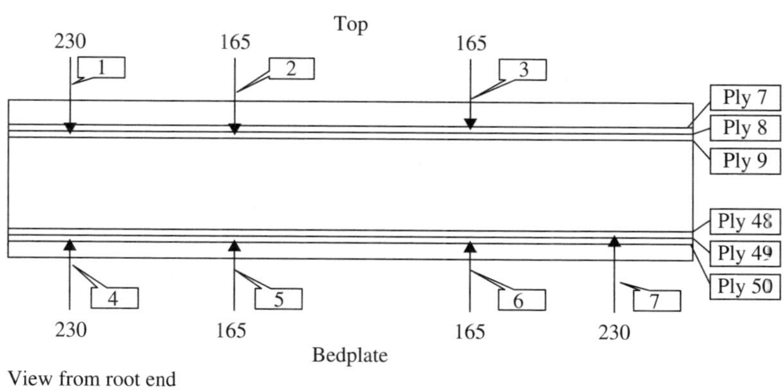

Figure 6.33 Location of sensors through the thickness of the wing-box skin

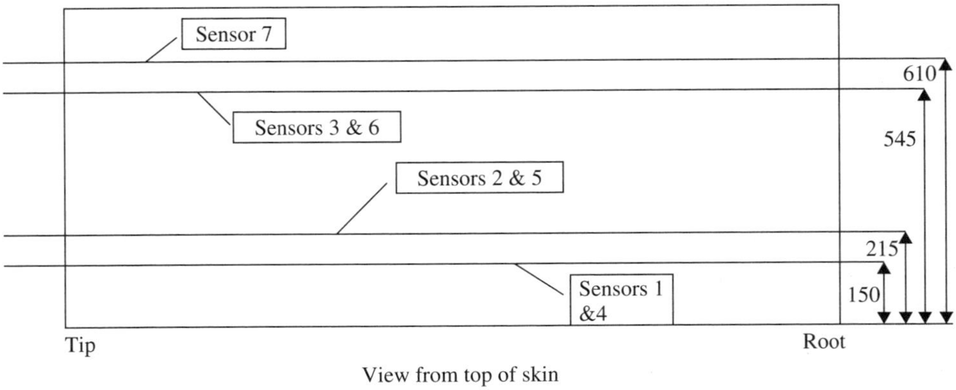

Figure 6.34 Sensor location in plane view

The evaluator test was conducted in accordance with the test schedule described in Section 6.3.2. Because the optical technique is comparative, a number of calibration scans were recorded so that any changes could be identified. Some examples of the results obtained are included here.

The outputs from the sensors are interferograms, which are plotted here as the modulus of the interferogram as a function of distance along the fibre. The distances along the fibre are given in mm, and are calibrated using the reference laser in the interferometer. The absolute magnitude of the interferogram is unimportant, and so is best regarded as being plotted in arbitrary units. Where traces are compared on the same graph, they are always scaled in the same way, so amplitude differences shown are real. It is important to obtain some appreciation of how much the signatures from each fibre vary even when damage is not introduced into the structure. An example is shown in Figure 6.35, where the response of the sensor is shown before and after two loading cycles of the wing-box to a maximum strain of over 2000 $\mu\varepsilon$. Some differences in the response can be seen, but these are relatively minor. Impact damage was introduced into the structure using an instrumented impactor. The impact sites are shown in Figure 6.36, along with the positions of the sensor fibres.

In Figure 6.37 the response of sensor 3 to impacts at site 3 is shown. The first trace shows the baseline trace before any damage was introduced. The second trace is the response after a 35 J impact, which introduced a damage area of approximately 157 mm^2 and the final trace was recorded after a 40 J impact, which increased the damage area to approximately 1731 mm^2. The location of the impact site along the length of the sensor is indicated on the figure. A number of changes in the traces can be seen, especially in the area surrounding the impact site. Sensor 3 is located near the top surface of the skin in between two 0° plies.

The response of sensor 6 to the same impacts, at site 3, is shown in Figure 6.38. This sensor is located close to the remote face of the wing-box skin, again between two 0°plies. This sensor shows a similar response to sensor 3, with perhaps a slightly greater sensitivity to the impact damage. Comparison of the results in Figures 6.37 and 6.38 suggests that the position in the depth of the skin has not has a significant influence on sensor performance.

The effect of embedding the sensor at the interface between a 0° ply and a 90° ply can be seen in Figure 6.39. Here the response of sensor 2 is shown to an impact at site 5.

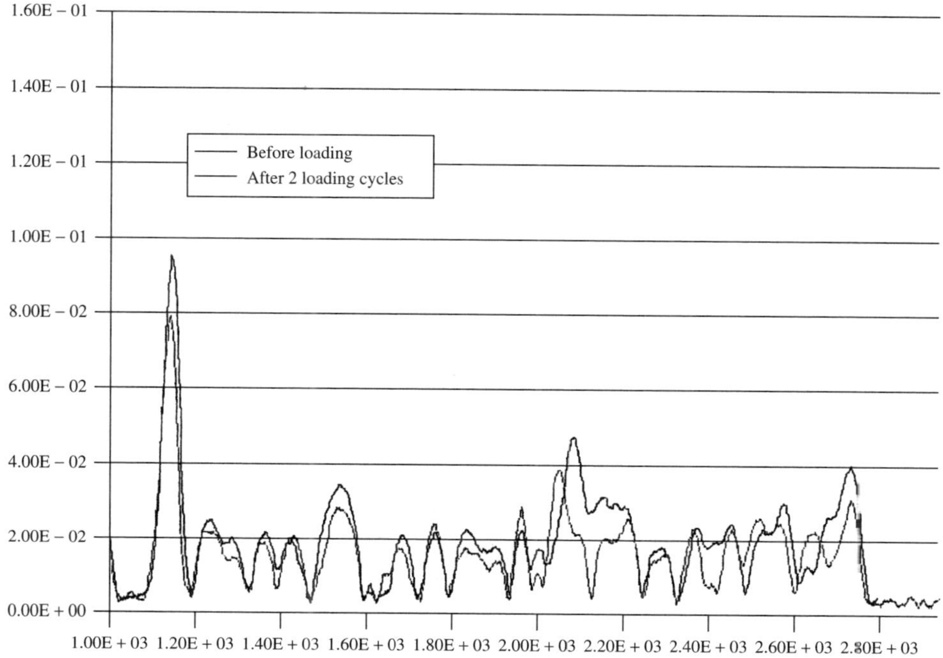

Figure 6.35 Comparison of the output from sensor 1 before and after two loading cycles

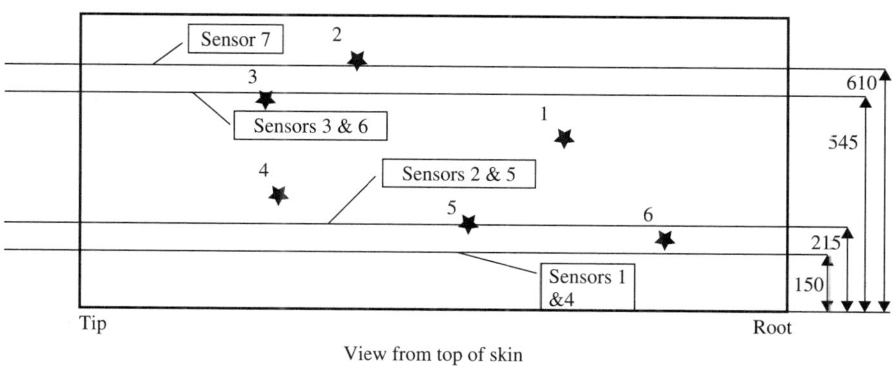

Figure 6.36 Impact sites relative to sensor fibres

In this case the indication due to the introduction is much stronger and unequivocal. The damage introduced by the 50 J impact at site 5 was approximately 1141 mm^2, which is less than the damage introduced at site 3 with the results shown in Figures 6.37 and 6.38.

In this section an overview of the performance of the optical damage detection system based on distributed sensors as assessed on the large composite evaluator has been presented. Conclusions drawn form the work can be summarised as follows:

- The optical fibre sensors survived the manufacturing process well. No failures were experienced either during manufacture of in subsequent rig fitting and testing.

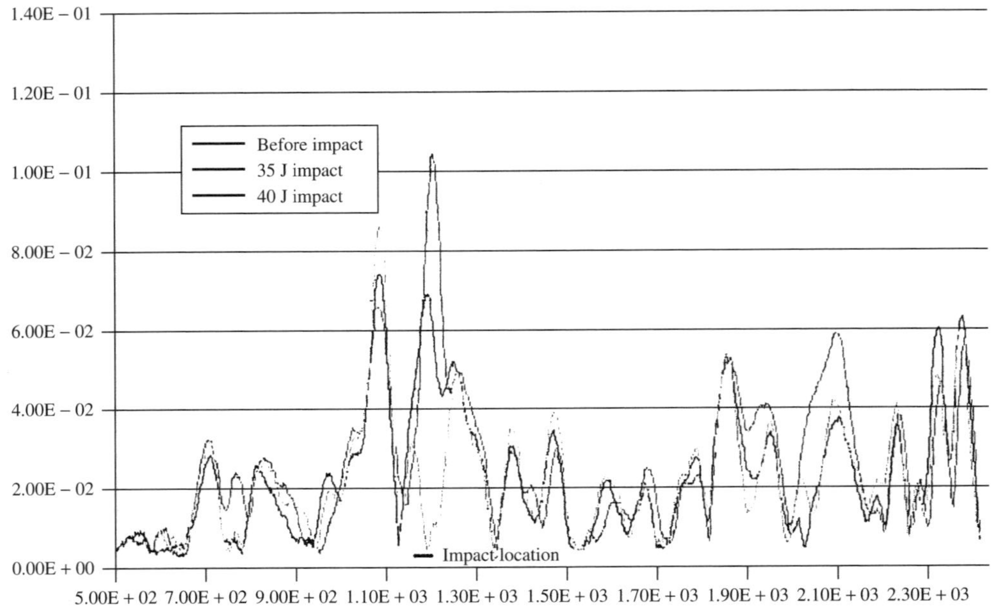

Figure 6.37 Response of sensor 3 to impact at site 3 before impact, after 37 J and 40 J impacts

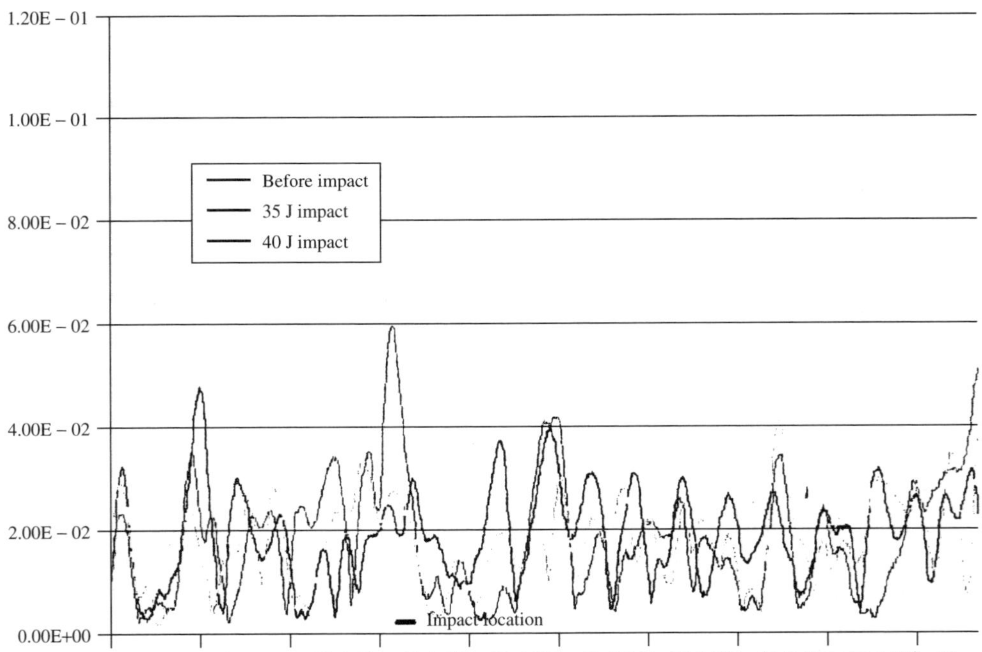

Figure 6.38 Response of sensor 6 to impact at site 3, after 37 J and 40 J impacts

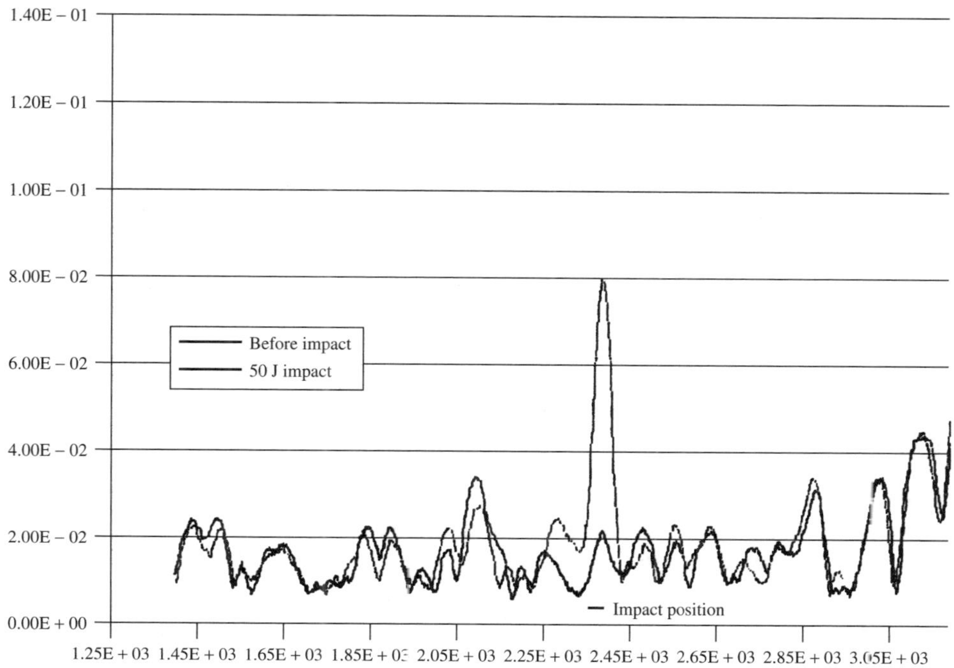

Figure 6.39 Response of sensor 2 to a 50 J impact at site 5

- The sensor is comparative. A baseline scan has to be recorded in the 'as manufactured' condition. This signature depends on a number of things, including the internal constraints imposed by the structure. Effects other than damage, such as mechanical loading, have relatively little effect on the baseline scan.
- Damage produces changes in the sensor response. These changes are permanent, so the system does not need to be running when the damage event takes place. The presence of damage can be detected and its position along the sensor fibre estimated with acceptable accuracy.
- The sensitivity of the sensor to damage depends slightly on the location of the fibre through the thickness of the skin, and much more on the interface at which it is located. Placing the sensor fibre between two 0° plies produces the minimum effect on structural performance, but limits sensitivity because the interface between the two plies effectively disappears after cure. Delaminations are unlikely to be formed at this interface. When the structure is impacted, delamination damage occurs first at interfaces with the largest angle between plies. For this reason the best location for the sensor is at such an interface. Sensor 2 was located at the interface between a 0° ply and a 90° ply. This configuration should provide the best compromise between minimal structural influence and damage detection sensitivity.
- The modifications made to the detection system have solved the problems experienced during the previous laboratory testing. During the testing schedule the technique proved to be reliable and rapid in operation. It took approximately five minutes to record each sensor scan, and this time could certainly be reduced if required.

6.3.5 Damage Detection Results – Bragg Grating Sensors

The optical fibre technology used in this section is complimentary with the operational load monitoring system used in the flight tests in Section 6.4. The technology for both systems is based on optical fibre Bragg grating sensors; both surface mounted and embedded sensors have been pursued. Although the details of the optoelectronic interface equipment will be different for the load monitoring and damage detection systems, synergy exists on the issues of structural implications of optical fibre embedment in carbon fibre composites and fibre sensor endurance to manufacturing and handling conditions.

The technology for impact detection is based on optical fibre Bragg grating sensors. These devices are fabricated within the cores of otherwise standard telecommunication optical fibre by an ultraviolet, laser exposure process. Details of the fabrication process, its variants and optical properties of these devices are described in Chapter 3. The nature of operation can be summarised as follows:

- Bragg gratings are localised regions in a length of optical fibre. Many gratings can be imprinted along a single length. Each grating is typically 2 or 3 mm in length. These gratings constitute the sensors.
- The gratings have the property of reflecting light that is shone down the fibre, in a pre-determined band of wavelength. The grating themselves are periodic ripples in the refractive index in the fibre's core.
- As the fibre, and hence the grating is strained, the band of wavelength at which the Bragg grating reflects is shifted. This strain can be induced quasi-statically, in which case the grating acts as a strain sensor, or can be dynamic such as the stress wave event caused by an impact.
- The sensor system operates by shining a broad range of wavelengths, simultaneously down the fibre. The wavelength of the reflected light is detected in an optoelectronic module where wavelength shift is converted to an electrical signal that constitutes the raw sensor signal. This can then be captured and analysed using standard data acquisition equipment.

The sensor system adopted for the trials is illustrated schematically in Figure 6.40. The sensor system comprised eight independent optical fibres, each containing a single chirped Bragg grating. These gratings were nominally identical and were produced in standard, single mode (1300 nm) telecommunications grade fibre. The eight sensors were split into two groups of four. Four sensors were embedded and four were surface mounted. Four gratings were embedded between ply layers 49 and 50 (0° plies). The sensor fibres were aligned lengthways with the local ply direction. The sensors were deployed in a triangular arrangement with a fourth, centrally placed. For the location of impact events within a nominally uniform plate, this arrangement provides the best coverage. Embedded sensors were labelled I2, I4, I6 and I7. After manufacture of the top skin but prior to installation on the test rig four more fibre optic sensors were bonded to the inner surface of the skin in a similar geometrical arrangement to those embedded. This installation should allow comparison of the respective performance of embedded and surface mounted fibre sensors. All embedded sensors were fabricated in a single mode optical fibre of standard 125-micron diameter. Jacket material was acrylate with Teflon protective sleeve at the

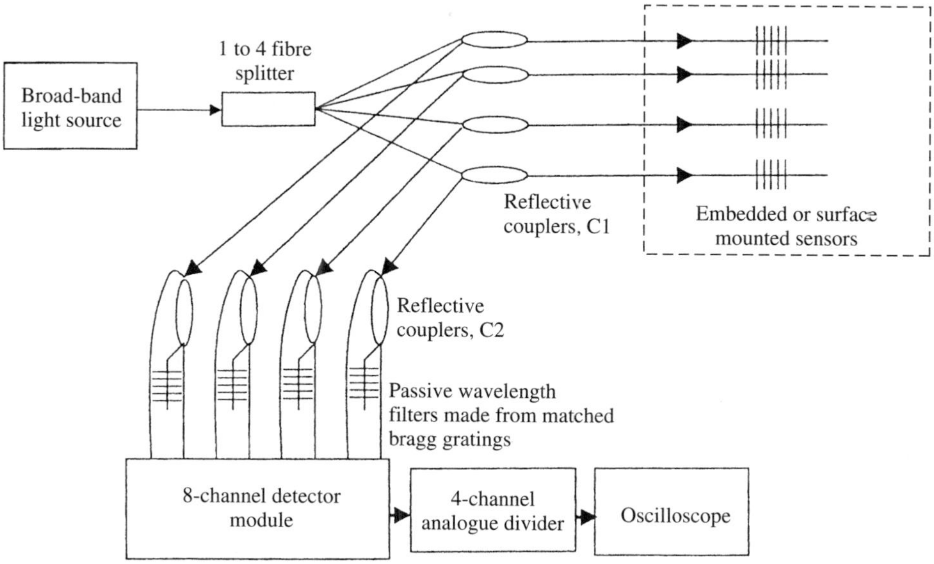

Figure 6.40 Schematic diagram of the optical fibre impact detection system using Bragg grating sensors

points of emergence of the fibres from the composite panel. The surface mounted sensors were laminated between films of polyimide (Kapton film), which were bonded with an adhesive layer. The fabrication process used was the same as for the fibre optic strain gauge rosettes described in Chapter 3. These sensors were bonded to the composite panel using strain gauge quality cyano-acrylate adhesive and surface preparation methods. The surface mounted sensors were labelled S2, S4, S6 and S7. Figure 6.41 shows the locations of the embedded and surface mounted sensors. All sensors were equipped with long fibre pigtails for easy connections to the optoelectronic system.

The assembled composite skin and wing-box were fitted to the test rig and the embedded and surface mounted sensors were tested. First, the sensing systems were calibrated. For the fibre optic vibration sensing system this involved recording a series of low energy impact events on the top skin. Calibration impacts were placed on a regular grid (see Figure 6.41). The calibration was performed for both the embedded and surface mounted sets of sensors.

After the initial calibration was completed the structure was loaded with hydraulic jacks at the tip end (root end fixed), as described in Section 6.3.2. Loading was monitored with an array of electrical strain gauges. Loading was limited to a maximum observed strain of $2000\,\mu\varepsilon$. This loading was repeated but the load was increased in $500\,\mu\varepsilon$ steps. Figure 6.42 shows the second loading cycle recorded using the fibre optic vibration sensing system, which can also be used to measure strain. The resolution was limited by the oscilloscope, which had to cope with a large DC offset. The data from the initial calibration grid were noisy, so 1 MHz bandwidth RC filters were fitted between the optoelectronic unit and the oscilloscope. After loading a second calibration grid was recorded. Comparison of the two calibration grids will highlight changes in the dynamic properties of the structure caused by initial loading.

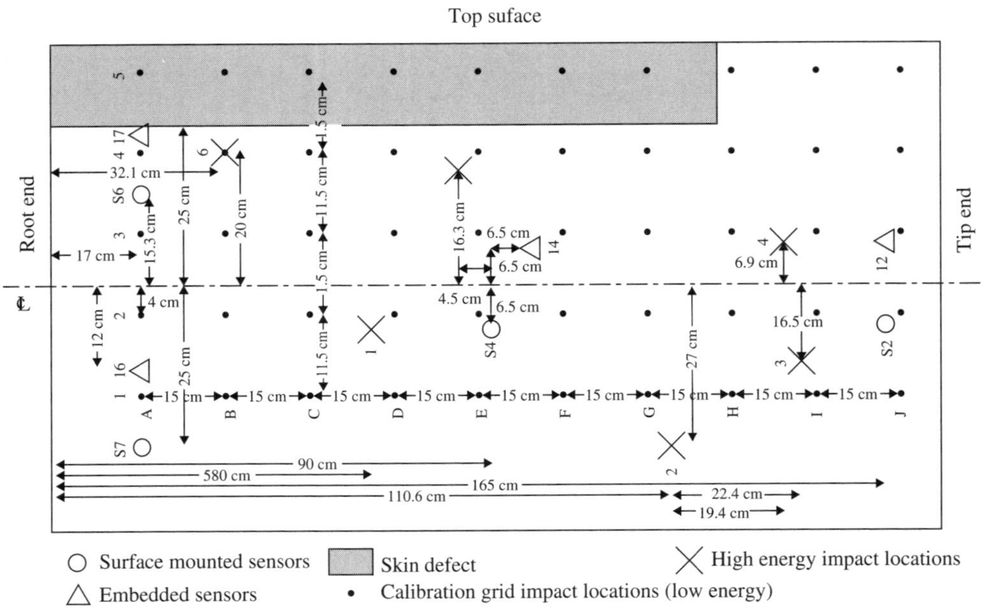

Figure 6.41 Composite evaluator top skin with locations of surface mounted and embedded optical fibre sensors

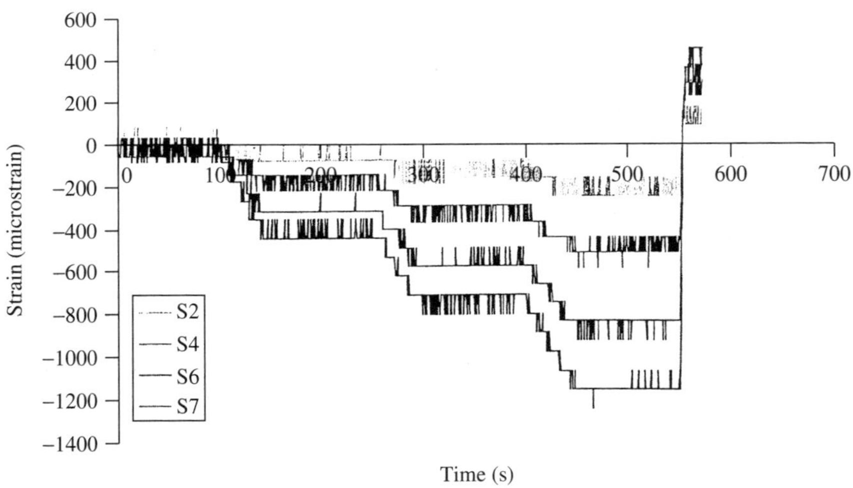

Figure 6.42 Second loading cycle of composite evaluator

Finally, a series of locations were chosen for high-energy impacts (see Figure 6.41). Figure 6.43 shows damaging and nondamaging impact records from impact at location 4. For clarity only data from sensor I4 is shown for both impacts. The amplitude of the signal has not altered significantly despite a ten-fold change in the impact energy. The shape of the curve is very different however. The damaging impact contains more high frequency

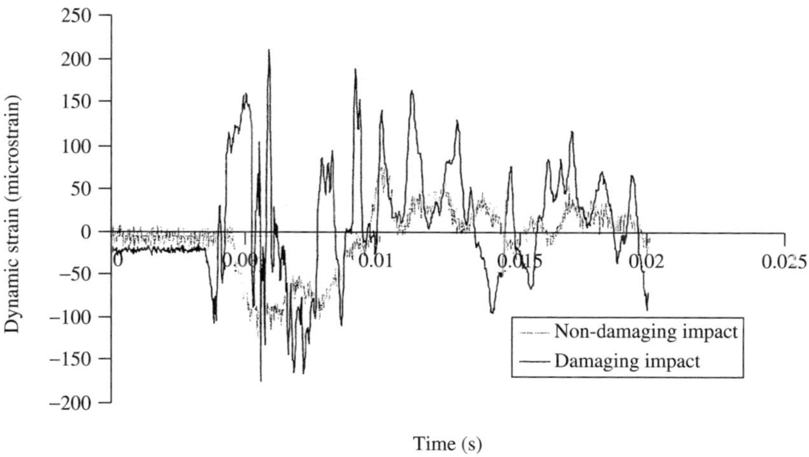

Figure 6.43 Damaging and non-damaging impact records

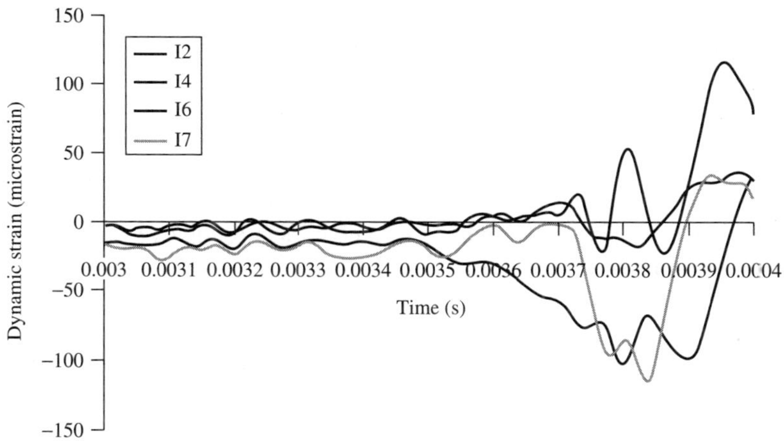

Figure 6.44 Initial response of sensors

components, which shows up as spikiness in the graph. It is possible that spikiness is indicative of damage. Figure 6.44 shows the leading edge of an impact shockwave as detected by the embedded sensor set. There is clearly a difference in the time of arrival of the shockwave at the individual sensors. This information could potentially be used to locate the impact positions. The data in Figures 6.48 and 6.49 have been filtered to remove noise at frequencies above 20 kHz.

In summary, the functioning of a four channel, optical fibre impact detection system has been verified. The system based on chirped fibre Bragg gratings, is capable of producing linear signal in response to static strain and dynamic stress wave events such as those resulting from impacts. The sensors were successfully embedded into a test panel manufactured in aerospace type composite material and underwent trial impacts up to and beyond damaging energies. The sensors functioned throughout. The work demonstrated the functionality of the sensor system and its ability to record impact event data to a resolution suitable for the location and identification of a damaging impact on a composite skin. No attempt has been made to analyse the sensor data other than to prove basic functionality and principle of operation.

6.3.6 Lamb Wave Damage Detection System

An active system based on Lamb wave testing was used to monitor the composite large-scale structure. A schematic diagram of the wing box composite specimen together with the positions of impact locations and piezoceramic transducer locations is shown in Figure 6.45.

Preliminary tests with wedge sensors were carried out in order to measure experimentally the attenuation of Lamb wave signals as function of the location in the plate and of the frequency. Attenuation measurements are essentials since they allow to know how far Lamb wave modes can be transmitted with a sufficient signal to noise ratio. Attenuation measurement results showed that for wave propagation parallel to the stringer and at a frequency of 250 kHz, which was the working frequency, Lamb waves could be propagated for a distance equal to 70 cm. This was a satisfactory result for a complex structure. According to these results, the piezoelectric elements were bonded on the surface of the composite panel.

The nondestructive testing using Lamb waves needs not only signal attenuation measurements, but also the experimental determination of the type of wave propagation, i.e. the various modes of vibrations and their velocities, which appear at any given frequency. The emphasis was placed on the sensitivity of Lamb wave modes to defects. This can be done only if the propagation modes are known. The technique used to perform the analysis of propagating multimode signals was based on a two-dimensional Fourier transformation described in (Alleyne and Cawley 1992). Figure 6.46 illustrates an example of these measurements performed near the transducer El. Six different Lamb wave modes have been identified between 0 and 800 kHz. Then, the experimental measurements of the phase velocities have been determined from these 3-D plots and compared with the computed phase velocities (see Figure 6.47). They were in good agreement for the first modes, which is sufficient since the chosen excitation frequency was 250 kHz. This comparative study was performed for each composite plate lay-up leading to satisfactory results.

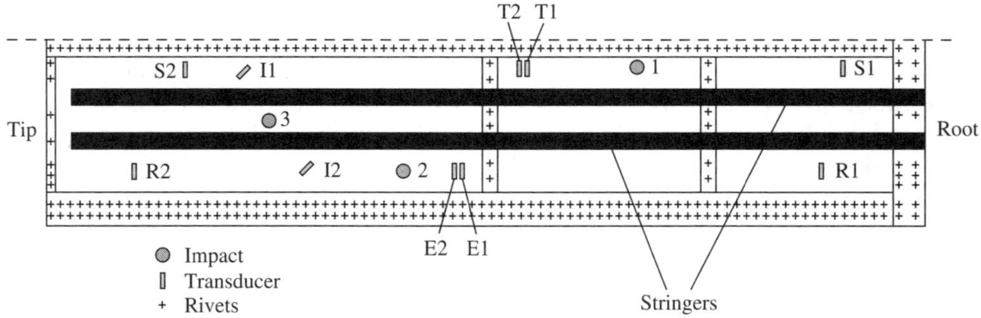

Figure 6.45 Locations of piezoceramic sensors – view from underside of the skin

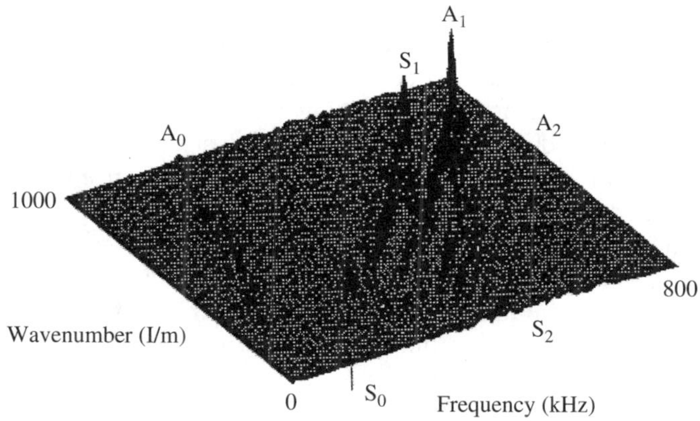

Figure 6.46 Measurements of the experimental Lamb wave wavenumbers

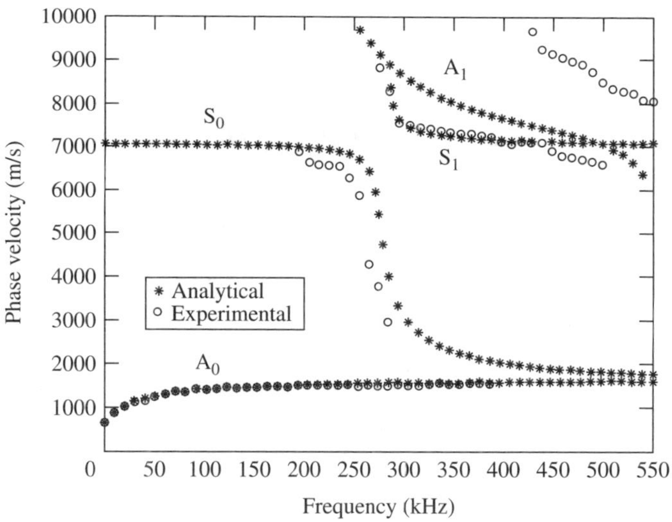

Figure 6.47 Comparison between the experimental and theoretical velocity measurements

Due to the thickness variations of the plate, the dominant component of the signal at 250 kHz was either the S_0 or S_1 mode. The sensitivity of excited modes depended on the location of the actuator. The dominant components propagating between the tip and the middle of the panel were the S_0 and A_0 modes, whereas the dominant component between the middle and the root of the plate was the S_1 mode.

The input signal chosen was a 250 kHz, 5-cycle tone burst modified by a Hanning window envelope. The advantage of this windowed tone burst was the elimination of significant side lobes allowing for the attenuation of the parasite modes.

Experimental analysis of Lamb waves with the use of two transducers as emitting elements was performed in order to get better control of the generated mode and to improve the signal to noise ratio. The first concern was the ability to receive the waves from different transducers in all directions. The study showed that it was possible to transmit Lamb waves not only parallel to the stringer over a distance of 70 cm but also through stiffeners and rivets. The second concern was the improvement of the responses induced by the use of several actuating elements with delay phase excitation. When suitable time delays were introduced between the excitation signals, significant improvements were observed, as shown in Figure 6.48. However, generation of a single mode was not possible due to the complexity of the monitored structure.

Figure 6.49 shows an example of Lamb wave responses obtained before and after damaging impacts. The responses were obtained from the R2 transducer using the E1 and E2 transducers as actuators. From measurements of the time of flight of the leading edge of the individual wave packets, the first wave packet was identified as the combination of the S_0 and S_1 modes. Since S_0 was excited in a dispersive frequency region, S_1 is the dominant mode. The 1 cm delamination due to a damaging impact in location 2, leads to shape changes of the response signal. The study shows that it is possible to detect impact damage using Lamb wave testing approach. However, more signal processing analysis is required to extract the information related to the severity of damage.

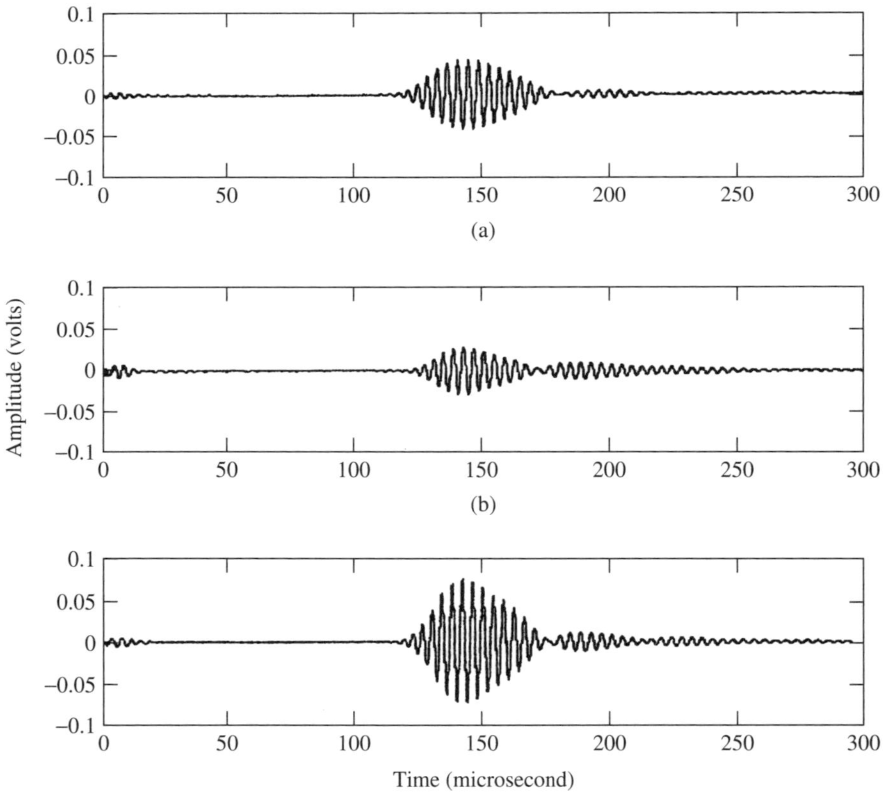

Figure 6.48 Examples of Lamb wave responses: (a) excitation with one piezoceramic element; (b) excitation with two piezoceramic elements in phase; (c) excitation with two piezoceramic elements with a phase delay

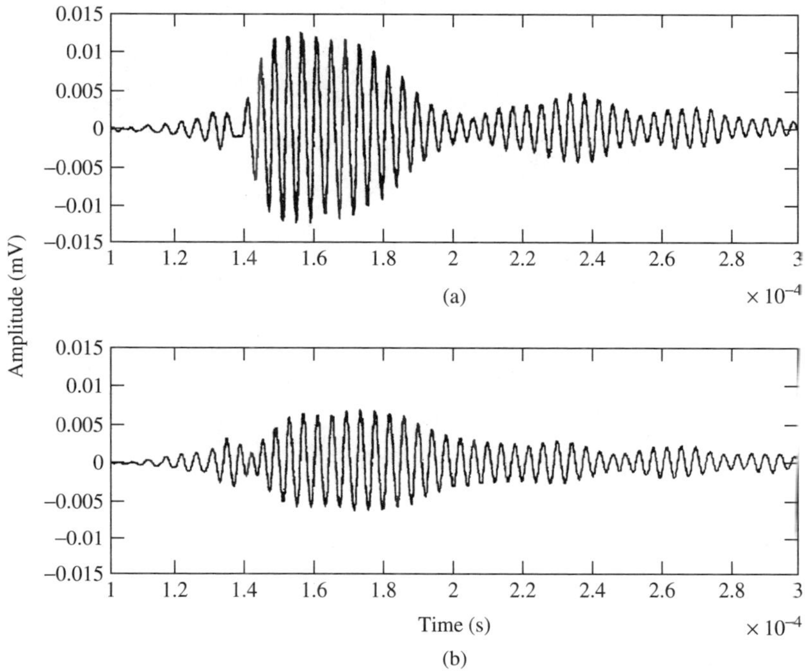

Figure 6.49 Damage detection using Lamb wave signals before (a) and after; (b) damaging impact

6.4 FLIGHT TESTS

Flight tests extend the laboratory results and evaluate load monitoring and noise levels for damage detection sensors under highly representative environmental conditions. This section demonstrates examples of flight test results from the optical fibre load monitoring and Acoustic Emission damage detection systems. Both monitoring technologies are described in detail in Chapters 3 and 4, respectively. The load measuring system was tested to see if strains could be reliably measured for various flight conditions. As mentioned earlier, the flight tests were not used for the detection of damage on primary of secondary components of the aircraft itself, because this would compromise structural integrity and would have further required clarification with the certification authorities. Instead the damage detection system was used to see if artificial crack-like signals could be differentiated from background noise during flight manoeuvres.

6.4.1 Flying Test-Bed

A *Jetstream 31* flying testbed aircraft, shown in Figure 6.50, was used in the tests performed at Warton Aerodrome over a two-week period in February and March 1999. The *Jetstream* is a twin-engine propeller driven aircraft. The cabin area, where passengers are normally carried, was adapted to accept equipment in 19-inch rack modules. Figure 6.51 shows the OLM system installed into an equipment rack. The sensor data were recorded

Figure 6.50 *Jetstream 31* flying testbed aircraft

Figure 6.51 Modified cabin area of the flight test aircraft

during the flights by engineers operating monitoring systems in the aircraft cabin. The operators sat in the seats opposite the equipment racks. Sensor data were transferred to ZIP disks so that it could be analysed on a ground-based computer. The ZIP disks also provided a convenient backup of the flight data.

Four types of flight were undertaken based on different flight configurations. The different configurations are:

- ground run at different speeds powered by one or two engines,
- flaps up and down on the ground,
- normal take-off and cruise climb,
- cruise flight, cruise climb and dive,
- maximum speed dive,
- maximum speed climb,
- 60° banked turns,
- cruise in turbulence,
- cruise with different thrust on the two engines,
- low speed with flaps down,
- low speed with undercarriage down,
- low speed with flaps and undercarriage down,
- touch and goes,
- land and taxi,
- reverse thrust.

There were two sets of operations, flight test and ground based manoeuvres; Tables 6.4 and 6.5 show the manoeuvres performed by the aeroplane within the flight test programme. Numbers indicate how many times a particular manoeuvre was performed. Banked turns and straight and level cruises were performed at different altitudes. At 20 000 feet the banking angle was limited to 45°.

Each flight involved verification of monitoring systems and data acquisition from various sensors. The flight test programme was designed to fulfil the requirements of the OLM and damage detection systems. This is illustrated in Table 6.6.

Table 6.4 Flight test programme – flight manoeuvres

Manoeuvre	Flight							
	1	2	3	4	5	6	7	8
Standard take-off	1	1	1	1	1	1	1	1
Maximum rate climb	1	2	3	–	1	3	2	2
Standard rate climb	2	3	3	3	4	5	3	5
60°/45° banked turns	6	–	6	–	–	6	–	6*
Standard rate decent	1	1	1	1	1	2	1	2
Powered shallow decent	–	1	1	–	1	1	1	1
Idle power rapid decent	–	1	1	–	1	1	1	1
Slow cruise – deploy flaps in stages	1	–	1	–	–	1	–	1
Slow cruise – landing gear down	1	–	1	–	–	1	–	1
Straight and level cruise	–	5	2	3	5	5	5	5
Straight and level cruise – asymmetric power	–	2	2	–	2	2	2	2
Roller landings	4	2	4	4	–	4	4	4
Full stop landing	1	1	1	1	1	1	1	1

Table 6.5 Flight test programme – ground based manoeuvres

Manoeuvre	On chocks	Taxi 1	Flight 4
Asymmetric power change throttle and RPM	Yes	2	2
Symmetric power change throttle and RPM	Yes	1	1
Deploy flaps in stages	Yes	1	1
Slow taxi – both engines	No	1	1
Slow taxi one engine	No	2	2
Fast taxi/aborted take-off	No	1	1

Table 6.6 Monitoring strategy for flight programmes

System	Operation	Flight segment	Comments
OLM	steady state transient	straight and level flight manoeuvres – dive banked turns	to flight limitations
		take off and landing	most extreme load transfers
Damage detection	ground run	flaps up and down	high noise level
		one engine/two engines	different rpm and pitch
	taxiing		different speeds
	transient	flight manoeuvres	different RPMs and pitch
		flaps up and down	cycling air turbulence
		take off and landing	reverse thrust
		Undercarriage up and down	cycling against locks and air turbulence

6.4.2 Acoustic Emission Optical Damage Detection System

Two linear optical sensor meshes were installed on the test bed. The first linear mesh (sensors 1 and 2) was placed on the fuselage near the wing. The distance between the sensors was approximately 820 mm. The second linear mesh (sensors 5 and 6) was placed on the lower part of the fuselage at the wing junction. The distance between the sensors was about 700 mm. The connection with the external sensors was achieved through an airtight connection panel in order to guarantee the cabin pressurisation. The preamplifiers of the external sensors were located inside the aircraft. The sensors 3 and 4 were located on the meshes 1 and 2, respectively and used as pulse generators in order to generate the sensitivity and the event localisation precision of the Acoustic Emission equipment in flight. Figure 6.52 shows the location of sensors 4, 5 and 6.

The Acoustic Emission equipment was installed in the racks inside the aircraft cabin (second rack from the right in Figure 6.51). The power supply used was 28 V DC. A *Thomson-CSF DD621* device was used for protection against power supply variations. A voltage converter 24 V DC to 220 V AC was connected to the protection device to power the computer, LCD and calibrator. The *PAC AECAL-2* Acoustic Emission calibrator was

Figure 6.52 Acoustic Emission sensor locations

used to simulate variable amplitude acoustic events. The control and acquisition computer equipped with six channels and the *MISTRAS* software. In order to minimise the risks of hard disk crashes and data losses the system used a flash disk during the flights. The mouse was replaced by a tracker ball for safety reasons. The keyboard was fitted in the rack. The six acoustic acquisition channels consisted of 1220*A* type preamplifiers with *AST* function, 'micro-dot' sensor cables and *R*15 type sensor (100–300 Hz). All the connections were achieved with 50 Ohms coaxial cables with *BNC* connectors. The bonding between the sensors and the structure was done with the *Silastic 732RTV* silicone glue vulcanised at ambient temperature. The integration and the fitting of the all the equipment were done by *Cranfield Aerospace*.

The coupling, performance and sensitivity of the Acoustic Emission sensors were very good. The effect of the temperature (from −23 °C to 0 °C) does not seem to modify the sensitivity of the sensors. However, the variation of acceleration (up to 2.75 G) slightly decreases the sensitivity. The different flight phases illustrate the good sensitivity and precision of the Acoustic Emission technique using simulated events. The localisation tests on the fuselage with the Hsu–Nilsen source (0.5 mm 2 H lead) have been done in level flight at different altitudes in order to take the temperature into account. All events were perfectly localised with amplitudes higher than 90 dB. The average precision of the localisation was about 2 % (14 mm on a 690 mm mesh). Considering the complexity of the structure on this mesh (riveted assembling, filler layer between structural elements) this shows good sensitivity of the Acoustic Emission System for the localisation of events. Except the artificial events a few events have been localised on the measurement mesh. The landings together with touch-and-go manoeuvres were by far the most active phases observed in the flight tests; counting these flight manoeuvres could be a good parameter to evaluate the ageing of an aircraft structure. The take-off phases exhibited the maximum load at the junction with the wings. The maximum activity was recorded at the thrust increase at take-off. Figure 6.53 shows an example of the Acoustic Emission activities

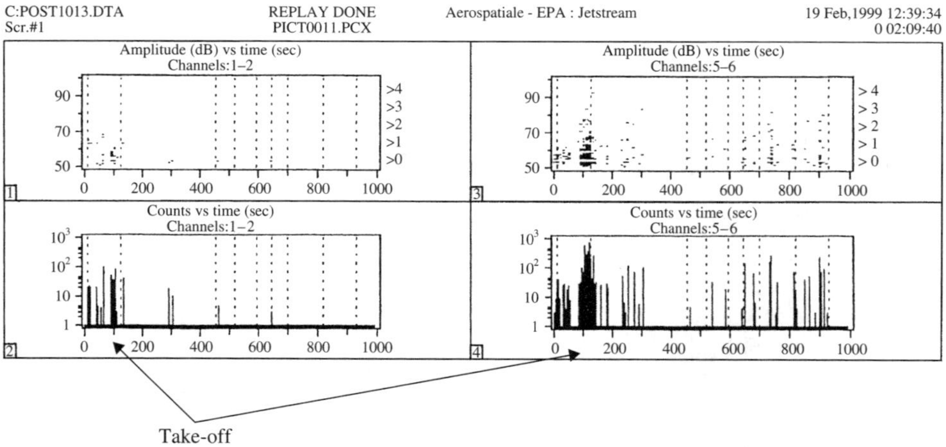

Figure 6.53 Example of Acoustic Emission events during take-off phase

recorded at take-off. During the dives the fuselage is very active and the activity was clearly linked with the dive speed. This reflects the capability of the aircraft to compensate the pressure differences.

The flight tests have proved the detection, filtering and selection capability of the Acoustic Emission system and the low rate of false alarms. The best application of the system could be in areas of the aircraft structure being difficult to access, where the maintenance operations are specifically more expensive. The tests have shown that system, sensors and wiring have to be optimised for a long-term use in flight conditions. Also, data processing has to be simplified to give only the few important parameters. A logic follow-up could be the in-flight damage monitoring on structural parts or on test specimens fitted on the structure in order to be coupled to the vibration environment.

6.4.3 Bragg Grating Optical Load Measurement System

As a part of the flight demonstrator programme a composite patch with embedded Bragg grating sensors has been developed and manufactured. The patch that was bonded to the aluminium wing skin of the aircraft with the objective to evaluate the performance of the Bragg grating sensors system in embedded sensory applications. The composite patch was bonded to lower port wing at outer skin, inboard of the engine pod.

The six ply thick patch was made from carbon fibre composite prepreg, *Ciba-Geigy HTA/6376C*. The stacking sequence was $[\pm 45/0_2/\text{of}]$. The patch had a dimension of 150 mm by 50 mm. A sketch of the patch is shown in Figure 6.54a. In the centre of the lay-up, two closely spaced optical fibres with sensors in two different optical fibres were embedded. These were protected with Teflon tubes at the ingress points at the two edges.

After curing in an autoclave at 180 °C for 2 hours, the quality of the patch and alignment of the optical fibres were checked. No damage was detected in the composite and alignment of the embedded optical fibres was good. The patch after curing is shown in Figure 6.54b.

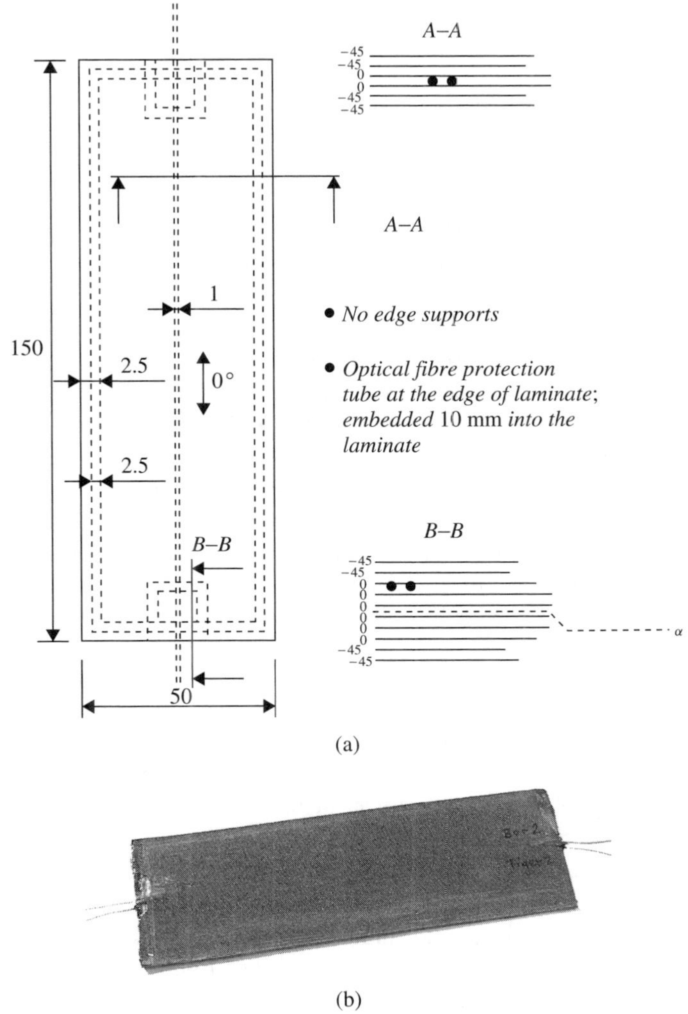

Figure 6.54 Composite patch with optical fibre sensors: (a) schematic diagram; (b) top view after curing

The approach to discriminate the temperature effect from the strain response of the Bragg grating sensors was based using sensors that have different thermal response. The sensing scheme consisted of seven pairs of closely spaced sensors. For each sensor pair, the two sensors are spatially embedded close to each other so those two sensors have experienced the same condition. Then, the sensor response can be expressed in terms of the Bragg wavelength, $\Delta\lambda$ axial strain ε_0, and temperature change, ΔT, as

$$\Delta\lambda_1 = (K_s \cdot \varepsilon_0 + K_{T,1} \cdot \Delta T) \tag{6.1}$$

for sensor 1 and

$$\Delta\lambda_2 = (K_s \cdot \varepsilon_0 + K_{T,2}, \cdot \Delta T) \tag{6.2}$$

for sensor 2. The approach was that the thermal sensitivity, K_T, is different in the two sensors, in this way, the temperature can be determined from

$$\Delta T = (\Delta\lambda_1 - \Delta\lambda_2)/(K_{T,1} - K_{T,2}) \tag{6.3}$$

To achieve as large as possible difference in thermal sensitivity a *Tiger* and a *Boron* doped optical fibre were selected. The thermal sensitivities have to be determined in advance to calibrate the system. The wavelength was measured in an oven at different temperatures for the patch. A linear thermal response was used to fit the data for each sensor. A large difference in thermal sensitivity of the *Tiger* and *Boron* fibre was observed when these were embedded in the $[\pm 45/0_2]$ patch. The thermal sensitivities were 0.0091 ± 0.00081 nm/°C and 0.0064 ± 0.00035 nm/°C in the *Tiger* and *Boron* fibres, respectively.

A finite element analysis including both the composite patch and the aluminium skin was performed to determine the longitudinal strain and thermal response when the patch was bonded to the aluminium wing skin as well as the free patch. Each ply in the composite patch was included in the model as well as the fibre optic sensors. The calculated strain and thermal sensitivity of the embedded sensors was used to evaluate the sensor response during flight. The finite element analysis was used to determine the thermal sensitivity for the sensors in a patch bonded to the aluminium skin. The results indicated a significant effect of the aluminium skin as well as the non-symmetrical geometry of the set-up on the sensor response.

A large set of temperature and strain data were generated from the flights. During the first five minutes the aircraft was climbing to 10 000 ft. A temperature of −1 °C was recorded from the aircraft instruments. The Bragg grating sensors indicated a decrease in temperature. Also, the temperature went back to a similar value at the end of the flight as that before the take-off. However, a nonlinear behaviour of the optical fibre sensor system resulted in the temperature values different from the values recorded by the instrument of the aircraft. The strain response was also evaluated. An example is given in Figure 6.55 when several approach, go-around and landing sequences were performed at the end of the flight. The strain was evaluated by subtracting the effect of time dependence in the wavelength output from the sensor system. The repeated peaks in Figure 6.55 show the landings and take-offs. The results clearly demonstrated the ability of the Bragg grating sensors to measure strain and temperature in flight conditions.

6.4.4 Fibre Optic Load Measurement Rosette System

The work described in this section uses the fibre optic strain monitoring system described in Chapter 3. It has been designed to be complementary with the vibration sensing system used in the composite evaluator described in Section 6.3.5. The technology for both systems is based on optical fibre Bragg grating sensors, both surface mounted and embedded approaches have been pursued. Although the details of the optoelectronic interface equipment will be different for the load monitoring and damage detection systems, synergy exists on the issues of structural implications of optical fibre embedment in carbon fibre composites and fibre sensor endurance to manufacturing and handling conditions.

In Chapter 3 the operational load monitoring system hardware was described, including the polyamide sensor patches as well as the optoelectronics. The work described

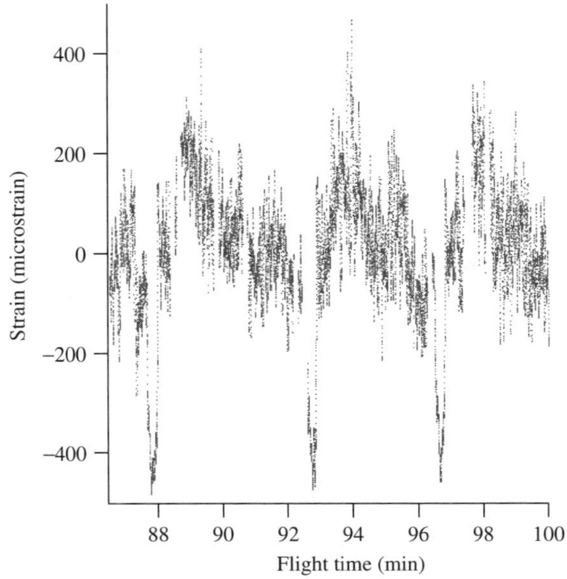

Figure 6.55 Flight test example of measured strain data

here utilises this hardware during full scale flight-testing. During flight the optoelectronic equipment in the cabin is subject to environmental conditions not encountered in the laboratory. These include:

- vibration; the cabin of the *Jetstream* aircraft lies between the two propellers which generate high levels of noise and vibration,
- nonstandard power supply; 110 V at 400 Hz,
- G-forces due to scheduled manoeuvres (e.g. 60° banked turns and roller landings),
- pressure changes; the cabin was pressurised but not at the equivalent of sea level pressure.

The OLM had to operate and be capable of recording flight data over the duration of each flight. Flights could last up to two and a half hours. This environment provides a significant test for components of the OLM including the Optical Signal Processor (OSP) and the single mode FC/PC connectors. The sensors and cabling had to endure an even harsher environment. Sensor patches were bonded to the external surface of the wing and protected with a layer of sealant. Cabling was then routed from the sensor patches through the leading edge of the wing into the cabin. As well as most of the conditions encountered by the optoelectronics, sensor patches and cabling also have to tolerate:

- large temperature changes/temperature cycling; ground temperatures at Warton in February were typically a few degrees above zero whereas at twenty thousand feet the temperature was below minus twenty degrees Celsius,
- large pressure changes; no pressurisation outside the cabin,

- airflow; as the sensors were positioned on the wing near the propeller they will have encountered a considerable wind-chill effect during flight.

Demonstration of the whole system operating in such unforgiving conditions can only increase confidence in the application of such a system to aerospace structures. The work described here is concerned only with the verification of the sensor apparatus in respect of the above requirements.

Details of the fabrication process, its variants and optical properties of these devices can be found in Chapter 3. The sensor system adopted for flight trials is shown in Figure 6.56. A driving signal from the processing electronics is used to scan the output wavelength from the OSP. When the scanning wavelength matches a Bragg grating reflection peak, light is reflected back to one of the photodiodes. During the wavelength scan the photodiodes see a series of light pulses corresponding to the Bragg grating sensors. The optical signal is converted into an electrical signal by the photodiodes. The output from the photodiodes is processed yielding the position of each light pulse along the wavelength scan. These encoded wavelength values are then stored on the computer's hard disk. A temperature controlled reference grating is stored within the OSP. The reference grating acts as an absolute wavelength reference. The characteristics of these Bragg gratings are distinct from the type used in the composite evaluator for vibration sensing. These gratings have a relatively narrow spectral profile. The spectral width of each grating's reflection spectrum was matched to the band of wavelengths output by the OSP for a given wavelength setting. Matching these wavelength profiles ensures the maximum reflected intensity and hence the best signal to noise ratio. Figure 6.57 shows a typical reflection spectrum from an array of sensor gratings. The difference in height of the reflection peaks is due to

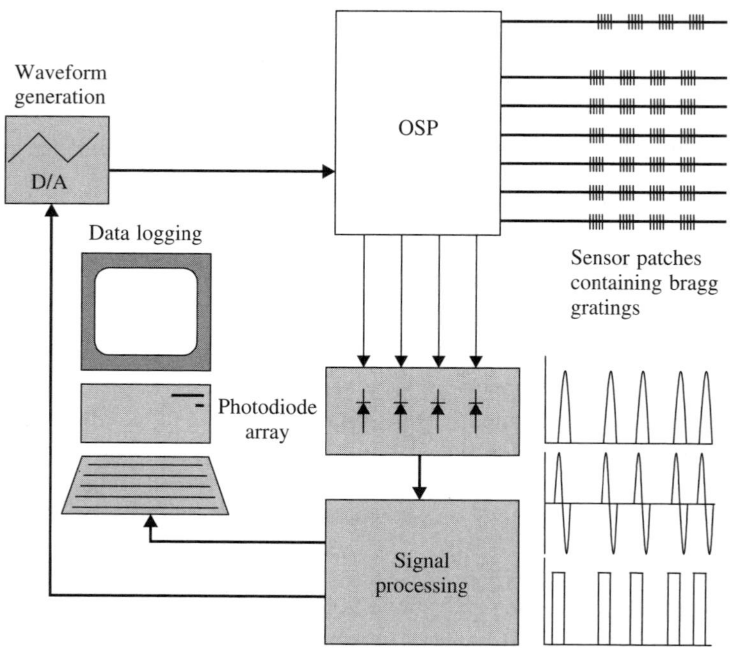

Figure 6.56 Schematic diagram of the OLM system using Bragg grating sensor patches

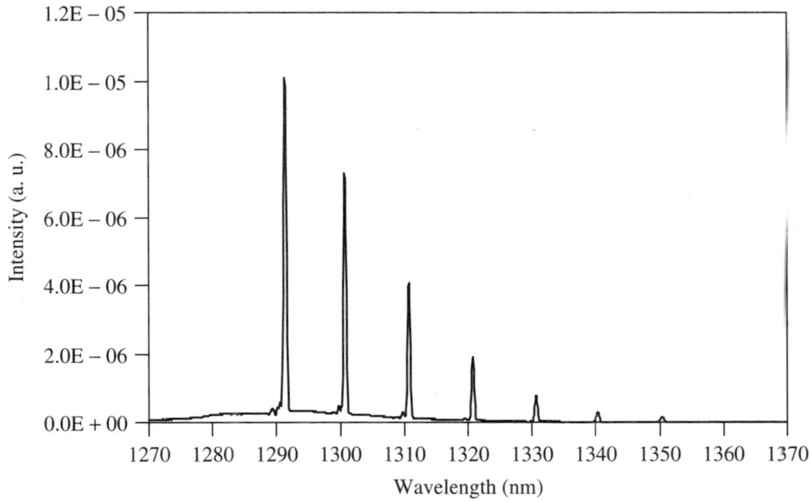

Figure 6.57 Reflection spectrum of Bragg grating sensor array system

the spectral profile of the illuminating source. The source within the OSP has a central wavelength of 1321 nm and will thus illuminate all the Bragg peaks more evenly. Each peak has a spectral width of approximately 0.5 nm.

A notable feature of this optical design is the single connecting fibre compared with the equivalent electrical device which requires nine cable connections, as described in Chapter 3. Vulnerability to failure in systems is usually correlated with the number of connectors present making this optical fibre solution potentially more reliable than the electrical equivalent. Each electrical connection needs screening or amplifying when conveying signals over significant distances (greater than a metre in aircraft applications). Optical fibres require no screening due to their inherent immunity to electromagnetic interference. The electrical gauges also require electrical power supply and are susceptible to moisture, which degrades their electrical properties.

Seven different patches with optical sensors were manufactured for flight tests. There were five plastic film patches (Figure 6.58), one metallic patch and the carbon fibre conformal patch. The bonding process differed for the different patches. Plastic patches were bonded using a hot bonding process developed for strain gauges. The metallic patch

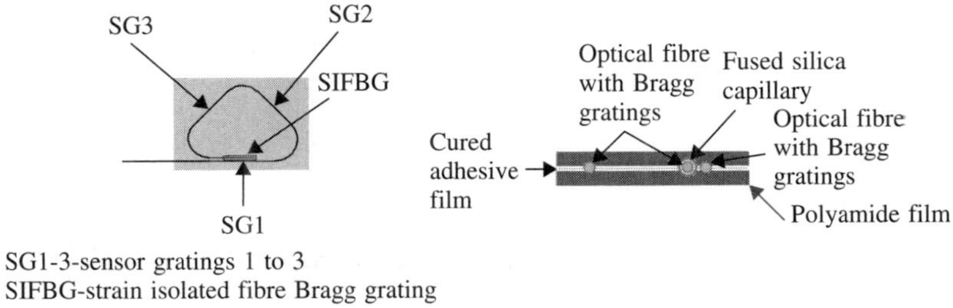

Figure 6.58 Plastic patch sensor design

was cold bonded using epoxy resin. The conformal carbon fibre patch was also cold bonded due to fears that a hot bonding process would buckle the wing skin. A resin impregnated glass fibre layer was sandwiched between the wing skin and the conformal patch during bonding. The glass fibre layer was used to prevent galvanic corrosion.

Installation on the aircraft required the patches to be connected to the OLM system by a rugged cable. The cabling would need to be long enough to connect the sensor on the wing to the OLM system in the cabin taking into account the available cabling routes. The cable would also have to be rugged enough to endure conditions on the wing surface. Figure 6.59 shows the cabling procedure. The optical fibre emerging from the plastic patches is protected only by its primary coating. The primary polyamide coating is only a few tens of microns thick; hence the fibre is quite fragile at this point. A Teflon tube is used to provide some protection. The edge of the patch has also been extended to accept the tubing and provide a gentle transition from patch to tube. This level of protection is suitable of a laboratory environment but is not sufficient for the flight trials.

Long connecterised pigtails were spliced to the patches to allow the patch to be connected and disconnected as required. The splice region is very fragile with the primary buffer removed. This region must be protected to be able to survive. However, normal splice protectors are rigid and bulky and may cause additional problems if the fibre has to be bent near the splice. A new method of protecting the splice was developed which would integrate with subsequent protective measures. The splice region lays between the Teflon tubing and the long pigtail whose central fibre had a secondary buffer coating. The outer diameter of the Teflon tubing and the secondary buffer were approximately the same. Clear heat shrink sleeving that would collapse to the same diameter was selected. A small amount of epoxy resin was applied to the splice region, which was then covered by the heat shrink sleeving. When heated the sleeving collapsed and the epoxy became more fluid. Thus by directing heat along the splice region the collapsing sleeving forced the epoxy resin to coat the exposed fibre. The length of heat shrink sleeving was chosen to be slightly longer than the splice region so that it overlapped on both ends.

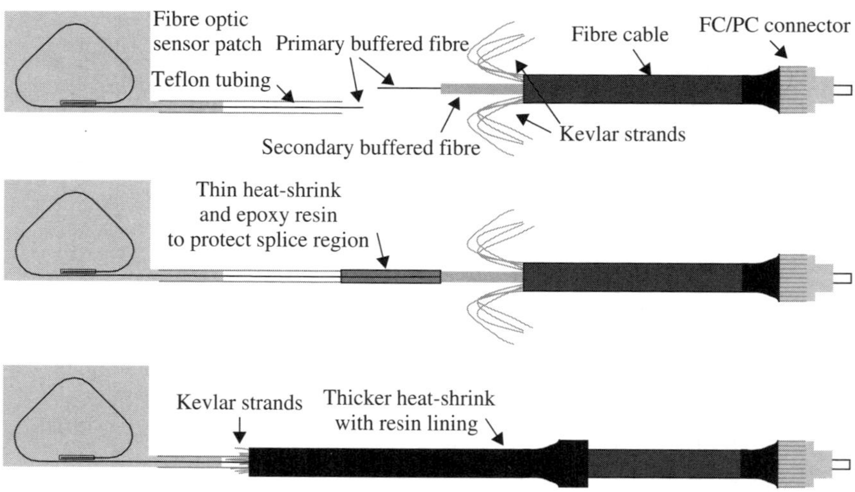

Figure 6.59 Adding rugged cables to patches

As well as the secondary buffer coating the long fibre pigtails had other protective features. Kevlar strands ran along side the optical fibre within a plastic sheath. The Kevlar strands are strong in tension and thus reduce tensile loading on the fibre. The additional plastic tube also offers some extra protection. It was advantageous to provide this level of security to the whole fibre length. Thus before splicing, the outer jackets of the long pigtails were cut back so that the resulting length of Kevlar strands was long enough to extend to the patch. After the splice region was protected the whole uncovered length of the Kevlar strands and Teflon tube were encapsulated by another layer of heat shrink. This secondary heat shrink was thicker than the first and was lined with resin so that it bonded to the Kevlar fibres. A small length of Teflon tubing was left uncovered so that the optical fibre would not be forced through tight bends when the patch was bonded to the aircraft.

Thus protected, all the patches were bonded onto the aeroplane's wing. The sensors alone are useless without some way of processing and recording the data that they produce. This function is performed by the OLM system. The heart of the OLM system is the OSP, which operates in conjunction with controlling electronics and a computer.

The OLM was calibrated using a low finesse Fabry–Perot cavity. The cavity consisted of a fibre end face and a mirror separated by a rigid spacer. The cavity was insulated to limit thermal effects. The reflection spectrum of the cavity was recorded using an optical spectrum analyser; it consisted of a series of spectral peaks. Figure 6.60 shows the normalised reflection spectrum of the cavity used to calibrate the OLM system. The OLM system finds peaks and assigns them a number that represents wavelength. The system is calibrated by determining the relationship of the OLM numbers to wavelength. Thus the wavelengths of the calibration peaks must be known accurately. The wavelength spectrum of the low finesse calibration cavity can be expressed as

$$y = A + B\cos\left(\frac{4\pi d}{\lambda}\right) \tag{6.4}$$

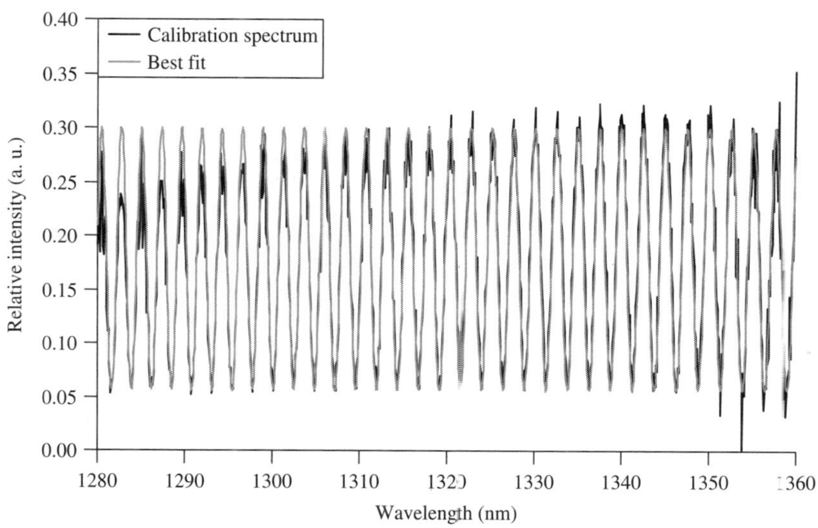

Figure 6.60 Reflection spectrum of the low finesse Fabry–Perot cavity

where y is the wavelength spectrum, A and B are constants, λ is the wavelength and d is the length of the cavity. By fitting this curve to the spectrum of the cavity a value for d can be determined. Figure 6.60 contains such a curve fit, with the value of d set to 0.3601331047 mm. Note that it is important that the fitted curve matches the data closely over a large range of wavelengths to obtain the correct value for d. Once the curve has been fitted to the data the positions of the spectral peaks can be determined.

The Fabry–Perot cavity was connected to each input channel of the OLM system in turn. The numbers produced by the OLM system were recorded by a computer. Thus, each channel and each scanning direction could be calibrated independently. Figure 6.61 shows a typical calibration curve. This calibration produces wavelength, which is required for the patches. To obtain strain and temperature, specific coefficients must be used.

The calibration curves display a slight nonlinearity, which is different for rising and falling sides of the driving waveform (scanning from short to long wavelengths and vice versa). Initially it was thought that these nonlinearities could be 'calibrated out' and data from the two scanning directions could be combined. However, the nonlinearities were found to vary over time, perhaps due to thermal effects. Also, it was not possible to track these changes with the single reference grating system used in the OLM. Thus eliminating the effects of the nonlinearities was not possible and a linear calibration was adopted.

The effect of the nonlinearities appears small in Figure 6.61 but can produce wavelength errors of up to 1 nm. Consider that 1 nm is equivalent to about 1000 $\mu\varepsilon$ and the required resolution is 10 $\mu\varepsilon$. Nonlinearities cause wavelength values of individual Bragg gratings to be significantly different for the two scanning directions; thus implying that the level of noise was higher than its actual value. Hence, the data for the scanning directions was separated. The update rate for each scanning direction was 25 Hz. After modification the OLM system could still address 26 sensor gratings and one reference grating, the data from each and every grating being updated at 25 Hz. The strain and temperature ranges of the gratings were unaffected by the modifications at $\pm 3000\,\mu\varepsilon$ and $-35\,^{\circ}$C to $+80\,^{\circ}$C respectively.

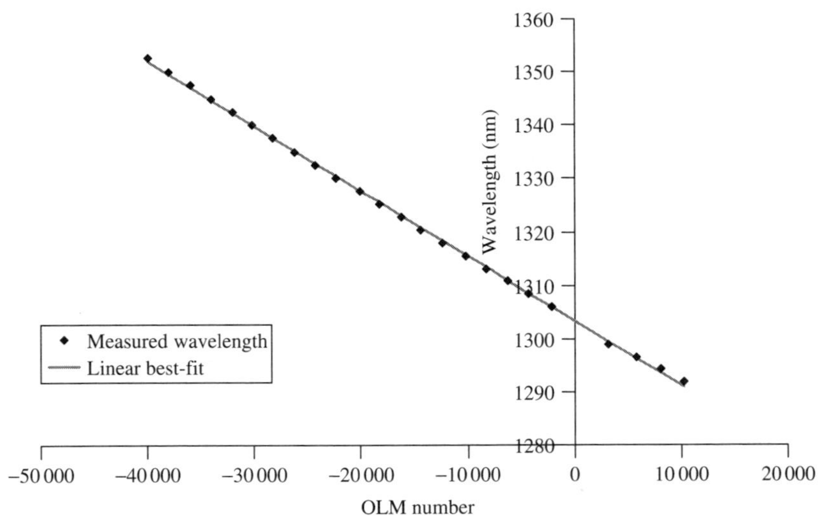

Figure 6.61 Example of typical calibration curve for the OLM system

The OLM system was installed in the rack inside the cabin – first rack from the right in Figure 6.51. The optical fibre leads connected the sensor patches on the wing to the OSP module within the OLM system. The components of the computer (screen, keyboard and computer) lie above and below the OSP module. The required length of optical fibre cable was not known initially and so the cables were left over long. Coils of excess fibre cable can be seen stowed on the side of the rack cabinet.

The plastic patches and the metallic patch contained four Bragg grating sensors and had one optical fibre connection each. The carbon fibre conformal patch contained fourteen Bragg grating sensors and had three fibre optic connections. These sensors have already been discussed in Section 6.3.5. There were only two optical fibres embedded into the conformal patch but both ends of one of the fibres were cabled so that if one was damaged the other would be useable. In fact all except one of the optical connections survived throughout the flight trials. This single failure was caused by accidental damage to a fibre cable during the bonding process. Figure 6.62 shows how the sensors were arranged; the plastic and metallic patches were placed in front of and behind the carbon fibre conformal patch. Fibre cables were looped round into the leading edge of the wing. A strain gauge was also bonded to the surface of the wing. The sensors were covered over with a protective layer of sealant. When the sealant was dry the whole area was repainted. During the flight trials the strain gauge bridge was found to be unconnected to the computer. At that stage it was too late to rectify the situation. Hence, no strain gauge data was recorded.

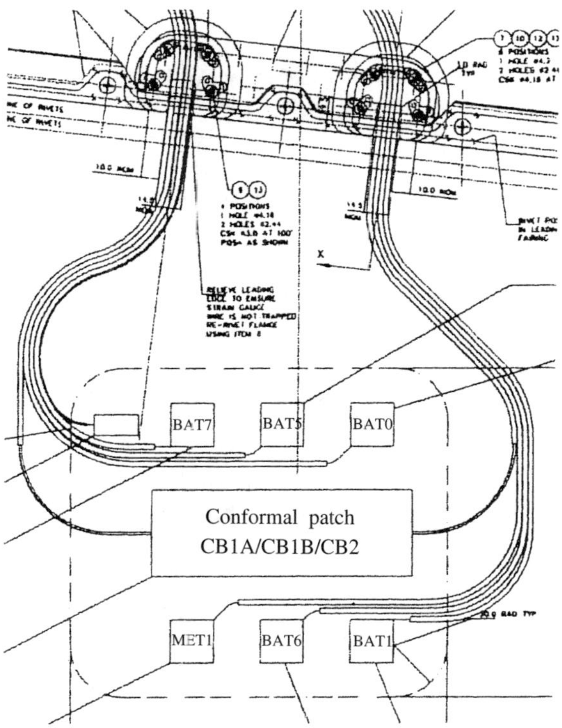

Figure 6.62 Optical fibre sensor layout seen from under the wing

The test procedures were designed to expose the sensing systems to the whole gamut of load and vibration possible within the flight envelope of the aircraft, as described in Section 6.4.1.

Figure 6.63 shows typical data recorded throughout the third flight test. The top graph shows the aircraft's altitude over the duration of the flight and the bottom graph shows the principle strains measured by one of the plastic patches. Flight test 3 is representative of all the flight tests as it contained all the previously described manoeuvres. These manoeuvres are indicated on the graphs. The altitude plot clearly indicates when the take-off, roller landings and the final full stop landing happened. Comparison of the strain and altitude graphs shows that there are significant strain changes at these take-off and landing points. The time of other manoeuvres was recorded manually; many of these events also show up on the strain record. There is a small gap in the strain record at 12:35 where data was down loaded to a zip disk in-flight. Analysis of the total volume of data after the flight indicated that in-flight downloading was unnecessary and this procedure was not repeated on subsequent flights.

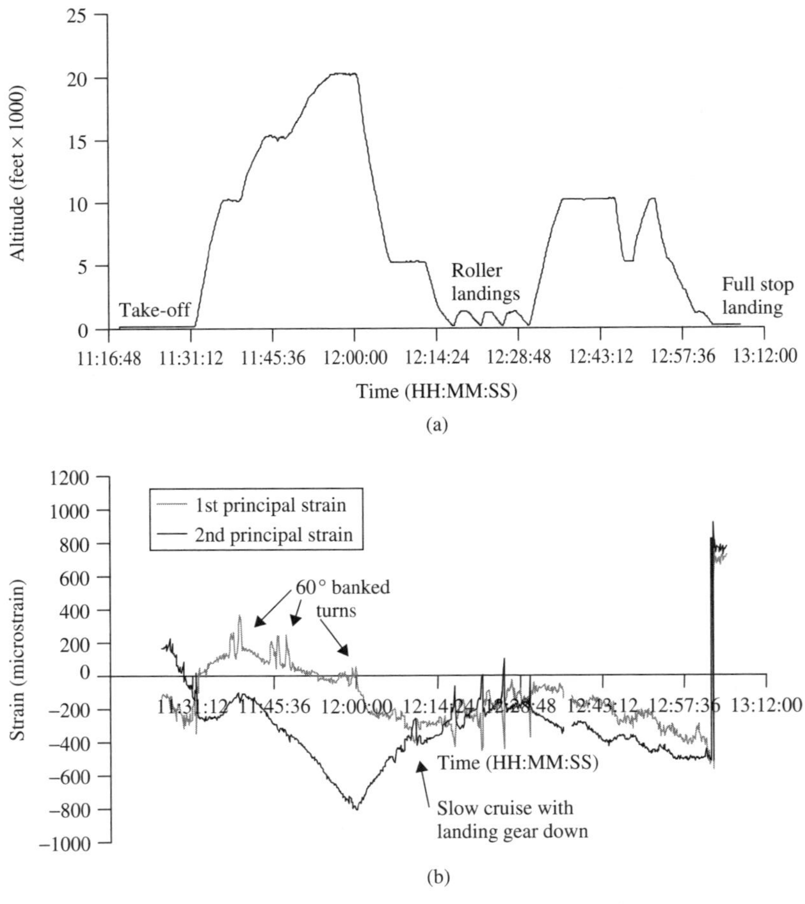

Figure 6.63 Data captured during flight test 3: (a) aircraft altitude; (b) strain from optical fibre sensors

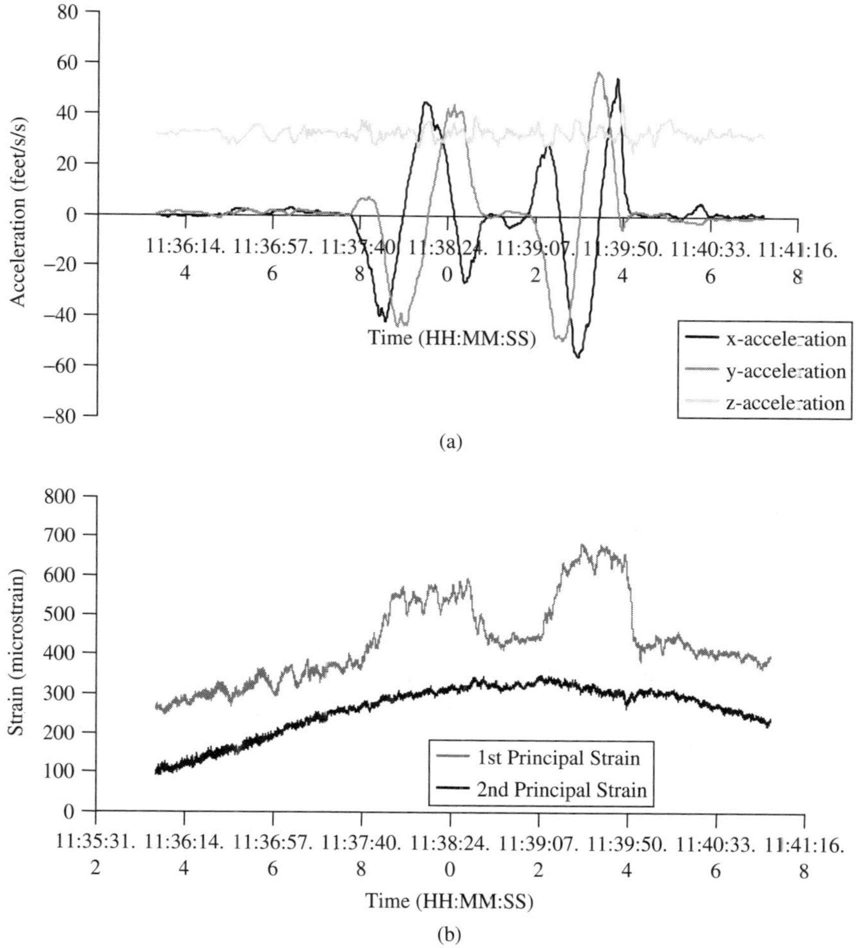

Figure 6.64 Data captured during flight test −60° banked turns anticlockwise and clockwise orbits: (a) acceleration; (b) strain from optical fibre sensors

The significant events can be investigated more closely. Figure 6.64 shows a five minute section of the third flight trial when 60° banked turns were being performed. The top graph shows accelerations measured by the test aircraft's own instrumentation and the bottom graph shows the principal strains measured by one of the patches. The x, y and z directions are defined relative to the aircraft, x- forward and back, y−right and left, z−up and down. The z−acceleration seems unaffected by the manoeuvre and further investigation indicates that the z-acceleration is insensitive to all manoeuvres indicating that this data are unreliable. The x- and y-accelerations clearly show the manoeuvres and the timing correlates well with the strain records. The 60° banked turns were designed to increase the G-force on the aeroplane and the strain record shows that loading on the wing also increases.

The angle of the first principal strain direction relative to the axis of the patch was also determined. Figure 6.65 shows this angle. While the aircraft is on the ground the strain

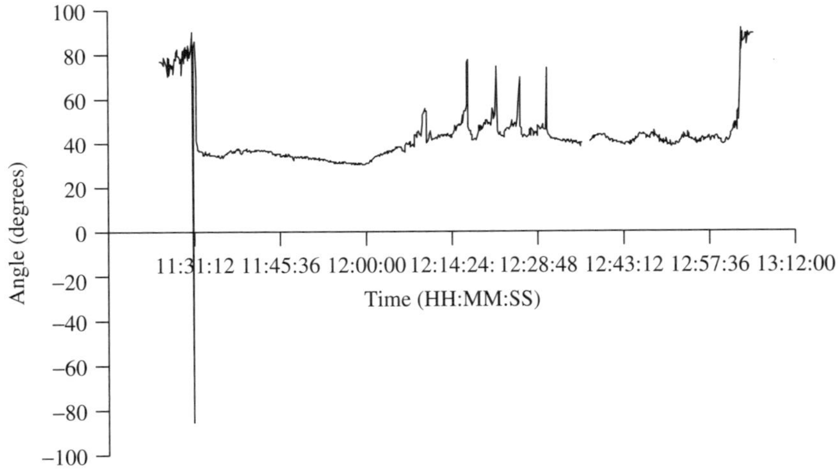

Figure 6.65 Data captured during flight test – strain angle from optical fibre sensors

level is low and the angle is not well defined. After take-off the strain levels become significant and the angle settles to a value of about 35°. The angle remains relatively constant over the duration of the flight indicating that the load path through the structure is also constant. The only significant variations in angle are during the roller landings and when the landing gear is lowered during a slow cruise. Roller landings briefly produce conditions similar to those experienced when the aircraft is on the ground. Thus, one would expect the strain angle to return to ground level values. One should note that lowering the landing gear while cruising is not a normal flight manoeuvre and would radically alter aerodynamic loads.

In order to measure the strain it is necessary to eliminate the effects of temperature on the Bragg grating sensors. The plastic patches contain a strain isolated Bragg grating to measure temperature. Output from this temperature sensor was used to compensate the other three sensors on the patch.

The data displayed in this section has been derived from the raw data files produced by the OLM system. The OLM system outputs data in a compact binary format. This format was chosen for two reasons:

- writing data from the optoelectronics directly to disk makes the amount of work that the computer has to do manageable and
- the file sizes remain relatively small despite the large amounts of data recorded.

For example – one hour of flight recording 27 Bragg gratings at 50 Hz produces 4.86 million data values. Conversion of the raw data files revealed an unforeseen problem. Occasionally, one or more replicas with a small wavelength offset would shadow the Bragg grating values. Thus instead of a single wavelength value, a Bragg grating may have several. These replica values were not consistent, appearing only for a limited time. Thus it was possible to track a grating wavelength over the duration of each flight. The shear size of the data files and the spurious wavelength values made the data processing a complex and difficult task.

6.5 SUMMARY

The aim of this chapter was to demonstrate the feasibility of load and damage monitoring technologies in representative component tests under realistic operational loads and environmental conditions. The components tested envisaged the use of three large-scale evaluators. The following two ground-based experiments and one flight test were performed:

- ground test on a large-scale metallic riveted structure to simulate damage monitoring on ageing aircraft;
- ground test on a composite wing-box structure to evaluate long term load and damaging obstacles in composite structures;
- flight test comprising a composite structure attached to the airframe together with a number of sensors attached alongside.

Additionally some results are reported being related to monitoring of an aircraft landing gear with Acoustic Emission.

The tests successfully demonstrated the function of fibre-optic load measurement both in embedded and as surface mounted devices. The use of piezoceramic and Acoustic Emission sensors was also successfully demonstrated. All these systems require signal processing and data logging systems to be carried in practice, which is definitely one of the key elements of any structural health and usage monitoring system.

REFERENCES

Alleyne, D.N. and Cawley, P.C. 1992. A two-dimensional Fourier transform method for the measurement of propagating multimode signals, *NDT&E International*, Vol. 25, pp. 11–22.

Hamonic, B.F., Debus, J.C. and Descarpigny, J.N. 1990. The finite element code ATILA, *Proceedings of the workshop held in Toulon, June*, ISEN, Lille, France.

Mondanos, M., Lloyd, P.A., Giles, I.P., Badcock, R.A. and Weir, K. 2000. 14th International Conference on Optical Fibre Sensors. October, Venice, Italy.

Index

Health Monitoring of Aerospace Structures – Smart Sensor Technologies and Signal Processing.
Edited by W.J. Staszewski, C. Boller and G.R. Tomlinson ISBN: 0-470-84340-3

Index compiled by Geraldine Begley